P9-AFV-205

Filtration Equipment for Wastewater Treatment

Prentice Hall Series in Process Pollution and Control Equipment

by Nicholas P. Cheremisinoff and Paul N. Cheremisinoff

Pumps and Pumping Operations

Compressors and Fans

Filtration Equipment for Wastewater Treatment

Filtration Equipment for Wastewater Treatment

Nicholas P. Cheremisinoff
Paul N. Cheremisinoff

Prentice Hall
Englewood Cliffs, New Jersey 07632

Library of Congress Cataloging-in-Publication Data

Cheremisinoff, Nicholas P.
Filtration equipment for wastewater treatment / Nicholas P. Cheremisinoff, Paul N. Cheremisinoff.
p. cm.
Includes bibliographical references and index.
ISBN 0-13-319559-7
1. Sewage--Purification--Filtration--Equipment and supplies.
I. Cheremisinoff, Paul N. II. Title.
TD753.C42 1993
628.3'52--dc20 92-6546
CIP

Editorial/production supervision
and interior design: *Brendan M. Stewart*
Prepress buyer: *Mary McCartney*
Manufacturing buyer: *Susan Brunke*
Acquisitions editor: *Michael Hays*

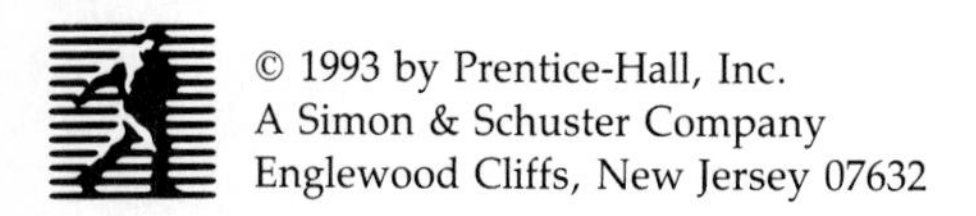

A Simon & Schuster Company
Englewood Cliffs, New Jersey 07632

The publisher offers discounts on this book when ordered in bulk quantities. For more information, write: Special Sales/Professional Marketing, Prentice Hall, Professional & Technical Reference Division, Englewood Cliffs, NJ 07632.

Printed in the United States of America
10 9 8 7 6 5 4 3 2 1

ISBN 0-13-319559-7

Prentice-Hall International (UK) Limited, *London*
Prentice-Hall of Australia Pty. Limited, *Sydney*
Prentice-Hall Canada Inc., *Toronto*
Prentice-Hall Hispanoamericana, S.A., *Mexico*
Prentice-Hall of India Private Limited, *New Delhi*
Prentice-Hall of Japan, Inc., *Tokyo*
Simon & Schuster Asia Pte. Ltd., *Singapore*
Editora Prentice-Hall do Brasil, Ltda., *Rio de Janeiro*

Contents

6 Selection and Sizing of Prefilter/Final Filter Systems 141

Preface

Filtration is a basic unit operation that has the objective of removing undissolved solids from liquids when the particles are of sufficient size. This is a physical separation technique whereby the diameter of the pores in the filter medium establishes the completeness of separation. The rate of flow of a given liquid is a direct function of the permeability or porosity of the filter medium. This means that the separation of particles in an extremely fine state of division can be accomplished only at the expense of capacity. Colloidal and many ultrafine suspensions cannot be filtered, nor even settled, or centrifuged out. To separate materials in this category the techniques of microfiltration, ultrafiltration, and reverse osmosis can be employed. Microfiltration separation generally involves removing particles from fluids based on size. In this operation, osmotic pressure is negligible. Ultrafiltration generally involves the separation of large molecules from smaller molecules, and overlaps somewhat with the porosity range of membranes used for reverse osmosis. Reverse osmosis can be thought of as a purification process, or the concentration of small molecules or ionic constituents in a solvent.

Filtration technology has been practiced for many years in the chemical process industries. Sometimes its objectives have been to recover valuable end or intermediate products, or to purify feedstocks and/or recyclable process streams. In environmental control the objectives are almost always to either concentrate waste products dispersed in water or process streams, or as an operation aimed at primary, secondary, and occasionally final treatment of water contaminated by particulates.

This third volume in the series is written to provide a working knowledge of the equipment and operational concepts of filtration. It has been prepared with the practitioner in mind, to enable a practical understanding of equipment configuration options, operating modes, and operating principles behind major types of filtration equipment. Selection criteria and performance data are provided to assist the engineer in the use of this operation.

—Nicholas P. Cheremisinoff
—Paul N. Cheremisinoff

Filtration Equipment for Wastewater Treatment

1

Filtration as a Unit Operation

Filtration is a fundamental unit operation aimed at the separation of suspended solid particles from a process fluid stream by passing the suspension through a porous substance referred to as a filter medium. In forcing the fluid through the voids of the filter medium, fluid alone flows, but solid particles are retained on the surface and in the medium's pores. The fluid discharging from the medium is called filtrate. The operation may be performed with either incompressible fluids (liquids) or slightly to highly compressible fluids (gases). The physical mechanisms controlling filtration, although similar, vary with the degree of fluid compressibility. In this book the unit operation is discussed specifically with reference to process liquid handling. Although there are marked similarities in the particle capture mechanisms between the two fluid types, mathematical description and design methodology presented strictly apply to liquid handling unless otherwise noted.

TYPES OF LIQUID FILTRATION

There are two major types of filtration: *cake* and *filter-medium* filtration. In the former, solid particulates generate a cake on the surface of the filter medium. In filter-medium filtration (sometimes called blocking or clarification), solid particulates become entrapped within the complex pore structure of the filter medium. The filter medium for the latter case consists of cartridges or granular media. Examples of granular materials are sand or anthracite coal. In design, to account for the parameters governing filtration and to select filtration equipment best suited for a particular application, two important parameters must be considered: (1) the method used for forcing liquid through the medium, and (2) the material that constitutes the filter medium.

When the resistance opposing fluid flow is small, gravity force effects fluid transport through a porous filter medium. Such a device is simply called a gravity filter. If gravity is insufficient to instigate flow, the pressure of the atmosphere is allowed to act on one side of the filtering medium, while a negative or suction pressure is applied on the discharge side. This type of filtering device is referred to as a vacuum filter. The application of vacuum filters is typically limited to 15 psi pressure. If greater force is required, a positive pressure in excess of atmospheric can be applied to the suspension by a pump. This motive force may be in the form of compressed air introduced in a montejus, or the suspension may be directly forced through a pump acting against the filter medium (as in the case of a filter press). Or centrifugal force may be used to drive the suspension through a filter medium as is done in screen centrifuges.

Filtration is a hydrodynamic process in which the fluid's volumetric rate is directly proportional to the existing pressure gradient across the filter medium, and inversely proportional to the flow resistance imposed by the connectivity, tortuosity, and size of the medium's pores, and generated filter cake. The pressure gradient constitutes the driving force responsible for the flow of fluid.

Regardless of how the pressure gradient is generated, the driving force increases proportionally. However, in most cases, the rate of filtration increases more slowly than the rate at which the pressure gradient rises. The explanation for this phenomenon is that as the gradient rises, the pores of the filter medium and cake are compressed and consequently the resistance to flow increases. For highly compressible cakes, both driving force and resistance increase nearly proportionally and any rise in the pressure drop has a minor effect on the filtration rate.

FILTER MEDIA

The filter medium represents the heart of any filtration device. Ideally, solids are collected on the feed side of the plate while filtrate is forced through the plate and carried away on the leeward side. A filter medium is, by nature, inhomogeneous with pores nonuniform in size, irregular in geometry, and unevenly distributed over the surface. Since flow through the medium takes place through the pores only, the microrate of liquid flow may result in large differences over the filter surface. This implies that the top layers of the generated filter cake are inhomogeneous and, furthermore, are established based on the structure and properties of the filter medium. Since the number of pore passages in the cake is large in comparison to the number in the filter medium, the cake's primary structure depends strongly on the structure of the initial layers. This means that the cake and filter medium influence each other.

Pores with passages extending all the way through the filter medium are capable of capturing solid particles that are smaller than the narrowest cross section of the passage. This is generally attributed to particle bridging or, in some cases, physical adsorption.

Depending on the particular filtration technique and intended application, different filter media are employed. Examples of common media are sand, diatomite, coal, cotton or wool fabrics, metallic wire cloth, porous plates of quartz, chamotte, sintered glass, metal powder, and powdered ebonite. The average pore size and configuration (including tortuosity and connectivity) are established from the size and form of individual elements from which the medium is manufactured. On the average, pore sizes are greater for larger medium elements. In addition, pore configuration tends to be more uniform with more uniform medium elements. The fabrication method of the filter medium also affects average pore size and form. For example, pore characteristics are altered when fibrous media are first pressed. Pore characteristics also depend on the properties of fibers in woven fabrics, as well as on the exact methods of sintering glass and metal powders. Some filter media, such as cloths (especially fibrous layers), undergo considerable compression when subjected to typical pressures employed in industrial filtration operations. Other filter media, such as ceramic, sintered plates of glass, and metal powders, are stable under the same operating conditions. In addition, pore characteristics are greatly influenced by the separation process occurring within the pore passages, as this leads to a decrease in effective pore size and consequently an increase in flow resistance. This results from particle penetration into the pores of the filter medium.

The separation of solid particles from a liquid via filtration is a complicated process. For practical reasons filter medium openings should be larger than the average size of the particles to be filtered. The filter medium chosen should be capable of retaining solids by adsorption. Furthermore, interparticle cohesive forces should be large enough to induce particle flocculation around the pore openings.

CAKE FORMATION

Filtration operations are capable of handling suspensions of varying characteristics ranging from granular, incompressible, free-filtering materials to slimes and colloidal suspensions in which cakes are incompressible. These latter materials tend to contaminate or foul the filter medium. The interaction between the particles in suspension and the filter medium determines to a large extent the specific mechanisms responsible for filtration.

In practice cake filtration is used more often than filter-medium filtration. On achieving a certain thickness, the cake is removed from the medium by various mechanical devices or by inverse filtrate flow. To prevent the formation of muddy filtrate at the beginning of the subsequent filtration cycle, a thin layer of residual particles is sometimes deposited onto the filter medium. For the same reason, the filtration cycle is initiated with a low but gradually increasing pressure gradient at an approximately constant flow rate. The process is then operated at a constant pressure gradient while experiencing a gradual decrease in process rate.

The structure of the cake formed and, consequently, its resistance to liquid flow depend on the properties of the solid particles and the liquid-phase suspension, as well as on the conditions of filtration. Cake structure is first established by hydrodynamic factors, cake porosity, mean particle size, size distribution and particle specific surface area and spherieity. It is also strongly influenced by some factors that can conditionally be denoted as physicochemical. These factors are:

1. the rate of coagulation or peptization of solid particles
2. the presence of tar and colloidal impurities clogging the pores
3. the influence of electrokinetic potentials at the interphase in the presence of ions, which decreases the effective pore cross section
4. the presence of solvate shells on the solid particles (as manifested at particle contact during cake formation)

Due to the combining effects of hydrodynamic and physicochemical factors, the study of cake structure and resistance is extremely complex, and any mathematical description based on theoretical considerations is at best only descriptive.

The influence of physicochemical factors is closely related to surface phenomena at the solid-liquid boundary. It is especially manifested by the presence of small particles in suspension. Large particle sizes result in an increase in the relative influence of hydrodynamic factors, while smaller sizes contribute to a more dramatic influence from physicochemical factors. No reliable methods exist to predict when the influence of physicochemical factors may be neglected. However, as a general rule, for rough evaluations their influence may be assumed to be most pronounced in the particle size range of 15 to 20 μ.

FILTRATION CONDITIONS

Two significant operating parameters influence the process: the pressure differential across the filtering place and the temperature of the suspension. Most cakes may be considered compressible and, in general, their rate of compressibility increases with decreasing particle size. The temperature of suspension influences the liquid-phase viscosity, which subsequently affects the ability of the filtrate to flow through the pores of the cake and filter medium.

In addition, the filtration process is aggravated by such factors as particle inhomogeneity, ability to undergo deformation when subjected to pressure, and settling characteristics under the influence of gravity. Inhomogeneity of particle sizes influences the geometry of the cake structure not only at the moment of its formation, but also during the filtration process. During filtration, small particles retained on the outer layers of the cake are often entrained by the liquid flow and transported to layers closer to the filter medium, or even into the pores themselves. This results in an increase in the resistances of the filter medium and cake formed.

Particles that undergo deformation when subjected to operating pressures are usually responsible for pore clogging. What nature has sometimes neglected in the filterability of suspensions, man can correct through the addition of coagulating and peptizing substances. These additives can drastically alter cake properties and, subsequently, resistances, ultimately increasing the filtration rate. Filter aids may be used to prevent the penetration of fine particles into the pores of a filter plate when processing low-concentration suspensions. Filter aids build up a porous, permeable, rigid lattice structure that retains solid particles on the filter-medium surface, while permitting liquid to pass through. They are often employed as precoats with the primary aim of protecting the filter medium. They may also be mixed with the suspension diatomaceous silica type (greater than 90 percent silica content). Cellulose and asbestos fiber pulps are also employed, although the latter is no longer widely used because of health reasons.

The brief discussions of the basic features of filtration given thus far illustrate that the unit operation involves some rather complicated hydrodynamics that depend strongly on the physical properties of both fluid and particles. The process is essentially influenced by two different groups of factors, which can be broadly grouped into macro- and microproperties. Macrofactors are related to variables such as the area of a filter medium, pressure differences, cake thickness, and the viscosity of the liquid phase. Such parameters are readily measured. Microfactors include the influences of the size and configuration of pores in the cake and filter medium, the thickness of the electrical double layer on the surface of solid particles, and other properties to be discussed later. At present, microproperties and their individual influence on the filtration process cannot be measured directly.

The inhomogeneous nature of both the suspension particles and the pores of the cake and filter medium prevent direct observation of microproperty influences on the filtration process. Measurements based on inference often lead to incorrect results. The uncertainty about these microfactors limits the development of reliable mathematical models.

WASHING AND DEWATERING STAGES

When objectionable or valuable suspension liquors are present, it becomes necessary to wash the filter cake to effect clean separation of solids from the mother liquor or to recover the mother liquor from the solids. Dewatering involves forcing a gas through the cake to recover residual liquid retained in the pores, directly after filtering or washing. Liquid is displaced from the pores by air (or other appropriate gas), which is often preheated to increase the flowability of entrapped liquid. Also, by preheating the gas, the hydrodynamic process is aided by diffusional drying.

Dewatering is a complex process on a microscale because it involves the hydrodynamics of two-phase flow. Although washing and dewatering are performed on a cake with an initially well-defined pore structure, the flows become greatly distorted and complex due to changing cake characteristics. The cake structure undergoes compression and disintegration during both operations, thus resulting in a dramatic alteration of the pore structure.

GENERAL CONSIDERATIONS IN FILTRATION INTENSIFICATION AND FILTER SELECTION

In specifying and designing filtration equipment, primary attention is given to options that will minimize high cake resistance. This resistance is responsible for losses in filtration capacity. One option for achieving a required filtration capacity is the use of a large number of filter modules. Increasing the physical size of equipment is feasible only within certain limitations as dictated by design considerations, allowable operation conditions, and economic constraints. A more flexible option from an operational viewpoint is the implementation of process-oriented enhancements that intensify particle separation. This can be achieved by two different methods. In the first method, the suspension to be separated is pretreated to obtain a cake with minimal resistance. This involves the addition of filter aids, flocculants, or electrolytes to the suspension.

In the second method the period during which suspensions are formed provides the opportunity to alter suspension properties or conditions that are more favorable to low-resistance cakes. For example, employing pure initial substances or performing a prefiltration operation under milder conditions tends to minimize the formation of tar and colloids. Similar results may be achieved through temperature control, by limiting the duration of certain operations immediately before filtering such as crystallization, or by controlling the rates and/or sequence of adding reagents.

Filtration equipment selection is often complex and sometimes confusing because of (1) the tremendous variations in suspension properties, (2) the sensitivities of suspension and cake properties to different process conditions, and (3) the variety of filtering equipment available. Generalities in selection criteria are, therefore, few. However, there are guidelines applicable to certain classes of filtration applications. One example is the choice of a filter whose flow orientation is in the same direction as gravity when handling polydisperse suspensions. Such an arrangement is more favorable than an upflow design, since larger particles will tend to settle first on the filter medium, thus preventing pores from clogging within the medium structure.

A further recommendation, depending on the application, is not to increase the pressure difference for the purpose of increasing the filtration rate. The cake may, for example, be highly compressible; thus, increased pressure would result in significant increases in the specific cake resistance. Design and selection generalities applicable to various classes of filtration problems are discussed in subsequent chapters. However, we may generalize the selection process to the extent of applying three rules to all filtration problems:

1. The objectives of a filtration operation should be defined.
2. Physical and/or chemical pretreatment options should be evaluated for the intended application based on their availability, cost, ease of implementation, and ability to provide optimum filterability.
3. Final filtration equipment selection should be based on the ability to meet all objectives of the application within economic constraints.

FILTRATION OBJECTIVES

The objectives for performing filtration usually fall into one of the following categories:

1. clarification for liquor purification
2. separation for solids recovery
3. separation for both liquid and solids recovery
4. separation aimed at facilitating or improving other plant operations

Clarification involves the removal of relatively small amounts of suspended solids from suspension (typically below 0.15 percent concentration). A first approach to considering any clarification option is to define the required degree of purification. That is, the maximum allowable percentage of solids in the filtrate must be established. Compared with other filter devices, clarifying filters are of lesser importance to pure chemical process work. They are primarily employed in beverage and water polishing, pharmaceutical filtration, fuel/lubricating oil clarification, electroplating solution conditioning, and dry cleaning solvent recovery. They are heavily employed in fiber spinning and film extrusion.

In filtration for solids recovery, concentration of solids suspension must be high enough to allow the formation of a sufficiently thick cake for discharge as a solid mass before the rate of flow is materially reduced. However, solids concentration alone is not the only criterion for adequate cake formation. For example, a 0.5 percent suspension of paper pulp may be readily cake forming, whereas a 10 percent concentration of certain chemicals may require thickening to produce a dischargeable cake.

Filtration for both solids and liquid recovery differs from filtration for solids recovery alone in the cake building, washing, and drying stages. If the filtrate is valuable liquor, maximum washing is necessary to prevent its loss, but if it is valueless, excess wash liquor can be applied without regard to quality.

Finally, filtration can be applied to facilitate other plant operations. Like other steps, filtration has the most immediate relationship to those operations immediately preceding and following it. Ahead of filtration, the step is often one of preparation for filtration. These prefiltration steps could include thickening, coagulating, heating, conditioning, pH adjustment, or the production of an unstable floc that must not be broken by rapid pumping or agitation before filtration. Such preparation stages are used to obtain more filterable material. This allows a continuous operating mode, smaller filter areas, or both. Figure 1.1 schematically summarizes the prefiltration and final processing steps.

Similarly, filtration is the preparatory step in the operation following it. The latter stages may be drying or incineration of solids, concentration or direct use of the filtrate. Filtration equipment must be selected on the basis of ability to deliver the best feed material to the next step. Dry, thin, porous,

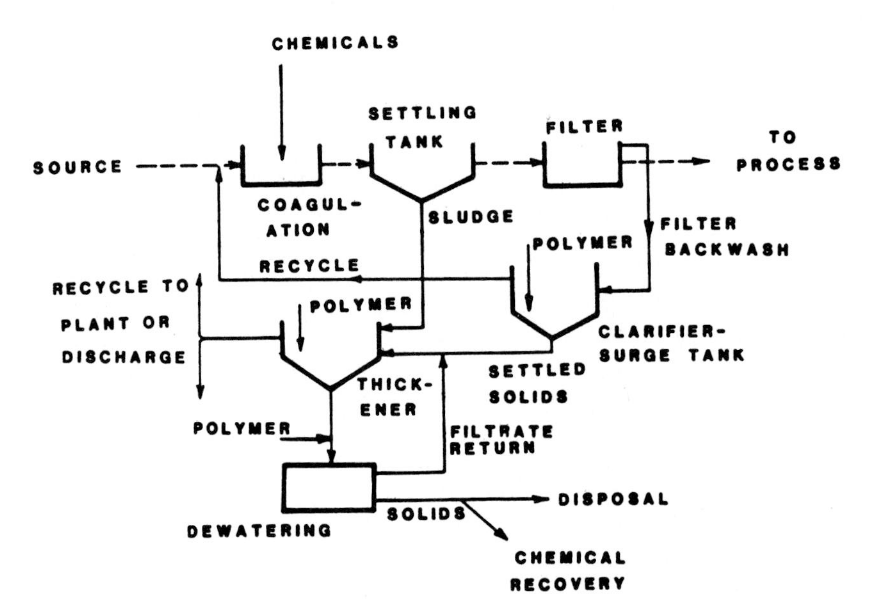

Figure 1.1 Summary of prefiltration and final processing steps in a filtering operation.

flaky cakes are best suited for drying where grinding operations are not employed. In such cases, the cake will not ball up, and quick drying can be achieved. A clear, concentrated filtrate often aids downstream treatment whereby the filter can be operated to increase the efficiency of the downstream equipment without affecting its own efficiency.

PREPARATION STAGES FOR FILTRATION

A number of preparation steps alluded to earlier assist in achieving optimum filterability. It is worthwhile to summarize the major ones.

Precoat and Filter Aids

Where particles of a colloidal nature are encountered in liquor clarification, a precoat and/or filter aid are often required to prevent deposited particles from being carried by stream-flow impact into the pores or the filter medium (or filter cake after formation), thus reducing capacity.

A precoat serves only as a protective covering over the filter medium to prevent the particles from reaching the pores, while the filter aid added to the influent assists in particle separation and cake formation. Filter aids serve as obstructions, intervening between the particles to prevent their compacting, and producing under the pressure-velocity impact a more or less impervious layer on the filter medium, or if a precoat is used, on it.

In some instances, precoats are used, not because of danger to filter cloth clogging, but to permit the use of a coarser filter medium such as metallic cloths. This can extend operating life or improve corrosion resistance.

Coagulation

Coagulation is another means of dealing with colloidal or semicolloidal particles. It applies particularly to clarification in water and sewage filtration and in the filtration of very fine solids. While flocculation often can be accompanied by agitation, the use of chemical additives results in alteration of the physical structure of the suspended solids to the extent of losing their colloidal nature and becoming more or less crystalline. This is usually accompanied by agglomeration. Clarification by settling may follow, if the specific gravity of the particles is sufficient to provide reasonably quick supernatant clarity. Direct filtration may be applied if the filter area is not excessive or if complete supernatant clarity is needed.

Temperature Control

Temperature has a direct impact on viscosity, which in turn affects the flow rate. It is an important factor in filtration, since lower viscosity leads to liquor penetration into smaller voids and in shorter times. Occasionally, temperature plays a part in altering the particle form or composition, and this in turn affects the clarification rate.

pH

Proper pH control can result in clarification that might otherwise not be feasible, since an increase in alkalinity or acidity may change soft, slimy solids into firm, free-filtering ones. In some cases precoats are employed, not because of the danger of filter cloth clogging, but to allow the use of a coarser filter medium, such as metallic cloth.

METHODOLOGY FOR EQUIPMENT SELECTION

Equipment selection is seldom based on rigorous equations. Where equations are used, they function as a directional guide in evaluating data or process arrangements. Projected results are derived most reliably from actual plant operational data where duplication is desired; from standards set up where there are few variations from plant to plant, so that results can be anticipated with an acceptable degree of confidence as in municipal water filtration or from pilot or laboratory tests of the actual material to be handled. Pilot runs are typically designed for short durations and to closely duplicate actual operations.

Based on earlier discussions it is possible to evaluate experimentally the optimum filter design and determine the filtration regime applicable to a given suspension.

Proper selection of equipment may be based on experiments performed in the manufacturer's laboratory, although this is not always feasible. Sometimes the material to be handled cannot readily be shipped. Its physical or chemical conditions change during the time lag between shipping and testing, or special conditions must be maintained during filtration that cannot be readily duplicated, such as refrigeration, solvent washing, and inert gas use. A filter manufacturer's laboratory has the advantage of having numerous types of filters and apparatus available with experienced filtration engineers to evaluate results during and after test runs.

The use of pilot-plant filter assemblies is becoming more common. These combine the filter with pumps, receivers, mixers, and so on, in a single compact unit and may be rented at a nominal fee from filter manufacturers, who supply operating instructions and sometimes an operator. Preliminary tests are often run at the filter manufacturer's laboratory. Rough tests indicate what filter type to try in the pilot plan.

Comparative calculations of specific capabilities of different filters or their specific filter areas should be made as part of the evaluation. Such calculations may be performed on the basis of experimental data obtained without using basic filtration equations. In designing a new filtration unit after equipment selection, calculations should be made to determine the specific capacity or specific filtration area. The basic filtration equation may be used for this purpose, with preliminary experimentation constants evaluated. These constants contain information on the specific cake resistance and the resistance of the filter medium.

However, the basic equations of filtration cannot always be used without introducing corresponding corrections. This arises from the fact that these equations describe the filtration process partially for ideal conditions when the influence of distorting factors is eliminated. Among these factors are the instability of the cake resistance during operation and the variable resistance of the filter medium, as well as the settling characteristics of solids. In these relationships, it is necessary to use statistically averaged values of both resistances and to introduce corrections to account for particle settling and/or other factors.

Within the subject of filtration, a distinction is made between micro- and macromodeling. The first one is related to modeling cake formation. The cake is assumed to have a well-defined structure, in which the hydrodynamic and physicochemical processes take place. Macromodeling presents few difficulties because the models are process oriented (i.e., they are specific to the particular operation or specific equipment). If distorting side effects are not important, the filtration process may be designed according to existing empirical correlations. In practice, filtration, washing, and dewatering often deviate substantially from theory. This occurs because of the distorting influences of filter features and the unaccounted for properties of the suspension and cake. These conditions are addressed in the chapters that follow.

Modeling of filtration, washing, and dewatering operations are of practical importance, and we shall endeavor to relate these to design practices. Existing statistical methods allow prediction of macroscopic results of the processes without complete description of the microscopic phenomena. They are helpful in establishing the hydrodynamic relations of liquor flow through porous bodies, the evaluation of filtration quality with pore clogging, description of particle distributions, and in obtaining geometrical parameters of random layers of solid particles.

2

Filtration Equipment

INTRODUCTION

A wide variety of liquid filtration equipment is commercially available, ranging from highly versatile units capable of handling different filtration applications to those restricted in use to specific fluids and process conditions. Proper selection must be based on detailed information of the slurry to be handled, cake properties, anticipated capacities, and process operating conditions. One may then select the preferred operational mode (batch, semibatch, or continuous), and choose a particular unit based on the preceding considerations and economic constraints.

Continuous filters are essentially a large number of elemental surfaces on which different operations are performed. To review, these operations performed in series are solids separation and cake formation, cake washing, cake dewatering and drying, cake removal, and filter media washing. The specific equipment used can be broken into two groups: (1) stationary components (which are the supporting devices such as the suspension vessel), and (2) scraping mechanisms and movable devices (which can be the filter medium, depending on the design).

Either continuous or batch filters can be employed in cake filtration. In filter-medium filtration, however, where particulates are retained within the framework of the filter medium, only batch systems are applicable. Batch filters may be operated in any filtration regime, whereas continuous filters are most often operated at constant pressure.

In an attempt to organize the almost overwhelming number of different types of filtration equipment, two classification schemes have evolved for continuous operations.

1. Continuous filtration equipment may be classified according to method of generating pressure difference. This classification scheme is:

	Pressure Differential (N/m^2)
Hydrostatic pressure of the suspension layer to be separated	Usually no more than 5
Action of compressors	5–9
Action of pumps	Up to 50 and higher

2. Classification may also be based on the relative directions between gravity force and the filtrate motion. Three orientations are possible: forces act in opposing directions (countercurrent), forces act in the same direction (cocurrent), and forces act perpendicular to each other (cross mode).

Because the influence of gravity is important to most filtration operations, the second classification basis is adopted. Descriptions of operating principles and important features of specific filtration equipment follow. The literature [1–7] provides further review of the operating performances and detailed design features of the following devices.

APPLICATIONS IN WATER REUSE

Although filtration is practiced in a variety of industrial applications and processes, it has become an even more critical unit operation due to large-scale efforts in more effective use of plant water. Water use in the chemical process industries (CPI) has been undergoing dramatic changes over the past decade. Interestingly, while gross water use has been rising, plant water intake and discharge have been falling. The reason for this has been greater water reuse in plant operations. The major force behind this water reuse trend is environmental protection. Federal, state, and local laws backed by stiff penalties set strict limits on the amount of effluents discharged by CPI.

Governmental regulations are not the only reason that CPI are turning toward water reuse. As fresh water becomes a scarce commodity, its cost is becoming prohibitive. Another contributing factor is the continually increasing cost of moving and heating fresh water, especially once-through water. The high cost of chemically treating fresh water cannot be overlooked either.

There is no doubt that a water-reuse system can be an asset to the processor, as long as it is functionally efficient in design and operation. However, a serious problem must be reckoned with, that is, recirculated water can have a much higher solids content. Where solids can cause a system problem, it is imperative that they be removed from the water. Although there is no clear-cut solution to the problem of solids buildup, filtration and/or chemical treatment can normally achieve the desired result.

Deposit formation on waterside surfaces of heat exchangers has always concerned cooling system and process operators. However, in the past the problem was considered relatively minor because the equipment was usually operated only six months (seldom longer than twelve months) between cleanings, or turnarounds. This is no longer possible. Many industrialists, in an effort to reduce overall operating costs, are working toward less frequent turnarounds, higher throughput with the same equipment, and lower maintenance costs. This is accomplished by removing the solids that cause deposit formation.

Cooling Towers

In cooling-tower operations, cascading water constantly "scrubs" airborne contaminants from the atmosphere and carries these particulates into the cooling system. Depending on geographic location and seasonal changes, these particulates can account for the highest concentration of suspended solids in the system. Other sources of fouling in cooling-tower operations can be traced to makeup water, scale from system piping, solids from the process, and dissolved solids from the atmosphere, which must be controlled chemically [8].

In the operation of heat-exchange equipment, corrosion control is of prime concern. Deposit control deserves at least equal attention because both problems occur jointly or cause one another. That is, as solids build up on the system surface, corrosion products begin forming, which in turn can produce a variety of undesirable operating conditions:

- reduced operating efficiency
- reduced or uneven heat transfer
- unexpected process equipment shutdown and associated maintenance
- shortened equipment life
- increased pumping costs

To avoid those solids-caused conditions, it is essential to remove the solids through filtration as soon as they enter the system.

In smaller systems with low flow rates, the filter can be placed directly on the tower outlet to handle the entire flow. However, full-stream filtration is not always necessary for large cooling systems with moderate or high flow rates.

An economical side-stream installation that continually filters a portion of the total system capacity will reduce tower blowdown and prevent loss of heat-exchange efficiency. This can be accomplished by installing a loop off the tower basin. Incorporated in this loop is a filter, a pump to pressurize the filter feed line, and a discharge line leading to the tower basin. At some point in the discharge line, a restriction (usually a control valve or orifice plate) is necessary to maintain pressure on the outlet header. This permits backwashing with filtered water.

For sizing a side-stream installation, recommendations will vary from user to user and even among filter manufacturers. Usually manufacturer suggestions are based on a percentage of the system flow rates. However, because almost every system and application are different, it is virtually impossible to give a standard formula that would apply to all.

Filter sizing requires careful and detailed analysis of many factors, such as flow rate, pressure, and system design. For example, if the makeup water is high in solids (from river or lake), it may be necessary to install a separate filter in the makeup system. An alternative is a central filter designed to handle both the makeup water and a portion of the recirculated water.

Choosing the right filter screen requires careful examination of the source of solids and nature particulates. Current operating systems on cooling towers are providing satisfactory performance with screen retention ratings ranging from 150 mesh to 5 to 10 microns.

For new cooling systems, initial installation of finer filter screens is recommended based on the anticipated solids size. On existing systems that have operated without filtering, coarse screens are generally used to reduce the solids concentration. After the backwash frequency has been stabilized, finer screens can be installed.

Because backwash filters use tower water for backwash cleaning, the frequency of tower blowdowns for removing the dissolved solids can be reduced. Also, savings are realized in the cost of chemicals used for treating makeup water.

Wet Scrubbers

Water for spray-type wet scrubbers can be recirculated with water from other processes. This water must be filtered as it is pumped to the scrubber. If not, solids can quickly clog small nozzle openings. This results in decreased scrubber efficiency. If the problem goes unchecked, the flow of water can eventually stop and downtime may be necessary to correct the situation.

If recirculated water is used in a scrubber system, it is unlikely that cartridge filters will be the primary equipment for solids removal because the water would form a sludge or slurry, which requires sophisticated clarification techniques for high solids removal. Generally, cartridge filters are used on the return line to the scrubber as a final filter for protecting spray nozzles.

Because of evaporation, scrubber systems using recirculated water still require some makeup water. Those systems are usually handled full-flow by cartridge filters. Filtering the makeup water ensures that further contamination is not introduced by this source. Automatic backwashing filters are best in that kind of application to prevent plugging of filter screens and restriction of water flow.

Boiler Feed Water

Boiler feed water can come from a variety of sources such as well water, city water, plant water, process water, and condensate. In boiler feed water systems, solids that can collect in boiler vessels and reduce boiler efficiencies must be removed. In this application, filtering the water down to 15 to 20 μm is common.

Food and Package Industry

Container rinse water can be recirculated water or water from other processes. As a rinse water, recirculated water is more advantageous than makeup water because it requires less energy for heating.

A major concern, especially to the food industry, is that no visible particles appear on the container surface. To prevent this, the rinse water can be clarified by filters with 25 mesh screens. Where smaller particulates such as dust and paper fibers are present, finer screens can be used.

Paper Industry

Throughout the paper industry, protection of machine showers is critical for product uniformity and consistency. Any filter installed to protect machine showers must satisfy three requirements:

1. It must have sufficient screen area to handle the required flow.
2. It must remove contaminants that may plug shower nozzles.
3. It must be easy to clean so that minimum operator attention is needed.

The desired operating pressure, location of the shower, and type of water used determine which filter will provide the best results.

Fresh-water filters are normally equipped with a wire screen; the mesh size is determined by the size of the nozzle orifice. A particle retention rating of five to six times smaller than the orifice should prevent plugging.

White-water showers pass a relatively high percentage of well-dispersed individual fibers and fines. However, because nozzles will plug with fiber bundles, pipe scale, and other contaminants, a white-water strainer is required. Pressure filters are widely used in the protection of these showers, normally after some type of primary clarification. This is generally accepted as the best method of assuming satisfactory operation of a shower system.

Steel Industry

The steel industry, one of the largest users of water in the United States, uses water for cooling purposes. One typical application is in rolling and drawing mills where recirculated water cools the equipment that forms the hot bars. As the water cascades over the rollers, it carries scale from the bars, airborne contaminants, and grease into a settling basin. At that point, most—but not all—contaminants and particulates settle in the bottom of the basin. Suspended solids that remain in the water must be removed before the water can be used. If not, the small water lines in the rollers and drawers would quickly plug and cause extensive damage to the equipment; also, the cooling-water distribution system would foul. Filters are usually installed immediately ahead of the rolling mills.

Other Industries

Many manufacturers use water to hydrostatically test their equipment. To prevent solids buildup in expensive testing equipment, the water is filtered between the collection sump and feed line to the test rack.

Another application of water reuse is for lubrication of pump seals. Again, this can be recirculated water or water from other processes. Most manufacturers of pump and seal equipment recommend filtration to 10 μm to remove sand or grit that can score the pump shaft or damage the scaling mechanism.

ROTARY DRUM FILTERS

Rotary drum filters constitute the countercurrent-mode type and are either vacuum operated or pressure operated. They are most frequently operated as vacuum filters. Although operated under pressure, they are rarely subjected to excessive pumping pressures. The principal advantage of these filters is the continuity of their operation. Unfortunately, total filtration cycles are limited to narrow time intervals. As such, it is necessary to maintain nearly constant slurry properties. Changing slurry properties can lead to wide variations in the required times for completing individual operations of the filtration process.

For separating low-concentration, stratified suspensions, rotary drum filters are normally specified at a submergence rate of 50 percent. Such slurries require only mild mixing to prevent particle settling. These filters are less useful in handling polydispersions containing particles with wide size ranges. Fouling by small solids is a frequent problem in these latter cases.

Drum Vacuum Filters with External Filtering Surfaces

These filters are characterized by the rate at which the drum is immersed in the suspension. These are perhaps the most widely employed countercurrent operated filters in industry, an example of which is shown in Figure 2.1. As shown, the design consists of hollow drum 1, with a slotted face, the outer periphery of which contains shallow tray-shaped compartments 2. The filter cloth is supported by a grid or a heavy screen which lies over these compartments. The drum rotates on a shaft with one end connected to drive 3, and the other to a hollow trunion adjoining to an automatic valve. The drum surface is partially immersed in the suspension contained in vessel 6. The cake formed on the outer surface of the drum is removed by scalper 7 as the drum rotates (as shown by the arrow).

Figure 2.2 shows a longitudinal view of the system. Compartments 2 of drum 1 are connected through pipe 3, passing through hollow trunion 4 of shaft 5, with automatic valve 6. Stirring device 7 is mounted under the drum to prevent particle settling.

A diagrammatic cross section of the filter is shown in Figure 2.3. As the drum rotates clockwise, each compartment is connected by pipe 2 with different chambers of immobile parts of automatic valve 4 and passes in series through the following operating zones: filtration, first dewatering, washing, second dewatering , cake removal, and cloth regeneration.

In the filtration zone, the compartment contacts the suspension in tank 11 and is connected to pipe 10 hooked up to a vacuum source. The filtrate is

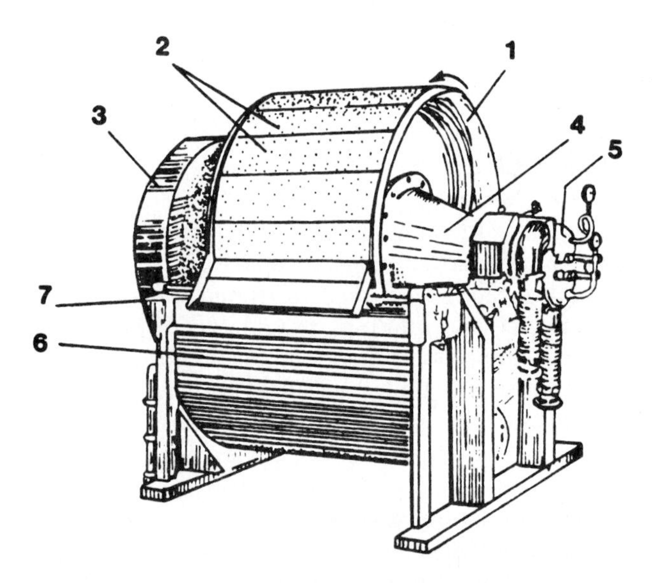

Figure 2.1 Rotary-drum vacuum filter with external filtration surface. 1 = hollow drum; 2 = filtration compartments; 3 = drive; 4 = hollow trunion, 5 = automatic valve; 6 = tank for suspension; 7 = knife for cake scraping.

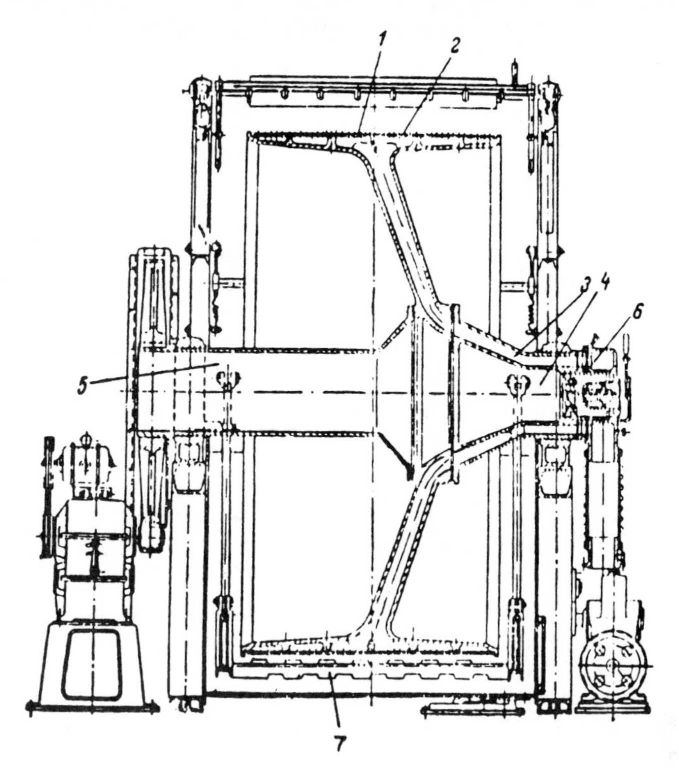

Figure 2.2 Longitudinal section of a rotary-drum vacuum filter with an external filtration surface 1 = drum; 2 = compartment; 3 = connecting pipe; 4 = hollow trunion; 5 = shaft; 6 = automatic valve; 7 = stirring device.

discharged through the pipe and space in the collector and the cake forms on the compartment's surface. In the first dewatering zone the cake comes in contact with the atmosphere, and the compartment is connected to space 10. Because of the vacuum, air is drawn through the cake, and for maximum filtrate recovery, the compartment remains connected to a collection port on the automatic valve.

In the washing zone the cake is washed by nozzles (or wash header) 8. The compartment is connected through port 6, which is also tied into a vacuum source. The wash liquor is removed in the other collector.

In the second dewatering zone, the cake is also in contact with the atmosphere, and the compartment is connected with port 6. Consequently, the washing liquid is displaced from the cake poles and delivered to the collector. To avoid cake cracking during washing and dewatering, endless belt 7 is provided, which moves over a set of guide rollers. In the discharge zone, the compartment is connected with port 5, which is supplied by a compressed air source. This reversal of pressure or "blow" loosens the cake from the filter medium, where it is removed by scraper or doctor blade 3.

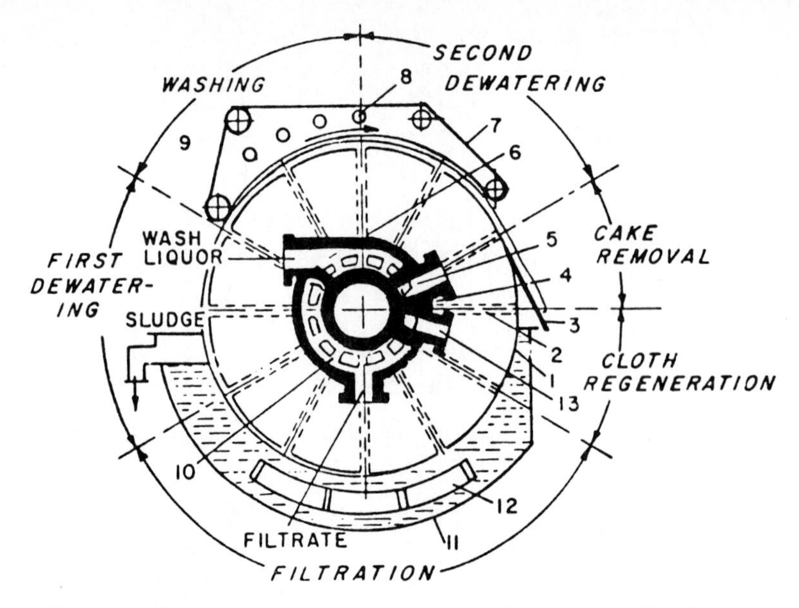

Figure 2.3 Diagrammatic cross section of rotary-drum vacuum filter; 1 = drum; 2 = connecting pipe; 3 = scraper; 4 = automatic valve; 5, 13 = chambers of automatic valve connected with a source of compressed air; 6, 10 = chambers of automatic valve connected with a source of vacuum; 7 = endless belt; 8 = wash header; 9 = guiding roll; 11 = tank for suspension; 12 = stirring device.

In the regeneration zone, compressed air is blown through the cloth; the air enters the compartment through the pipe from port 13. The automatic valve serves to activate the filtering, washing, and cake discharge function of the filter sections. It provides separate outlets for the filtrate and wash liquid, and a connection by which the compressed air blowback can be applied.

COCURRENT FILTERS

Cocurrent or top-feed filters employ flat and cylindrical filtering media. In flat designs, the angle between the directions of gravity force and filtrate motion is 0° but may vary to larger angles. In this class of filter, the directions of gravity force action and filtrate flow coincide. Filter designs in this class are quite different from the counterflow type. They include sophisticated rotary drum filters, continuous belt filters, Nutsch batch filters, and filter presses with horizontal chambers. These filters are most often used for separating stratified slurries. Separation by filters of the first subgroup is based on intensive slurry mixing by agitators. It is especially advisable to use these filters for the separation of polydispersed systems. In this case, the cake formed is properly stratified with large particles adjacent to the filter medium.

Flat filtering surfaces form a cake of uniform thickness and homogeneous structure at any horizontal plane. This permits highly effective washing.

Internal Rotary Drum Filters

The internal rotary drum filter is illustrated in Figure 2.4. The filter medium is contained on the inner periphery. This design is ideal for rapid settling slurries that do not require a high degree of washing.

Tankless filters of this design consist of multiple-compartment drum vacuum filters. One end is closed and contains an automatic valve with pipe connections to individual compartments. The other end is open for feed entrance. The drum is supported on a tire with rigid rollers to effect cake removal. The drum is driven by a motor and speed reducer connected to a riding roll shaft.

The feed slurry is discharged to the bottom of the inside of the drum from the distributor and is maintained as a pool by a baffle ring located around the open end and the closed portion of the outer end. As a drum revolves, the compartments successively pass through the slurry pool, where a vacuum is applied as each compartment becomes submerged. Slurry discharge is accomplished at the top center where the vacuum is cut off and gravity (usually assisted by blowback) allows the solids to drop off onto a trough. From there,

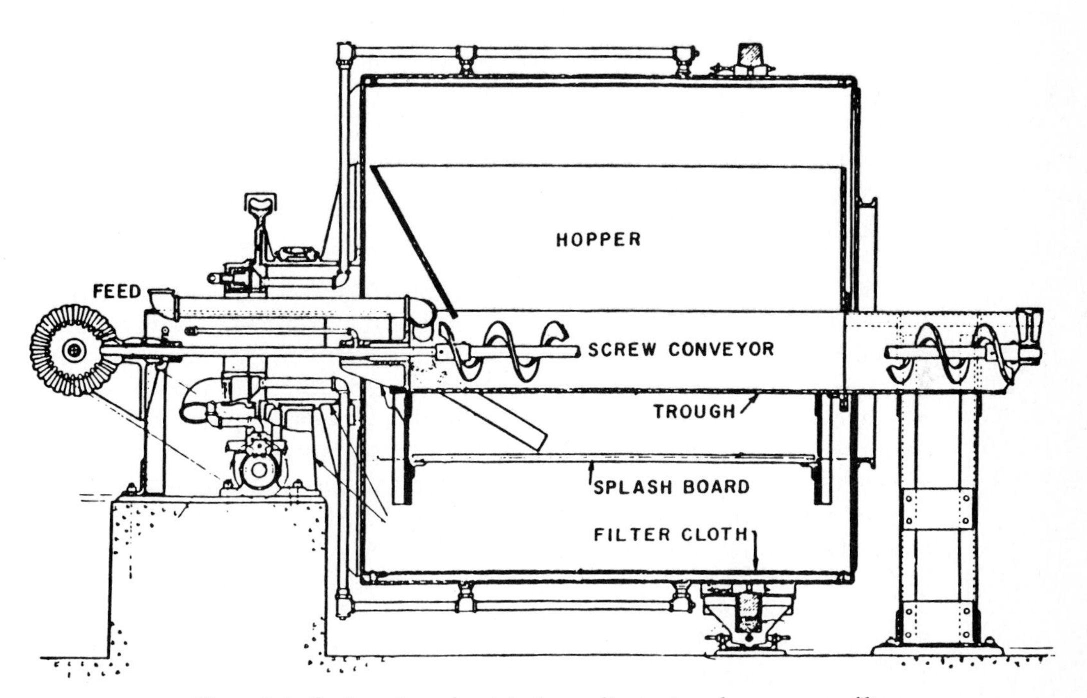

Figure 2.4 Section view of an interior medium rotary-drum vacuum filter.

a screw or bell conveyor removes the solids from the drum. This filter is capable of handling heavy, quick-settling materials. Dickey [9] notes that variations in feed consistency cause few or no difficulties.

Nutsch Filters

Nutsch filters are one design type with a flat filtering plate. This configuration basically consists of a large false-bottomed tank with a loose filter medium. Older designs employ sand or other loose, inert materials as the filtering medium and are still employed in water clarification operations. In vacuum filtration, these false-bottom tanks are of the same general design as the vessels employed for gravity filtration. They are, however, less widely used being confined for the most part to rather small units, particularly for acid work. Greater strength and more careful construction are necessary to withstand the higher pressure differentials of vacuum over gravity. This naturally increases construction costs. However, when high filtering capacity or rapid handling is required with the use of vacuum, the advantages may more than offset higher costs.

Construction of the vacuum false-bottom tank is relatively simple; a single vessel is divided into two chambers by a perforated section. The upper chamber operates under atmospheric pressure and retains the unfiltered slurry. The perforated false bottom supports the filter medium. The lower chamber is designed for negative pressure, and to hold the filtrate.

Nutsch filters are capable of providing frequent and uniform washings. A type of continuous filter that essentially consists of a series of Nutsch filters is the rotating-tray horizontal filter. Berline [10] provides details of this system.

Horizontal Rotary Filters

The horizontal rotary filter shown in Figure 2.5 is well adapted to filtering quick-draining crystalline solids. Due to its horizontal surface, solids are prevented from falling off or from being washed off by the wash water. As such, an unusually heavy layer of solids can be tolerated. The basic design consists of a circular horizontal table that rotates about a center axis. The table is comprised of a number of hollow pie-shaped segments with perforated or woven metal tops. Each of the sections is covered with a suitable filter medium and is connected to a central valve mechanism that appropriately times the removal of filtrate and wash liquids and the dewatering of the cake during each revolution. Each segment receives the slurry in succession. Wash liquor is sprayed onto each section in two applications. Then the cake is dewatered by passing dry air through it. Finally, the cake is scooped off of the surface by a discharge scroll.

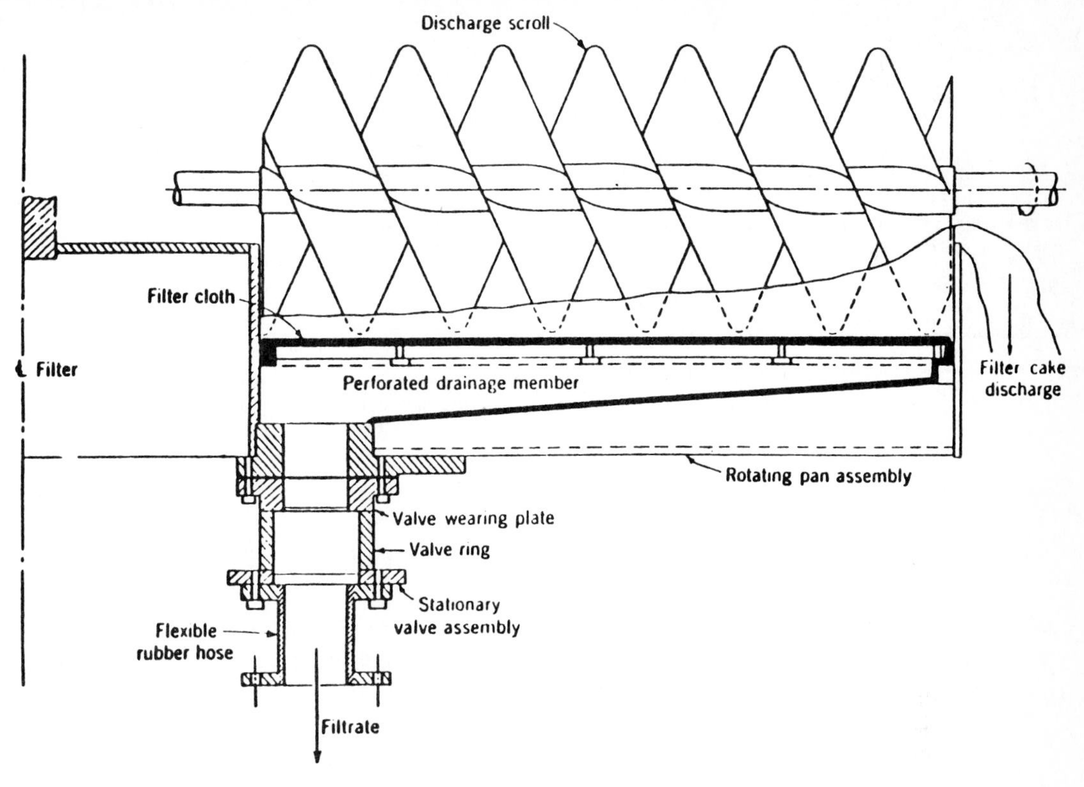

Figure 2.5 Cross section of a rotary horizontal vacuum filter showing filtrate removal system, filter cloth and discharge scroll.

Belt Filters

Belt filters consist of a series of Nutsch filters moving along a closed path. Nutsch filters are connected as a long chain so that the longitudinal edge of each unit has the shape of a baffle plate overlapping the edge of the neighboring unit. Each unit is displaced by driving and tensioning drums.

Nutsch filters are equipped with supporting perforated partitions covered with the filtering cloth. The washed cake is removed by turning each unit over. Sometimes a shaker mechanism is included to ensure more complete cake removal.

In contrast, a belt filter consists of an endless supporting perforated rubber belt covered with the filtering cloth. The basic design is illustrated in Figure 2.6. Supporting and filtering partitions 1 are displaced by driving drum 2 and maintained in a stretched condition by tensioning drum 3, which rotates due to friction against the rubber belt. Belt edges (at the upper part of their path) slide over two parallel horizontal guide planks. Elongated chamber 4

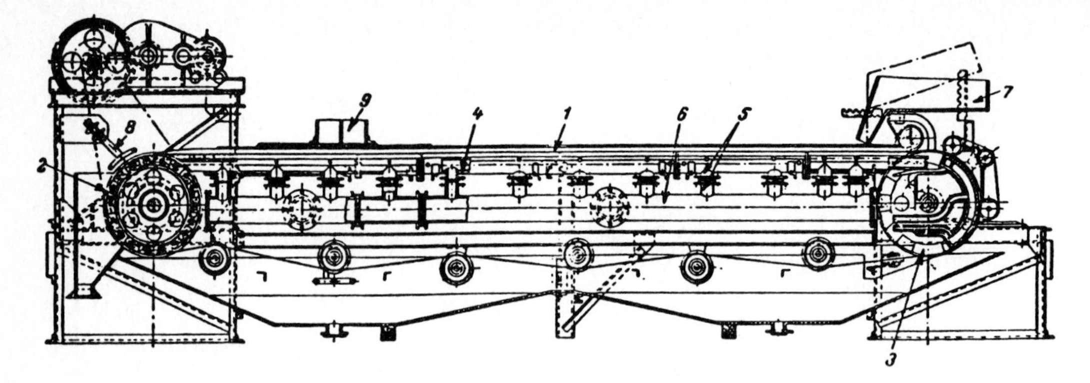

Figure 2.6 Belt filter, 1 = supporting and filtering partition; 2 = driving drum; 3 = tensioning drum; 4 = elongated chamber; 5 = nozzles; 6 = collector for filtrate; 7 = trough for sludge feed; 8 = device for cake washing; 9 = tank for washing liquid.

is located between the guide planks. The chamber in the upper part has grids with flanges adjoining the lower surface of the rubber belt. The region under the belt is connected by nozzles 5 to the filtrate collector, which is attached to a vacuum source. The chamber and collector are divided into sections from which filtrate and washing liquid may be discharged. The sludge is fed by trough 7. The cake is removed from drum 2 by gravity or blowing, or sometimes it is washed off by liquid from the distributor nozzle. The washing liquid is supplied from tank 9, which can move along the filtering partition. It can be washed during the belt's motion along the lower path.

The filtering partition, shown in Figure 2.7, consists of riffled rubber belt 1 with slots 2, grooves 3, and filter cloth 4, which is fixed in a set of grooves by cords 5. Slots 2, through which the filtrate passes, are located over the grids of the elongated chamber. The edges of the rubber belt are bent upward by guides forming a gutter on the upper path of the belt.

The velocity of the filtering partition depends on the physical properties of the sludge and the filter length. The cake thickness may range from up to

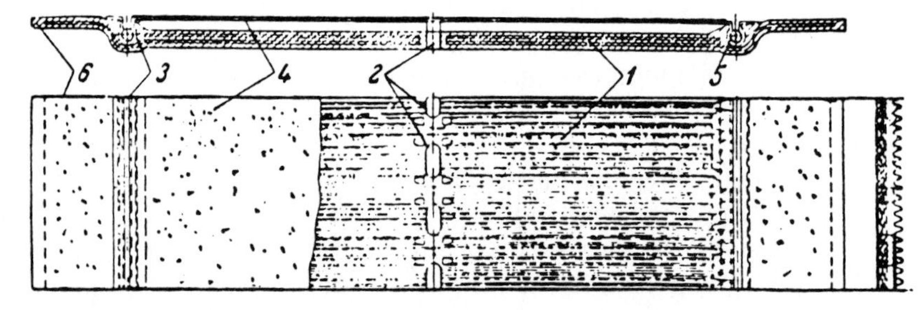

Figure 2.7 Filtering partition for a belt filter, 1 = rubber belt; 2 = slots; 3 = grooves; 4 = filtering cloth; 5 = cord; 6 = edges of rubber belt.

25 mm. The advantages of belt filters are their simplicity in design compared to filters with automatic valves, and their abilities to provide countercurrent cake washing and removal of thin layers of cake. Their disadvantages include large area requirements, inefficient use of the total available filter area, and poor washing at the belt edges.

CROSS-MODE FILTERS

Filters of this group have a vertical flat or cylindrical filtering partition. In this case, filtrate may move inside the channels of the filtering elements along the surface of the filtering partition downward under gravity force action, or rise along this partition upward under the action of a pressure differential. In the separation of heterogeneous suspensions, nonuniform cake formation along the height can occur because larger particles tend to settle out first. This often results in poor cake washing due to different specific resistances over the partition height. The cake may creep down along the partition due to gravity; this is almost inevitable in the absence of a pressure gradient across the filtering partition. The vertical filtering partition makes these filters especially useful as thickeners, since it is convenient to remove cake by reverse filtrate flow.

Filter Presses

The common filter press is the plate-and-frame design, consisting of a metal frame made up of two end supports rigidly held together by two horizontal steel bars. Varying numbers of flat plates containing cloth filter media are positioned on these bars. The number of plates depends on the desired capacity and cake thickness. The plates are clamped together so that their frames are flush against each other, forming a series of hollow chambers. Figures 2.8 and 2.9 show the principal feature of a plate-and-frame filter press. The faces of the plates are grooved, either pyramided or ribbed. The entire plate is covered with cloth which forms the filtering surface. The filter cloth has holes that register with the connections on the plates and frames, so that when the press is assembled these openings form a continuous channel over the entire length of the press and register with the corresponding connections on the fixed head. The channel opens only into the interior of the frames and has no openings on the plates. At the bottom of the plates, holes are cored so that they connect the faces of the plates to the outlet cocks. As the filterable slurry is pumped through the feed channel, it first fills all of the frames. As the feed pump continues to supply fluid and build up pressure, the filtrate passes through the cloth, runs down the face of the plate, and passes out through the discharge cock. When the press is full, it is opened and dumped. Cake cannot be washed in these units and therefore the discharge contains

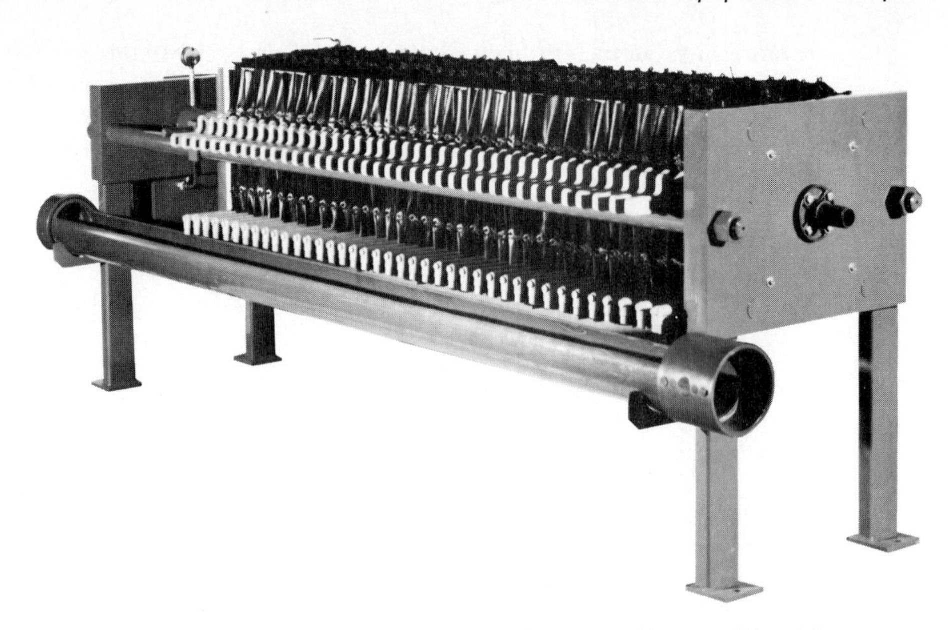

Figure 2.8 Manual recessed chamber plate filter press. *(Courtesy of Netzsch Inc., 119 Pickering Way, Exton, PA)*

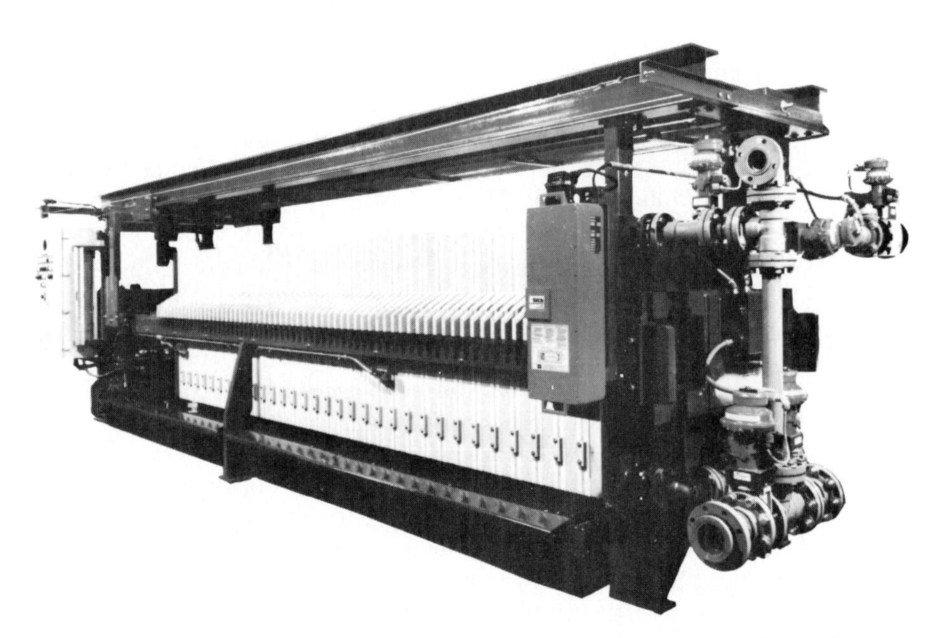

Figure 2.9 Fully automatic recessed chamberplate press. *(Courtesy of Netzsch Inc., 119 Pickering Way, Exton, PA)*

a certain amount of cake with whatever valuable or undesirable material it may contain.

Each plate discharges a visible stream of filtrate into the collecting launder. Hence, if any cloth breaks or runs cloudy, that plate can be shut off without spoiling the entire batch.

If the solids are to be recovered, the cake is usually washed. In this case, the filter has a separate wash feed line and the plates consist of washing and nonwashing types arranged alternately, starting with the head plate as the first nonwashing plate. The wash liquor moves down the channels along the side of each washing plate, and moves across the filter cake to the opposite plate and drains toward the outlet. This is illustrated in Figure 2.10. To simplify assembly, the nonwashing plates are marked with one button and the washing plates with three buttons. The frames carry two buttons.

In open-delivery filters the cocks on the one-button plates remain open and those on the three-button plates are closed. In closed-delivery filters a separate wash outlet conduit is provided. Figure 2.11 illustrates the basic design of a frame, a nonwashing plate and a washing plate. These plates and frames in closed-delivery filters are shown in Figure 2.12.

In terms of initial investment and floor area requirements, plate-and-frame filters are inexpensive in comparison to other filters. They can be operated at full capacity (all frames in use) or at reduced capacity by blanking off some of the frames by dummy plates. They can deliver reasonably well

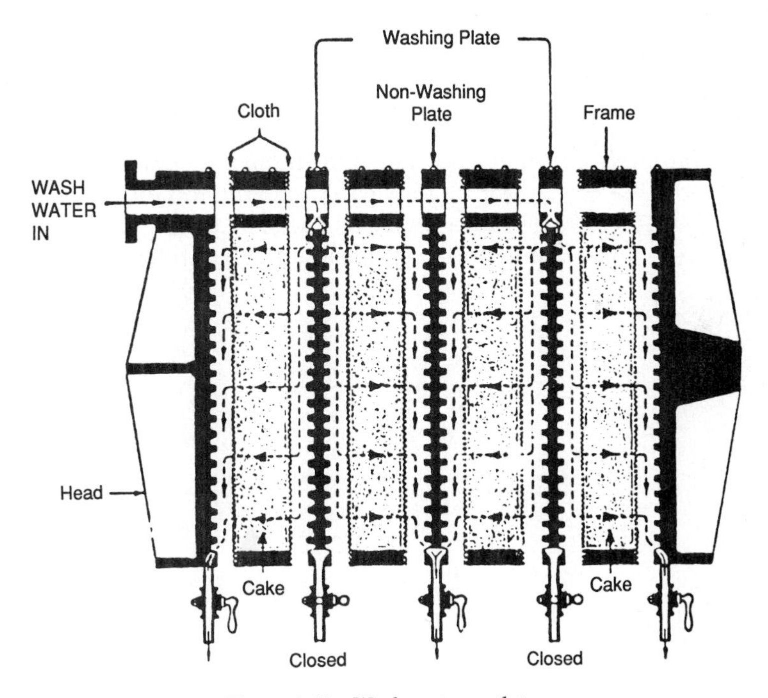

Figure 2.10 Wash water outlet.

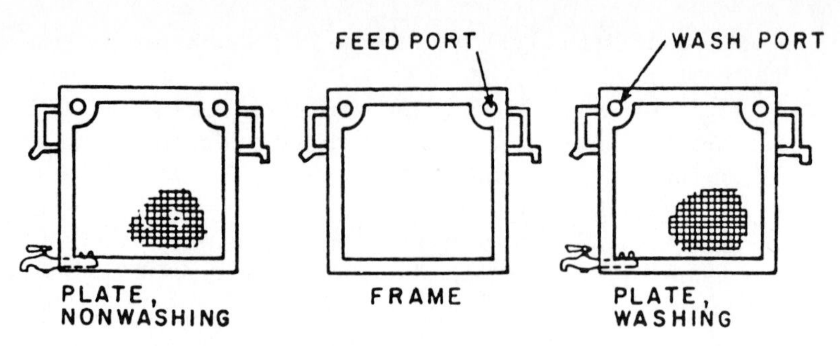

Figure 2.11 Plates and frames of an open-delivery through-washing filter.

washed and relatively dry cakes. However, the combination of labor charges for removing the cakes and fixed charges for downtime may constitute a high percentage of the total cost per operating cycle.

Figures 2.13 and 2.14 illustrate other common plate-and-frame press designs. Figure 2.13 shows a two-stage roll-over model, offering either a filter with top or bottom inlet which can be reversed in position for completely draining liquid and completely drying cake. This unit is also excellent for thoroughly washing cake while preventing channeling. The unit is designed to be placed in an autoclave for complete sterilization.

Figure 2.14 shows a square two-stage machine for submicrometer sterilization of liquids. This unit can also be used as two separate single-stage machines with an auxiliary pump, but no blank divider is necessary.

Leaf Filters

Leaf filters are similar to plate-and-frame filters in that a cake is deposited on each side of the leaf (Figure 2.15), and the filtrate flows to the outlet in the channels provided by a coarse drainage screen in the leaf between the cakes. The leaves are immersed in the sludge when filtering and in the wash liquid

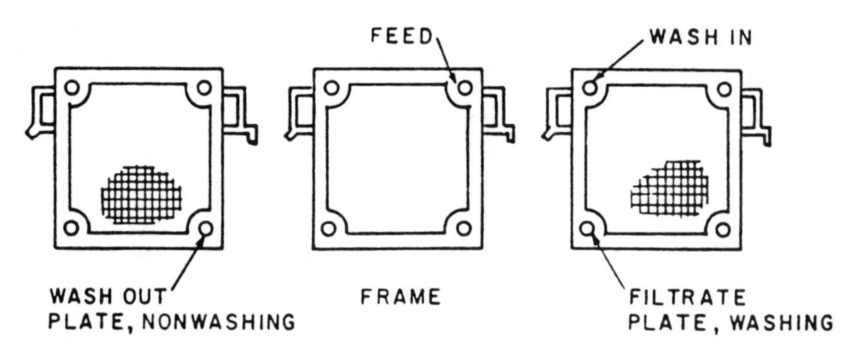

Figure 2.12 Plates and frames of a closed-delivery through-washing filter.

Figure 2.13 Two-stage roll-over plate-and-frame filter press. *(Courtesy of SciTech Publishers, Inc., Morganville, N.J.)*

when washing. Therefore, the leaf assembly may be enclosed in a shell, as in pressure filtration, or simply immersed in sludge contained in open tanks, as in vacuum filtration.

In operating a pressure leaf filter, the sludge is fed under pressure from the bottom and equally distributed. The clear filtrate from each leaf is collected in a common manifold and carried away. In filters with an external filtrate manifold (Figure 2.16), the filtrate from each leaf is visible through a respective sightglass. This not possible when the leaves are mounted on a hollow shaft that serves as an internal filtrate collecting manifold (such as the Vallez rotary leaf filter). The filter cakes are built on each side of the leaves and filtration is continued until the required cake thickness is achieved. For washing the excess sludge is usually drained, simultaneously admitting compressed air (3 to 5 lb pressure), which serves mainly to prevent the cake from peeling off the leaves.

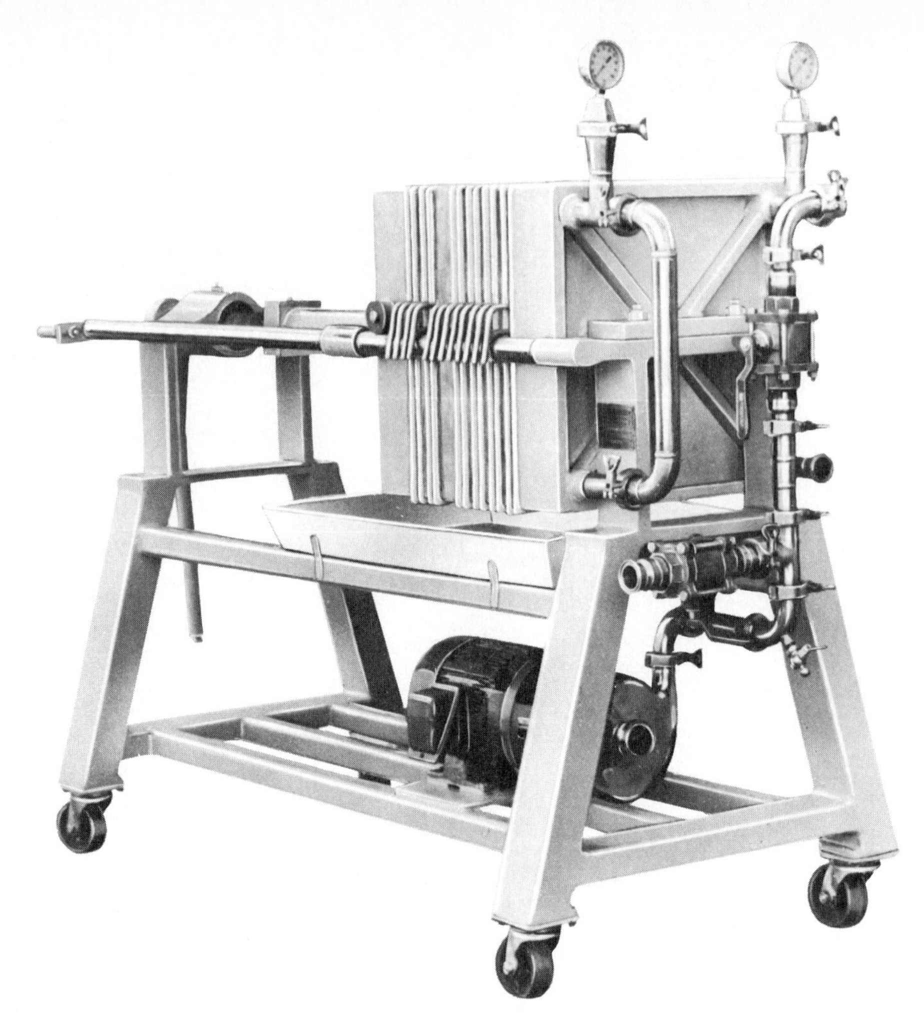

Figure 2.14 Square two-stage plate-and-frame filter press. *(Courtesy of SciTech Publishers, Inc., Morganville, N.J.)*

Disk Filters

Disk filters consist of a number of concentric disks mounted on a horizontal rotary shaft and operate on the same principle as rotary drum vacuum filters. The basic design is illustrated in Figure 2.17. The disks are formed by using V-shaped hollow sectors assembled radially around the shaft. Each sector is covered with filter cloth and has an outlet nipple connected to a manifold extending along the length of the shaft and leading to a port on the filter valve. Each row of sectors is connected to a separate manifold. The sludge

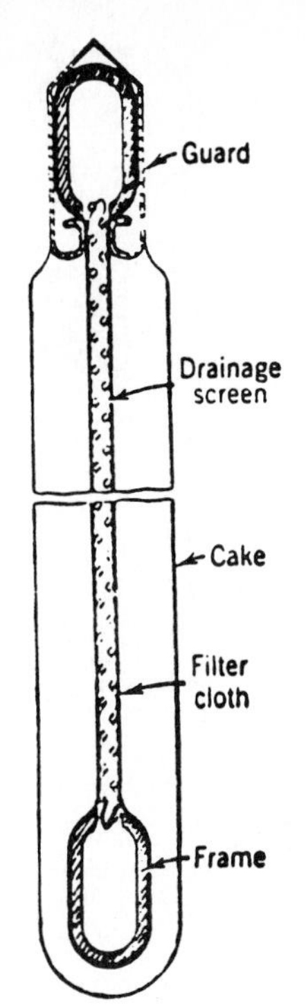

Figure 2.15 Sectional view of a filter leaf showing construction and approximate location of cake.

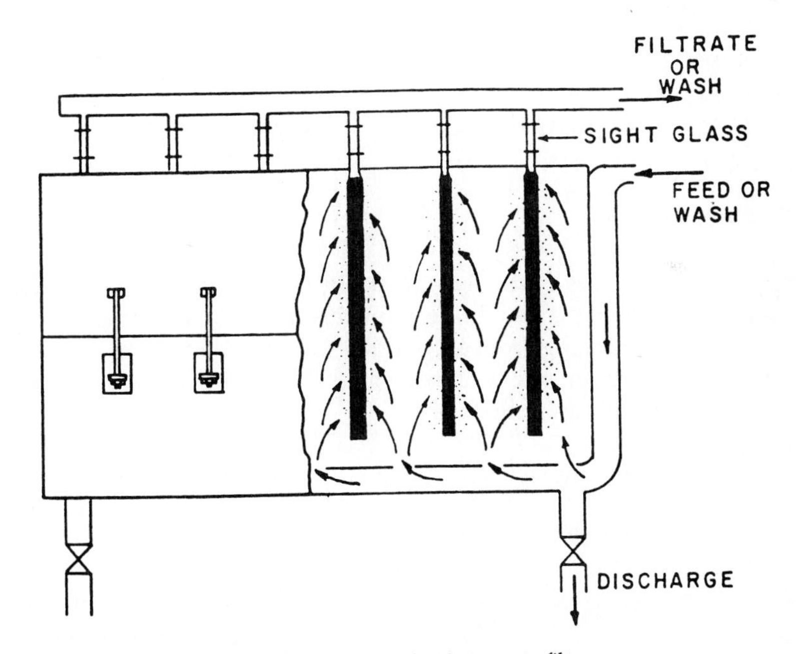

Figure 2.16 Sweetland pressure filter.

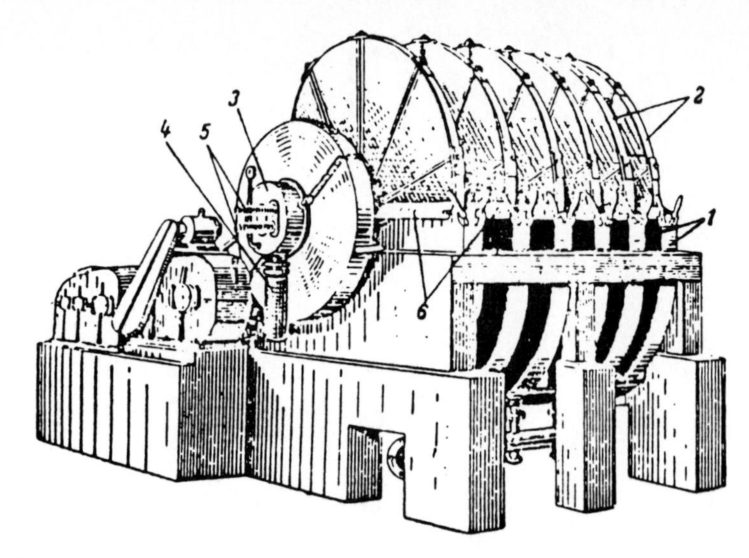

Figure 2.17 Rotary-disk vacuum filter, 1 = section; 2 = filtering disks; 3 = automatic valve; 4 = manifold for vacuum and filtrate discharge; 5 = piping for compressed air; 6 = doctor's knives for cake removal.

level in the tank should provide complete submergence to the lowest sector of the disks.

Compared to drum vacuum filters, the greatest advantage of the disk filter is that for the same filtering area, it occupies considerably less floor space. However, because of vertical filtering surfaces, cake washing is not as efficient as when a drum filter is used. The disk filter is ideal when the cake is not washed and floor space is at a premium.

CARTRIDGE FILTERS

Cartridge filters normally operate in the countercurrent mode, but because of their extensive use throughout the chemical and process industries in applications ranging from laboratory-scale to commercial operations for flows extending to an excess of 5,000 gal/min, a separate discussion is warranted.

Typical applications are:

- remove undispersed solids
- remove precipitated solids
- protect catalyst beds
- protect instruments
- remove DE filter carryover
- filter poultry and meat wash water
- remove oversize particles from slurries
- clean electrolytic solutions
- filter waste oil for reuse
- remove plastic fines from water

- keep spray nozzles open
- filter recirculating water
- remove particles from coatings
- filter cooling-tower water
- remove char particles
- filter condensate
- filter bottle and can wash water
- filter scrubber water
- filter boiler feed water
- filter pump seal water
- protect reverse osmosis systems
- protect chiller and air conditioners
- remove pulp from juices

Industrial applications of cartridge filters follow. The chemical industry uses cartridge filters to handle:

- acetic acid
- calcium carbonate
- brine
- ethylene glycol
- herbicides
- hydrochloric acid
- latexes
- resins
- polymers
- sulfuric acid
- cooling-tower water
- pelletizer water

The food industry uses cartridge filters to handle:

- corn syrup
- dextrose
- lard
- jelly
- juices
- milk sugar
- edible oils
- tea liquor
- city and well water
- extracts
- chocolates
- soybean concentrate
- peanut butter

The paper industry uses cartridge filters to handle:

- pigmented coatings
- white water
- fresh water
- size
- starch
- TiO_2 slurry
- mill water
- dyes
- cooling water
- pump seal water
- decker shower water
- wet end additives
- clay slurry

The petroleum industry uses cartridge filters to handle:

- amine
- feed stocks
- reduced crudes
- naphtha
- hydraulic oil
- injection fluids
- completion fluids
- cooling-tower water

- fuel oil
- motor oil
- pump seal water
- synthetic lubricants

Miscellaneous industrial uses of cartridge filters include:

- adhesives
- resins
- solvents
- paints
- shampoo
- dyes
- cooling water
- pharmaceuticals
- beverages
- toothpaste
- liquors
- beer

Table 2.1 gives typical filtration ranges encountered by industry. Early designs, still widely used, consist of a series of thin metal disks that are 3-in to 10-in diameter, set in a vertical stack with very narrow uniform spaces between them. The disks are supported on a vertical hollow shaft and fit into a closed cylindrical casing. Liquid is fed to the casing under pressure, where it flows inward between the disks to openings in the central shaft and out through the top of the casing. Solids are captured between the disks and remain in the filter. Since most of the solids are removed at the periphery of the disks, the unit is referred to as an edge filter. The accumulated solids are periodically removed from the cartridge.

More recent designs are simpler, experience lower pressure drop, and have fewer maintenance problems. Figure 2.18 illustrates the operating principle behind one type of design. Unfiltered liquid enters the inlet (bottom) port. It flows upward, around, and through the filter medium, which is a stainless steel or fabric screen reinforced by a perforated stainless steel backing. Filtered liquid discharges through the outlet (top) port. Because of the outside-to-inside flow path, solids deposit onto the outside of the element, so screens are easy to clean. This external gasketing design prevents solids from bypassing the filter and contaminating the process downstream. Principal advantages are no Q-ring seals that can crack, no channeling media that can fail, and no cartridges that can collapse or allow bypassing.

As with any filter, careful media selection is essential. Media that are too coarse, for example, will not provide the needed protection. However, specifying finer media than necessary can add substantially to both equipment and operating costs. Factors to be considered in media selection include solids content, type of contaminant, particle size and shape, amount of contaminant to be removed, viscosity, corrosiveness, abrasiveness, adhesive qualities, liquid temperature, and required flow rate. Typical filter media are wire mesh (typically 10 to 700 mesh), fabric (30 mesh to 1 μm), slotted screens (10 mesh to 25 μm) and perforated stainless steel screens (10 to 30 mesh). Table 2.2 provides typical particle retention sizes for different media.

Figure 2.19 shows single-unit filters. Single filters may be piped directly into systems requiring batch or intermittent service. Using quick-coupling

TABLE 2.1 TYPICAL FILTRATION RANGES

Industry and liquid	Typical filtration range
Chemical Industry	
Alum	60 mesh–60 μm
Brine	100–400 mesh
Ethyl Alcohol	5–10 μm
Ferric Chloride	30–250 mesh
Herbicides/Pesticides	100–700 mesh
Hydrochloric Acid	100 mesh to 5–10 μm
Mineral Oil	400 mesh
Nitric Acid	40 mesh to 5–10 μm
Phosphoric Acid	100 mesh to 5–10 μm
Sodium Hydroxide	1–3 to 5–10 μm
Sodium Hypochlorite	1–3 to 5–10 μm
Sodium Sulfate	5–10 μm
Sulfuric Acid	250 mesh to 1–3 μm
Synthetic Oils	25–30 μm
Drugs and Cosmetics	
Acetic Acid	40–150 mesh
Aerosol	60–200 mesh
Bath Oil	400–700 mesh
Citric Acid	60 mesh to 1–3 μm
Glycerine	5–10 μm
Lipstick	60–150 mesh
Shampoo	100–250 mesh
Soap	10–250 mesh
Suntan Lotion	15–20 μm
Tallow	700 mesh to 25–30 μm
Toothpaste	100 mesh
Food and Beverage	
Apple Juice	5–10 μm
Beer	250–400 mesh
Brine	400 mesh to 15–20 μm
Chocolate	10–400 mesh
Corn Syrup	80 mesh to 5–10 μm
Fructose Syrup	5–10 to 25–30 μm
Fruit Juices with Pulp	10–100 mesh
Jelly	700 mesh
Lard	500 mesh to 5–10 μm
Lemon Effluent	60–150 mesh
Liquors	700 mesh to 15–20 μm
Vegetable Oil	150 mesh to 5–10 μm
Wash Water	20–250 mesh
Petroleum Industry	
Atmospheric Reduced Crude	25–75 μm
Completion Fluids	200 mesh to 1–3 μm
DEA	250 mesh to 5–10 μm
Deasphalted Oil	200 mesh
Decant Oil	60 mesh
Diesel Fuel	100 mesh

(*continued*)

TABLE 2.1 TYPICAL FILTRATION RANGES (*continued*)

Industry and liquid	Typical filtration range
Gas Oil	25–75 μm
Gasoline	1–3 μm
Hydrocarbon Wax	25–30 μm
Isobutane	250 mesh
MEA	200 mesh to 5–10 μm
Naphtha	25–30 μm
Produced Water for Injection	1–3 to 15–20 μm
Residual Oil	25–50 μm
Seawater	5–10 μm
Steam Injection	5–10 μm
Vacuum Gas Oil	25–75 μm
Pulp and Paper	
Calcium Carbonate	30–100 mesh
Clarified White Water	30–100 mesh
Dye	60–400 mesh
Fresh Water	30–200 mesh
Groundwood Decker Recycle	20–60 mesh
Hot Melt Adhesives	40–100 mesh
Latex	40–100 mesh
Mill Water	60–100 mesh
Paper Coating	30–250 mesh
River Water	20–400 mesh
Starch Size	20–100 mesh
Titanium Dioxide	100–200 mesh
All Industries	
Adhesives	30–150 mesh
Boiler Feed Water	5–10 μm
Caustic Soda	250 mesh
Chiller Water	200 mesh
City Water	500 mesh to 1–3 μm
Clay Slip (ceramic and china)	20–700 mesh
Coal-Based Synfuel	60 mesh
Condensate	200 mesh to 5–10 μm
Coolant Water	500 mesh
Cooling-Tower Water	150–250 mesh
Deionized Water	100–250 mesh
Ethylene Glycol	100 mesh to 1–3 μm
Floor Polish	250 mesh
Glycerine	5–10 μm
Inks	40–150 mesh
Liquid Detergent	40 mesh
Machine Oil	150 mesh
Pelletizer Water	250 mesh
Phenolic Resin Binder	60 mesh
Photographic Chemicals	25–30 μm
Pump Seal Water	200 mesh to 5–10 μm
Quench Water	250 mesh
Resins	30–150 mesh
Scrubber Water	40–100 mesh
Wax	20–200 mesh
Well Water	60 mesh to 1–3 μm

Figure 2.18 Operating principle behind a single-cartridge filter.

TABLE 2.2 TYPICAL FILTER RETENTIONS

	Mesh or mesh equivalent	Nominal particle retention		Percentage of open area
		(in.)	(μg)	
Wire Mesh	10	0.065	1,650	56
	20	0.035	890	46
	30	0.023	585	41
	40	0.015	380	36
	60	0.009	230	27
	80	0.007	180	32
	100	0.0055	140	30
	150	0.0046	115	37
	200	0.0033	84	33
	250	0.0024	60	36
	400	0.0018	45	36
	700	0.0012	30	25
Perforated	10	0.063	1,575	15
	20	0.045	1,125	18
	30	0.024	600	12
Slotted	10	0.063	1,600	50
	15	0.045	1,140	43
	20	0.035	890	36
	30	0.024	610	30
	40	0.015	380	20
	60	0.009	230	18
	80	0.007	180	25
	100	0.006	150	13
	120	0.005	125	11
	150	0.004	75	9
	200	0.003	50	7
	325	0.002	25	5
		0.001		3
Fabric	60	0.009	230	NA[a]
	80	0.007	180	NA
	100	0.0055	140	NA
	150	0.0046	115	NA
	250	0.0024	60	NA
	500	0.0016	40	NA
		0.0010–0.0012	25–30	NA
		0.0006–0.0008	15–20	NA
		0.0002–0.0004	5–10	NA
		0.00004–0.00012	1–3	NA

[a] NA = percentage of open area not applicable to fabric media.

connectors, the media can be removed from the housing, inspected, or cleaned. Also, filtering elements are interchangeable. Hence, while one is being cleaned, another can be placed in service. Table 2.3 gives typical dimensions for the units shown in Figure 2.19. Figure 2.20 shows a plot of differential pressure versus flow for several models of one manufacturer.

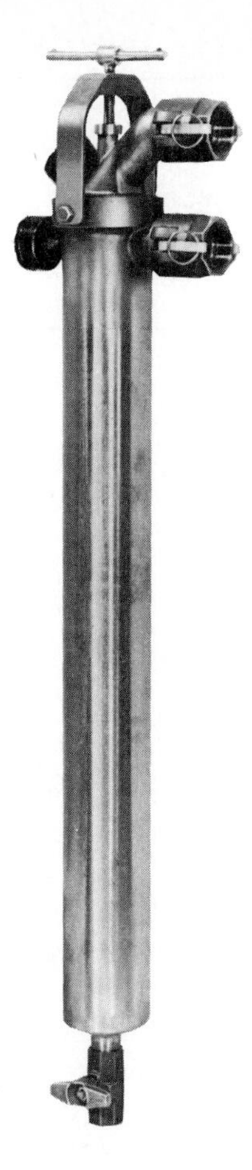

Figure 2.19 Typical single-cartridge filter units.

Multiple filters are also common, consisting of two or more single-filter units valved in parallel to common headers, as shown in Figure 2.21. The distinguishing feature of these filters is the ability to sequentially backwash each unit in place while the others remain on stream. Hence, these systems are essentially continuous filters. These units can be fully automated to eliminate manual backwashing. Backwashing can be controlled by changes in differential pressure between the inlet and outlet headers. One possible arrangement consists of a controller and solenoid valves that supply air signals to pneumatic valve actuators on each individual filter unit. As solids collect on the filter elements, flow resistance increases. This increases the pressure differential across the elements and thus between inlet and outlet headers on the system. When the pressure drop reaches a preset level, an adjustable differential pressure switch relays information through a programmer to a set of solenoid valves, which in turn sends an air signal to the pneumatic valve actuator. This rotates the necessary valve(s) to backwash the first filter element. When the first element is cleaned and back on stream, each suc-

TABLE 2.3 TYPICAL DIMENSIONS OF SINGLE-FILTER UNITS

Body size (in.)	Inlet/outlet diameter NPTI	Filter element		Straight-through filter dimensions (in.)		Standard filter dimensions (in.)	
		Dimensions (in.)	Area (in.^{2})	A	B	C	D
$2\frac{1}{2}$	1	2×12	75	$24\frac{1}{16}$	$19\frac{7}{16}$	$24\frac{11}{16}$	4
$2\frac{1}{2}$	1	2×18	112	$30\frac{1}{16}$	$25\frac{7}{16}$	$30\frac{11}{16}$	4
3	$1\frac{1}{2}$	$2\frac{1}{4} \times 24$	168	$38\frac{5}{16}$	$32\frac{15}{16}$	$37\frac{7}{8}$	$4\frac{5}{8}$
3	$1\frac{1}{2}$	$2\frac{1}{4} \times 36$	256	$50\frac{5}{16}$	$44\frac{15}{16}$	$49\frac{7}{8}$	$4\frac{5}{8}$
4	2	$3\frac{1}{4} \times 18$	182	$32\frac{7}{8}$	$26\frac{1}{2}$	$34\frac{3}{16}$	$5\frac{3}{16}$
4	2	$3\frac{1}{4} \times 36$	364	$51\frac{5}{16}$	$44\frac{15}{16}$	$52\frac{13}{16}$	$5\frac{3}{16}$
4	3	$3\frac{1}{4} \times 18$	182	$33\frac{9}{16}$	$26\frac{1}{2}$		
4	3	$3\frac{1}{4} \times 36$	364	52	$44\frac{15}{16}$		

cessive filter element is backwashed in sequence until they are all cleaned. The programmer is then automatically reset until the rising differential pressure again initiates the backwashing cycle.

Filter cartridges or tubes are made from a variety of materials. Common designs are natural or synthetic fiber wound over a perforated plastic or metal core. A precision winding pattern covers the entire depth of the filter tube with hundreds of funnel-shaped tunnels, which become gradually finer from the outer surface to the center of the tube and trap progressively finer particles as the fluid travels to the center. This provides greater solids retention capacity than is associated with surface filter media of the same dimensions. Typical cartridge materials are cotton, Dynel, polypropylene, acetate, porous stone, and porous carbon filter tubes. Supported perforated cores for cotton, Dynel,

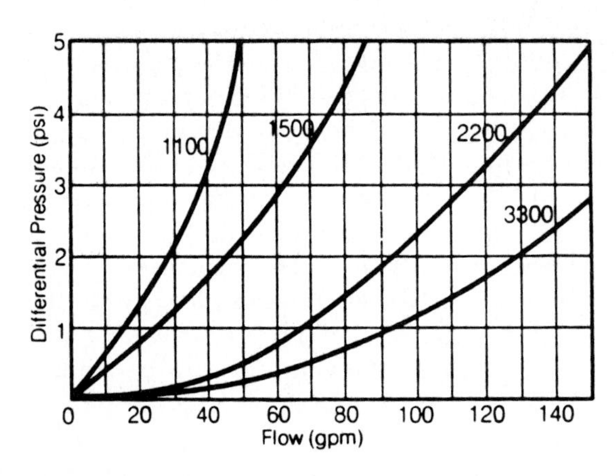

Figure 2.20 Differential pressure vs flow for several single-filter models. *(Courtesy of SciTech Publishers, Inc., Morganville, N.J.)*

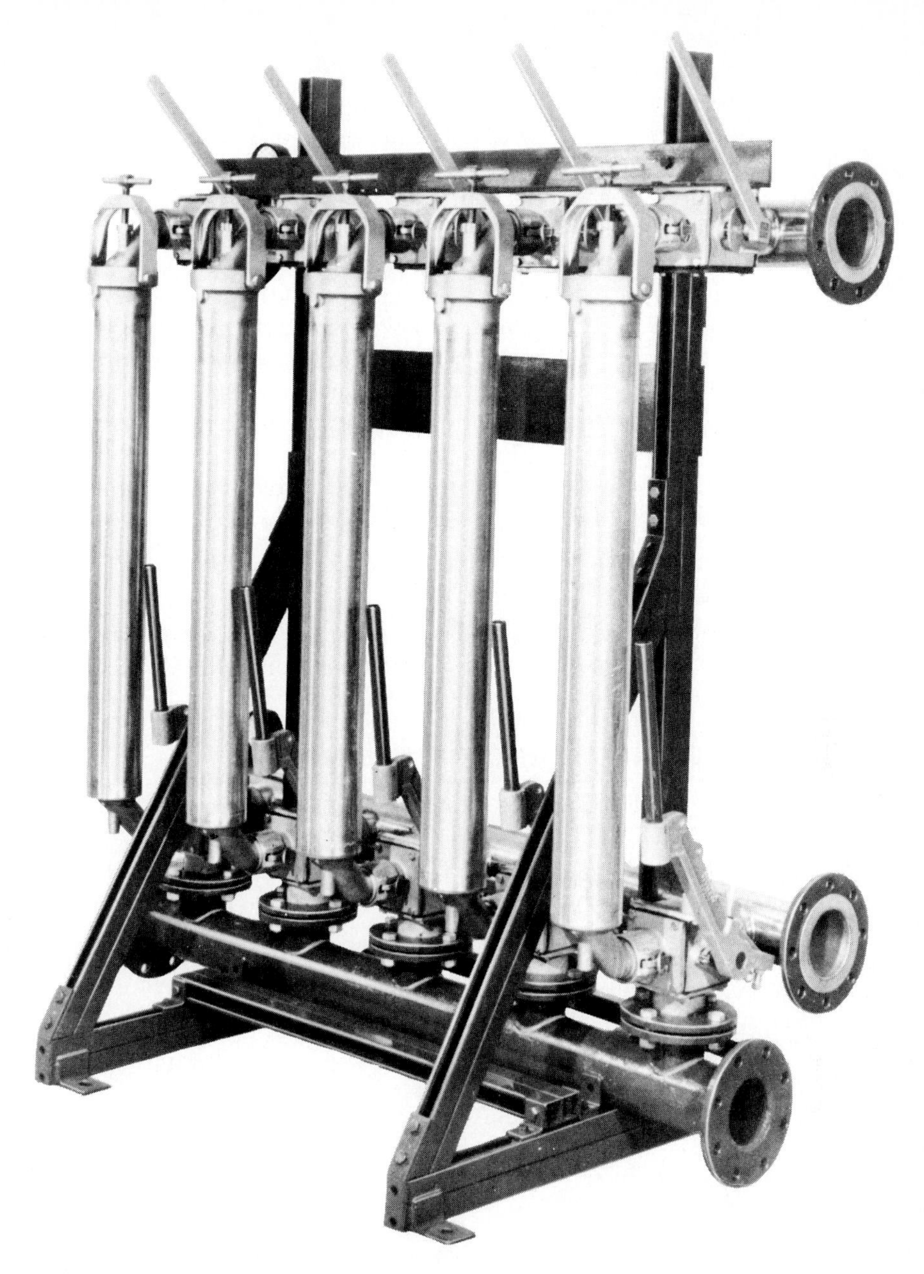

Figure 2.21 Internal and external-backwashing multiplex filtering units.

or polypropylene are stainless steel, polypropylene, or steel. Supporting cores for acetate tubes are tin-plated copper with voile liner. Porous stone and porous carbon filter tubes do not require supporting cores. Stainless steel cores are recommended for mildly acid and all alkaline solutions, pH 4 to 14. Polypropylene cores are used where all metal contact must be eliminated or where stainless steel is attacked, such as high chloride and sulfuric acid solutions. It is recommended for all acid and alkaline solutions, pH 4 to 14. Two types of polypropylene cores are available: mesh polypropylene and rigid perforated polypropylene. Mesh polypropylene is satisfactory for temperatures below 140°F. The more expensive rigid polypropylene cores are used for temperature applications over 140°F, and for double- and triple-tiered filter chambers because their greater strength is needed here. Perforated steel cores are used for dilute alkaline solutions, solvents, lacquers, oils, emulsions, and so on.

Table 2.4 can serve as a rough guide to filter cartridge selection. The following general guidelines are useful.

- Cotton filter tubes are recommended for moderately acid and alkaline solutions in the pH range 3 to 11.
- Polypropylene, Dynel, and porous carbon filter tubes are recommended for concentrated acid and alkaline solutions and for all fluoborate solutions over the entire pH range (0 to 14).

TABLE 2.4 GUIDE TO FILTER CARTRIDGE SELECTIONS

	Electroplating solutions	Filter tube (Material/core)
Acid		
Fluoborates	Cu, Fe, Pb, Sn	Polypropylene (PP) or Dynel/PP
Nonfluoborates	Cu, Sn, Zn: <6 oz/gal H_2SO_4	PP or cotton/PP
	Cu, Sn, Zn: >6 oz/gal H_2SO_4	PP or Dynel/PP
	Cr	PP or Dynel/PP
	Au, In, Rh, Pd	PP or Dynel/PP
	$FeCl_2$ (190°F)	PP/rigid PP (RPP) or porous stone
	Ni (Woods)	PP or Dynel/PP
	Ni (Watts type & bright)	PP or cotton/PP
	Ni (high-chloride)	PP or cotton/PP
	Ni (sulfamate)	PP or cotton/PP
	Electrotype Cu and Ni	PP or cotton/PP
Alkaline	Sn (stannate)	Cotton/stainless steel (SS)
Cyanide	Brass, Cd, Cu, Zn[a]	Cotton/SS, PP or Dynel/PP
	Au, In, Pt, Ag	Cotton/SS, PP/PP
Pyrophosphate	Cu, Fe, Sn, etc.	Cotton/SS or PP
Electroless	Ni plating: <140°F	Cotton/SS or PP
	Ni plating: >140°F	PP/RPP, cotton/SS
	Cu: <140°F	PP/PP
	Cu: >140°F	PP/RPP

(*continued*)

TABLE 2.4 GUIDE TO FILTER CARTRIDGE SELECTIONS (*continued*)

	Chemicals	Filter tube (material/core)
Acids	Acetic: dilute	Cotton/SS, PP/PP
	Acetic: concentrated	PP or Dynel/PP
	Boric	Cotton/SS, PP/PP
	Chromic, hydrocloric, nitric, phosphoric, sulfuric	PP or Dynel/PP, porous stone[b]
	Hydrofluoric, fluoboric	PP or Dynel/PP
Alkalies	NaOH or KOH	PP/PP
	NH_4OH: dilute	cotton/SS, PP/PP
	NH_4OH: concentrated	PP or Dynel/PP
Misc. Chemicals	Biological solutions	Cotton/SS, PP/PP, porous stone[b]
	Electropolishing solutions	Porous stone, PP/PP
	Pharmaceutical solutions	Cotton/SS, PP/PP, porous stone[b]
	Photographic solutions	Cotton/SS, PP/PP
	Radioactive solutions	Cotton/SS, porous stone[b]
	Ultrasonic cleaning solutions	Cotton special B compound/SS
	Nickel acetate (190°F)	Cotton/SS
	Food products	Cotton/SS, PP/PP
Organic Liquids	CCl_4	Cotton/steel or SS
	Dichloroethylene	Cotton/steel or SS
	Hydraulic fluids	Cotton/steel or SS
	Lacquers	Cotton/steel or SS
	Per- and trichloroethylene	Cotton/steel or SS
	Solvents	Cotton/steel or SS
Petroleum Products	Fuel oil, diesel, kerosene, gasoline, lube oil	Cotton/steel or SS

[a] When operated as high-speed baths at high temperatures (>140°F) or with high alkali content, use PP or Dynel/PP.

[b] Porous stone is recommended for all acids except hydrofluoric and fluoboric.

- Polypropylene filter tubes are also recommended for electropolishing solutions, as well as certain other highly corrosive solutions.
- Porous stone filter tubes are recommended for concentrated acid solutions.
- Acetate filter tubes are recommended for water.

STRAINER AND FILTER BAG BASKETS

Strainer filter baskets and filter bag baskets are used as prefiltering devices. This prestraining or prefiltering stops the larger contaminated particles and thus extends the life of the entire system.

Single-stage strainers and bag filters differ only in the basket design. Strainer baskets have solid flat bottoms, and baskets for filter bags have perforated bottoms to accept standard-sized filter bags.

Dual-stage straining/filtering action is achieved by insertion of a second inner basket. It is supported on the top flange of the outer basket. Both baskets can be strainers (with or without wire mesh linings) or both can be baskets for filter bags. They may also be a combination: one a strainer basket, the other a filter bag basket. Dual-stage action increases strainer or filter life and reduces servicing needs.

Figure 2.22 shows details of the basket seal, which prevents unfiltered liquid from bypassing the strainer or filter bag basket. The seal is maintained during operation by a hinged basket bail handle being held down under the closed cover, which holds the basket down against a positive stop in the housing.

There are a variety of strainer and filter bag basket arrangements. Figure 2.23 shows different single-stage and double-stage basket units.

Fabric bag filter baskets are capable of providing removal ratings from 20 mesh to nominal 1 μm, for both Newtonian and viscous liquids. Wire mesh or fabric baskets can be cleaned and reused in many applications, or are disposable when cleaning is not feasible. Side-entry models feature permanent flanged connections for line pressures to 150 psi. These filters are fabricated to American Society of Mechanical Engineers (ASME) codes for applications that must comply with piping standards established in many processing plants.

Top-entry models feature the inlet connection as an integral part of the lid. The inlet can be equipped with different types of quick disconnects for fast basket removal.

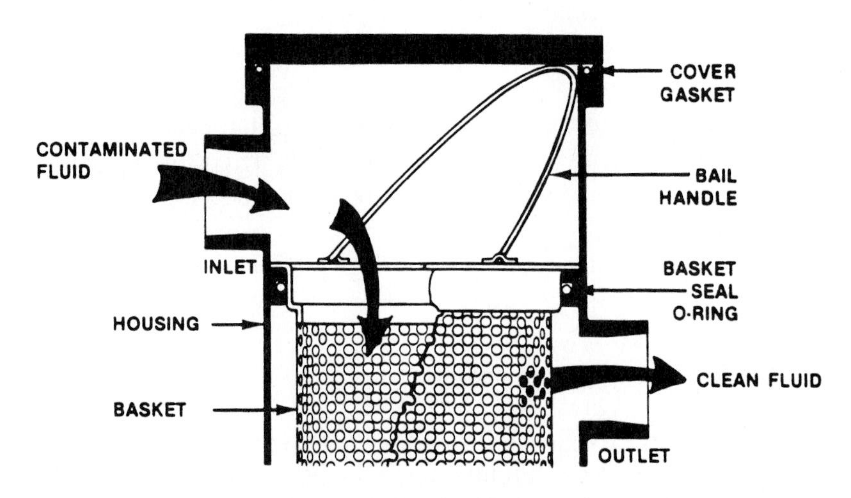

Figure 2.22 Details of basket seal.

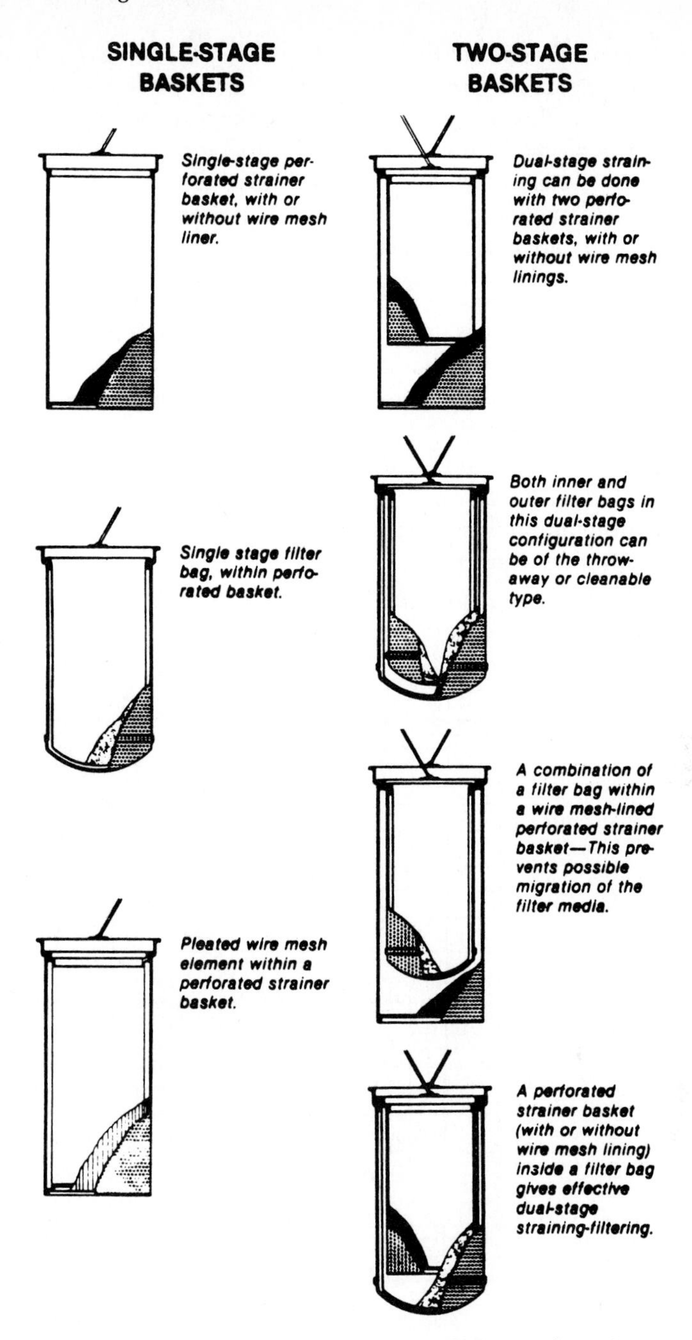

Figure 2.23 Different one- and two-stage basket configurations.

Strainers should be selected so that the pressure drop incurred does not exceed a specified limit with a clean strainer basket (typically 2 psi). Pressure drop versus flow capacity curves for basket strainers are given in Figure 2.24. This plot provides gross pressure drop for different capacities of water flow at suitable strainer pipe sizes. The value obtained must be corrected on the basis of the actual fluid viscosity and strainer opening size to be used. These corrections are given in Table 2.5 and the procedure is as follows:

1. Enter the pressure drop value from the bottom scale with the specified flow rate. Read up to where its vertical line intersects the diagonal representing a strainer pipe size that gives a reasonable pressure drop, which is found by following the horizontal line to the pressure drop scale at the left.
2. To correct this figure to match the actual fluid viscosity (and strainer media choice), use Table 2.5. Read down the appropriate viscosity column and across from the appropriate strainer media description to find the correction factor.
3. Multiply the pressure drop figure found in step 1 by the factor found in step 2 to obtain the adjusted pressure drop.

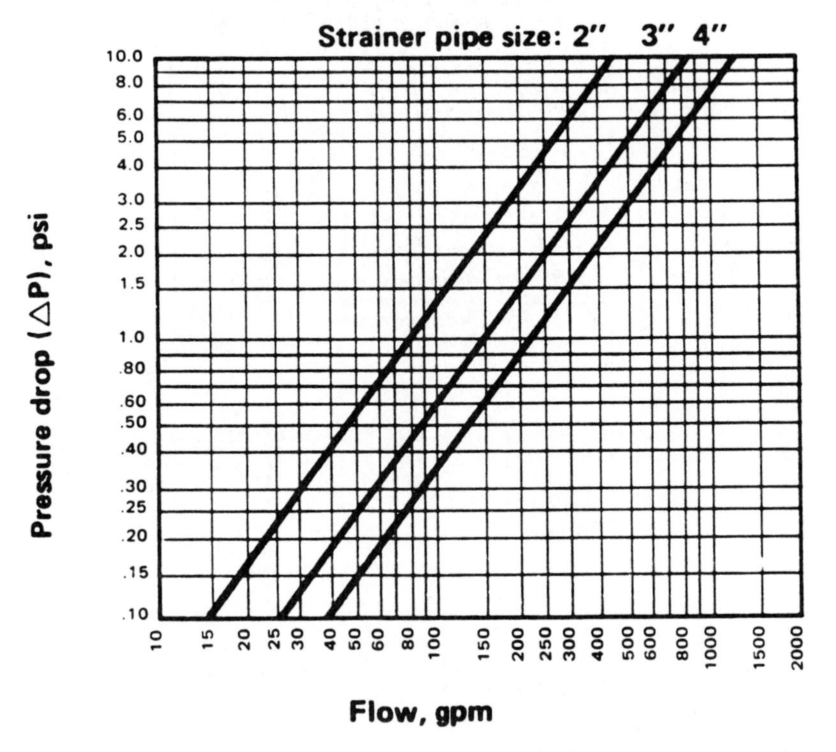

Figure 2.24 Pressure drop vs. flow capacity for basket strainers.

TABLE 2.5 VISCOSITY CORRECTION FACTORS TO BE USED WITH FIGURE 2.24

	Viscosity (cP)								
	1[a]	50	100	200	400	600	800	1,000	2,000
All unlined baskets, with or without pleated inserts	0.65	0.85	1.00	1.10	1.20	1.40	1.50	1.60	1.80
40-mesh lined	0.73	0.95	1.20	1.40	1.50	1.80	1.90	2.00	2.30
60-mesh lined	0.77	1.00	1.30	1.60	1.70	2.10	2.20	2.30	2.80
80-mesh lined	0.93	1.20	1.50	1.90	2.10	2.40	2.60	2.80	3.50
100-mesh lined	1.00	1.30	1.60	2.20	2.40	2.70	3.00	3.30	4.40

[a] H_2O.

4. If you are using a mesh-lined (not pleated) strainer basket, the pressure drop can be lowered by using a basket 30 in. deep instead of the one that is 15 in. deep. Divide the pressure drop figure from step 3 by 15.

DIAPHRAGM FILTERS

Diaphragm filters are specially designed filter presses. They have the ability to reduce sludge dewatering costs by a squeezing cycle using a diaphragm. Instead of the conventional plate-and-frame unit in which constant pumping pressure is used to force the filtrate through the cloth, diaphragm filters combine an initial pumping followed by a squeezing cycle that can reduce the process cycle time by as much as 80 percent [11].

The operating cycles for this design are illustrated in Figure 2.25. During the filtration cycle, sludge is fed at approximately 100 psi into each chamber through an inlet pipe in the bottom portion of the filter plate. The number of chambers can range from a few dozen to more than 100. The sludge feed pump continues to feed sludge into the chamber until a predetermined filtering time has been achieved. Filtrate passes through the cloth on both sides of the chamber. The filtration cycle is completed independently in each chamber. Short filtration cycles produce cake thicknesses of 0.5 in. to 0.75 in. (12.7 mm to 19.1 mm).

Once the filtration cycle is complete, the sludge pump is stopped and a diaphragm in the chamber is expanded by water pressurized up to 213 psi. This compresses the sludge on both sides of the chamber into a thin, uniform cake with a solids content of more than 35 percent. The uniform water content of the thin cake (no wet cores) results in easier shredding and conveying and makes it much more adaptable to self-sustained thermal destruction or land-

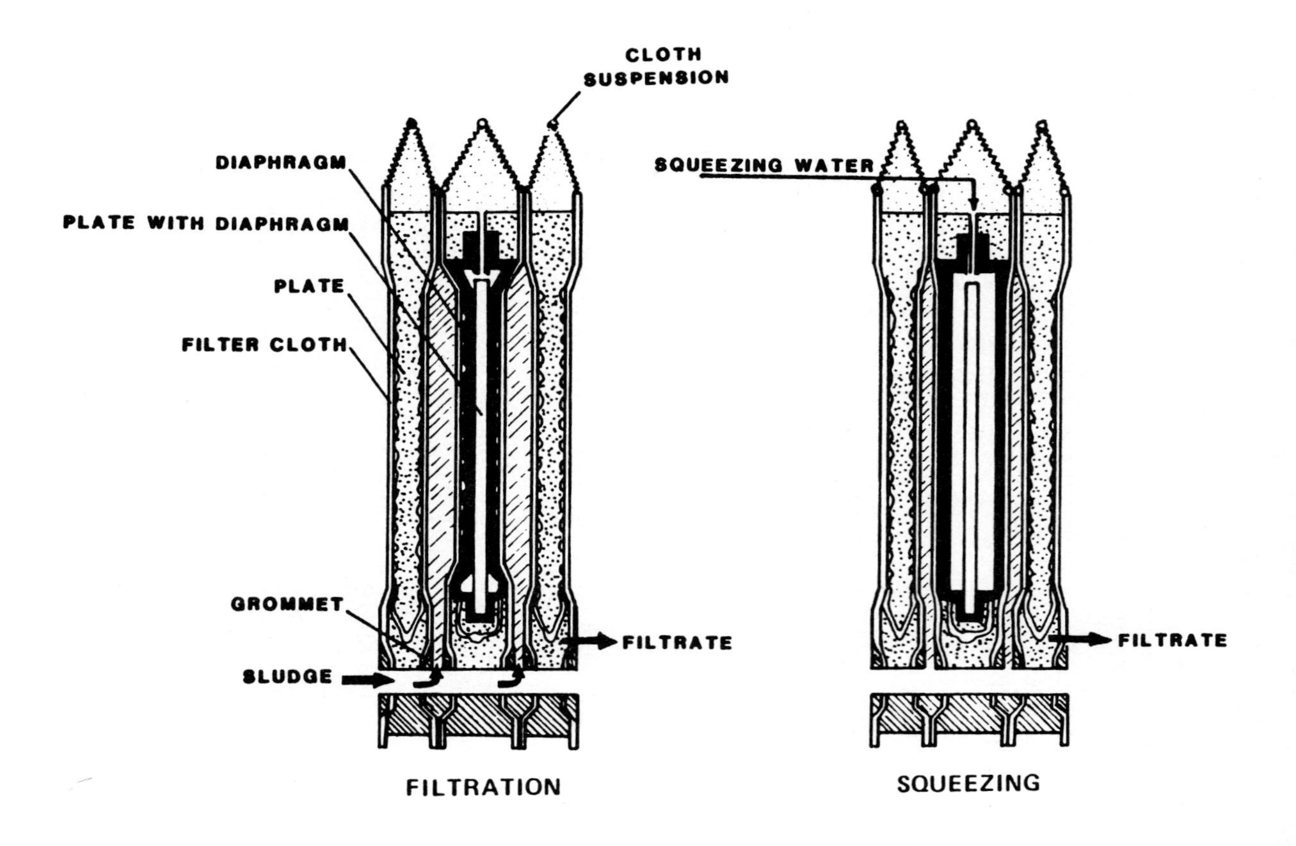
CLOTH SUSPENSION
DIAPHRAGM
PLATE WITH DIAPHRAGM
PLATE
FILTER CLOTH
GROMMET
SLUDGE
FILTRATE
SQUEEZING WATER
FILTRATE
FILTRATION
SQUEEZING

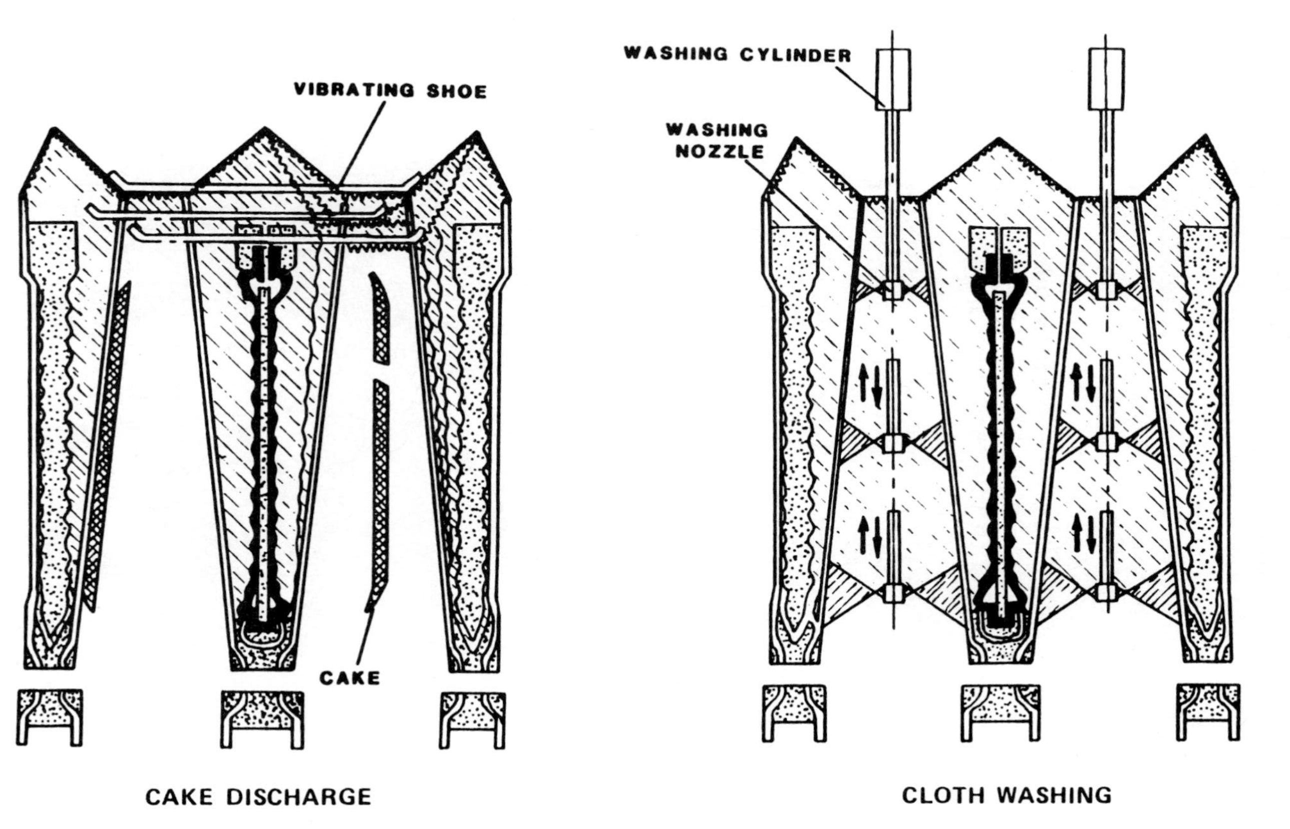

Figure 2.25 Operating cycles of a diaphragm press.

fill. Optimum filtering and squeezing time cycles vary, depending on the type of sludge, and can be determined accurately by bench tests. Squeezing water is recycled. A hydraulic ram keeps the chambers in position during both cycles.

On completion of the filtration and squeeze cycles, the chambers are automatically opened and the cakes are discharged, usually onto a belt conveyor. No precoating is required. Two chambers are normally opened at a time in sequence. This reduces the impact loading on the belt conveyor. Any sludge or filtrate remaining in the feed and filtrate lines is automatically purged by high-pressure (100 psi) air before the next cycle begins. This purging prevents wet sludge from discharging and keeps sludge lines from plugging.

Cake discharge from filter presses is fast. After a number of cycles (depending on the sludge type), the filter cloth will require cleaning. This can be accomplished manually or can be performed automatically at preset frequencies with an automatic cloth washer using a jet of 1,000-psi wash water.

Where even faster cake discharge is desired or where sludge cakes may tend to be sticky, automatic cloth vibrators can be provided. These units help speed mechanical discharge and help remove cakes where poor sludge conditioning causes excessive sticking. This reduces the need for continuous monitoring by operations personnel. Cloth vibrators also simplify cloth selection, since cloths can be selected to assure clearer filtrate or better filtering qualities rather than sacrificing these advantages for a cloth that allows for better discharge characteristics. Cake discharge is illustrated in Figure 2.26.

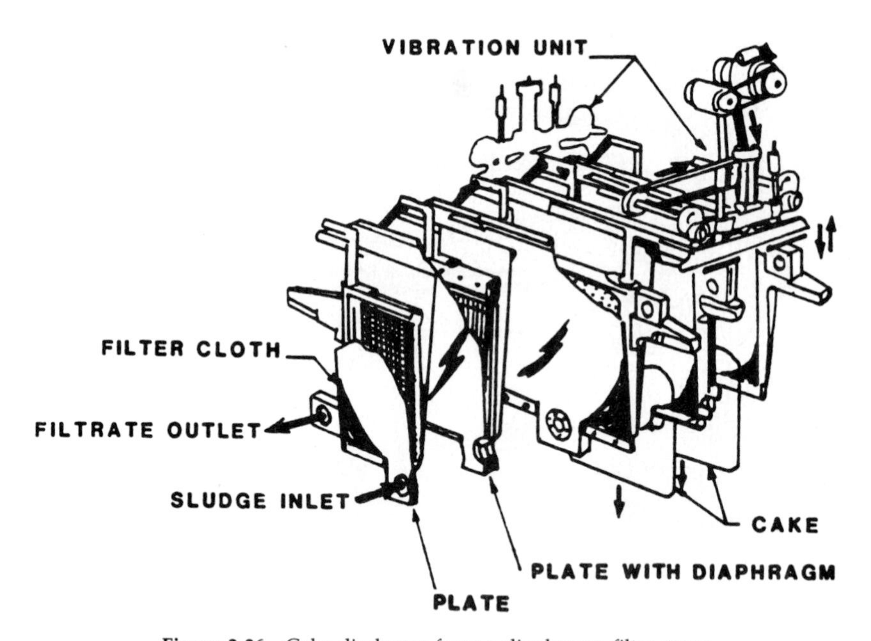

Figure 2.26 Cake discharge from a diaphragm filter press.

TABLE 2.6 CAPACITIES AND DIMENSIONS FOR DIAPHRAGM PRESSES

Number of chambers[a]	Filter area[b]		Cake volume at 0.75-in. thickness		Weight[c]		Length[d]	
	(m^2)	(ft^2)	(liters)	(ft)	(kg)	(tons)	(mm)	(ft-in.)
30	115	1,237	1,002	36	53,200	58	7,355	24-1
40	154	1,656	1,337	47	59,400	65	7,975	26-2
52	200	2,152	1,736	61	67,800	75	8,720	28-7
66	254	2,732	2,211	78	79,600	88	10,035	32-11
78	300	3,228	2,611	92	88,700	98	10,780	35-4
92	354	3,808	3,079	109	99,000	109	11,650	38-2
104	400	4,304	3,479	123	110,400	121	12,895	42-3
118	454	4,884	3,947	139	122,700	135	13,760	45-1
130	500	5,380	4,353	154	130,200	143	14,505	47-7

[a] Presses available from 30 to 130 chambers in two-chamber increments.
[b] Nominal plate size is 1,500 × 1,500 mm (4 ft-11 in. square).
[c] Weight of the press only (without sludge).
[d] Overall length. All presses have an overall width of 3,000 mm and an overall height of 4,200 mm (9 ft-10 in. × 13 ft-9 in.).

Typical capacities and dimensions for diaphragm presses are given in Table 2.6.

HIGH-PRESSURE, THIN-CAKE FILTERS

Thin-cake staged filters have been used effectively at high flow rates per unit area for many years in both Eastern and Western Europe. Use of filtration ultrathin cakes is a useful technique for increasing flow rates. The operating characteristics of filters with thin cakes are discussed in the literature [12–16].

The basic elements of the filter are shown in Figure 2.27. Filtration surfaces are recessed plates equipped with rotating turbines that maintain permanent precoat-type thin cakes throughout the filter. Cake thickness is prevented from growing beyond the in situ precoat formed during the first few minutes of the operation by blades on a rotating shaft passing through the axis of the filter.

Slurry flows into the first stage and then flows around the turbines and through the clearances between the shaft and active filter surfaces. As liquid is removed, the thickening slurry moves from stage to stage. The unit acts as a filter thickener, producing a continuous extrudate that may contain a higher solids content than is normally encountered in conventional filters. The turbine plates sweep close to the filter cloths, leaving a thin, permanent cake on each stationary plate (see Figure 2.27). Even in the last stage of the filter where the slurry is highly non-Newtonian, a thin, easily identifiable, hard cake is maintained.

At low turbine velocities, the blades serve as scrapers that limit the cake thickness to the dimensions of the clearance. At higher velocities, the cake thickness is reduced and can be as thin as 1.0 mm with 3-mm clearance. For filters that depend entirely on fluid action, shear forces at the cake surface depend on fluid properties and velocity distributions.

The combination of high pressure (300 psi) and thin cakes produces high rates. Washing is accomplished either cocurrently or countercurrently. Separate filters in series can be employed in a manner similar to conventional thickeners. Washing may also be performed within a single unit whereby an initial portion of the filter must be used to remove liquid. In this case, the final stages are used for concentration. Clean wash liquid may be injected after the initial filtering at one or several intermediate stations. Injection wash tends to increase the overall filtrate rate but decreases the cake output rate.

Tiller [17] developed design methodology for staged filters. An important consideration is changing feed rates at constant pressure and rotational speed of the turbines. Following Bagdasarian and Tiller [15], Figure 2.28 shows the material balance for filter stage $n + 1$. A balance for the liquid is:

$$GY_n = GY_{n+1} + A\rho q_{n+1} \tag{2.1}$$

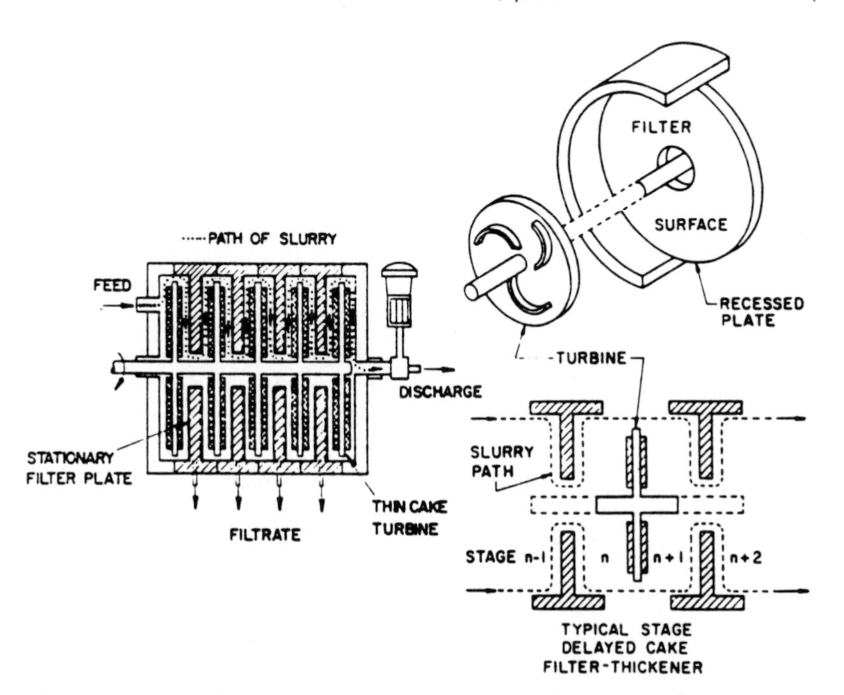

Figure 2.27 Principal components of a staged, thin-cake filter thickener.

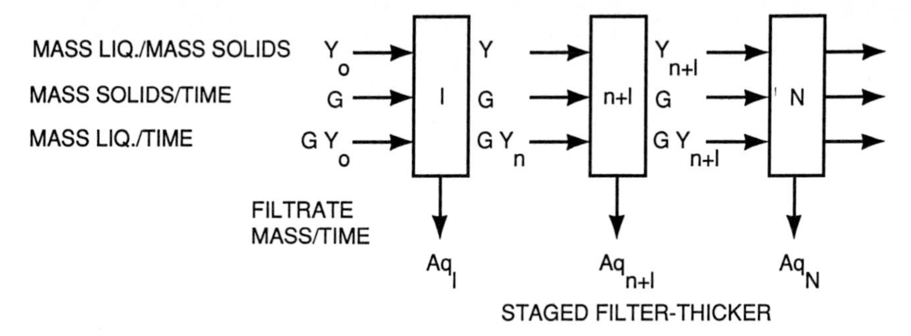

Figure 2.28 Material balance about stage n + 1.

where

G = mass of dry solids per time
Y = mass ratio of liquid to inert solids
ρ = density
q = filtration rate $(V/L^2 - t)$

Solving for the mass ratio:

$$Y_n = Y_{n+1} + \frac{\rho q_{n+1}}{G/A} \tag{2.2}$$

$q_n + 1$ is the mass flow rate per unit area of filtrate. Figure 2.29 gives a plot of ρq_{n+1} versus Y. Also shown (using right axis of Figure 2.29) is a plot of $\rho Q/(G/A)$, for different flow rates of dry solids. This term denotes the change in Y from stage to stage. Data are based on filtration of calcium carbonate in water.

Curve B represents the change in concentration from stage to stage based on data obtained from a full-scale filter. The conditions for these data were: area = 5.95 m², plates = 13, stages = 26 (the number of stages is twice the number of plates; see Figure 2.27), dry flow rate G = 2,396 kg/hr. Based on data for a single run, Bagdasarian and Tiller explain the following technique to estimate results at other conditions: Using a stage as the basic unit, A = 5.95126 = 0.288 m². This represents the one-sided area of a filter surface. Then

$$G/A = 2{,}396/0.2288 = 10{,}470 \text{ kg/hr} - \text{m}^2$$

Dividing this into ρq of curve A (Figure 2.29) yields curve B. The other curves represent other values of ρq (G/A at values of G = 50, 75, and 125 percent of the actual run).

The curves can be used for direct stage-to-stage calculations, or a graphical procedure can be used similar to that used in stagewise operations involving distillation, absorption, extraction, and washing. The latter method is illustrated in Figure 2.30, where 26 stages are required to change the inlet

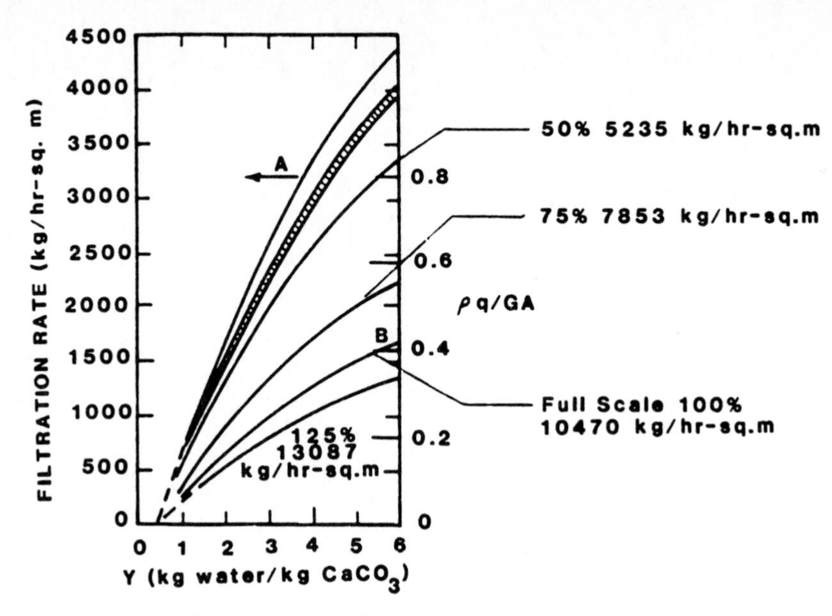

Figure 2.29 Filtrate rate and ρQ/(G/A) vs. mass ratio of water to calcium carbonate.

concentration of 0.13 mass fraction ($Y_o = 0.87/0.13 = 6.69$ kg water/kg solid) to 0.50 mass fraction ($Y_n = 1$).

The plot of Y_n versus Y_{n+1} shown in Figure 2.30 is based on Equation 2.2. For example, at $Y = 4$ in Figure 2.28, a value of $\rho q(G/A) = 0.315$ is found from curve B. Then the value $Y_n = 4.315$. $Y_{n+1} = 4$ becomes one point on the operating line. The entire curve is constructed on a similar basis.

At $Y = 0.46$, $q = 0$, and the operating line intersects the 45° line. An infinite number of stages would be required to reach that point.

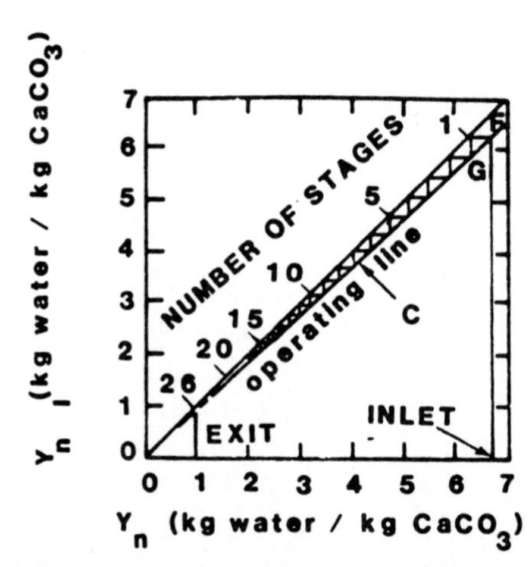

Figure 2.30 Graphical evaluation of the required number of stages.

To obtain the number of required stages, $Y_n = 6.69$ is located as the inlet concentration, and the point $(Y_1, Y_{0'})$ is found at point F on the operating line. The first stage, represented by (Y_1, Y_1), is obtained by passing horizontally from F to point 1. Having Y_1 the point (Y_1, Y_2) is located at G. The process is continued until 26 stages locate the specified exit concentration [15].

Concentration changes are small in the final stages. Use of a logarithmic plot, as shown in Figure 2.31, helps to improve accuracy. Two operating lines are shown for 75 percent and 125 percent of the solid rate in the experimental run. The exit concentration for the experimental run is shown as an open circle at $Y = 1.0$. For 125 percent of G, 26 stages yield $Y = 1.38$ or 0.42 mass fraction of carbonate. For 75 percent of G, the exit concentration is $Y = 0.755$, or 0.57 mass fraction. At the right side of Figure 2.31 is a plot of the exit concentration versus the ratio of the solid rates with the experimental value as a base.

Exit concentration is sensitive to throughput solid rate. Filtration pressure can be used to maintain desired specifications with changing flow rates. For the standard run with G = 2,396 kg/hr, liquid removed in increasing concentration from 0.13 to 0.50 mass fraction is given by overall filtration rate

$$= \frac{2396}{0.13} - \frac{2396}{0.50} = 13{,}639 \text{ kg/hr.}$$

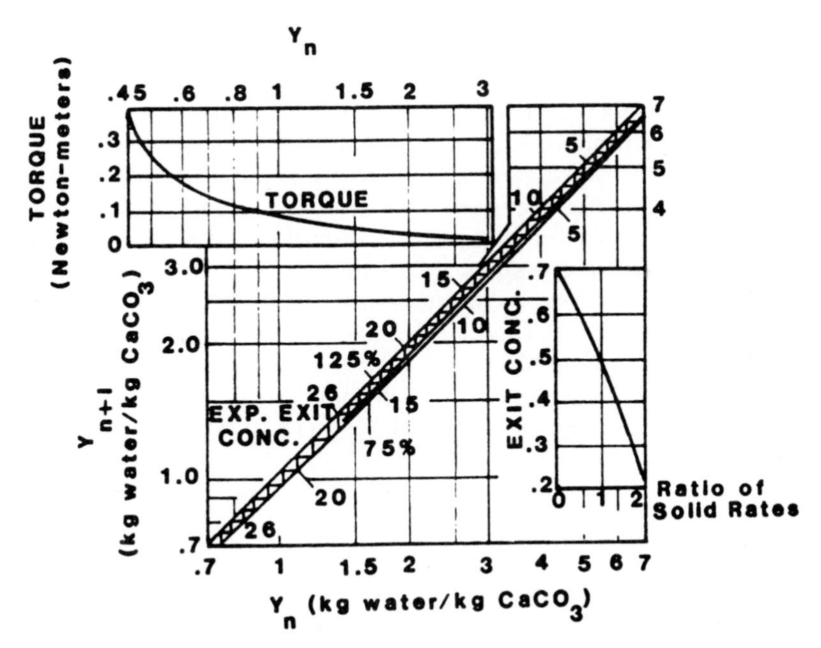

Figure 2.31 Graphical determination of the required number of stages on expanded log scales.

Figure 2.32 Continuous thin cake filter unit. *(Courtesy of SciTech Publishers, Inc., Morganville, N.J.)*

If the rate of solid flow is increased by 50 percent, the exit concentration obtained from Figure 2.31 would be 0.36 mass fraction. To restore the original 0.50 concentration, the filtration rate must be increased to (1.5)(13,639) = 20,459 kg/hr. Assuming the rate varies as the 0.6 power of the pressure, the pressure must be increased 1.97 times the original pressure of 827 $kN^5 - m^2$ or to 1,625 $kN^5 - m^2$ (236 psi). Curves of the exit concentration versus solid rate at different pressures are needed.

Figure 2.32 shows a commercial unit manufactured in the United States. This is a continuous solid/liquid separator, since the feed slurry rate is flow controlled and the filtrate and solids are removed in a steady-state operating mode. The thickened cake or extrudate flows out from the machine through a modulated, air-operated diaphragm-type control valve.

These filters can effectively process slurries from which the solid phase produces a cake that (1) is nonbinding, (2) has specific rheological properties so that a continuous extrudate or paste can be generated, and (3) is not affected by the shearing action of the rotating wiper blades. Examples of materials successfully handled by this type filter are dyes, polymer slurries, precipitated chemicals, organic pigments, kaolins/clays, pharmaceuticals, solvent slurries, inorganic pigments, and metal carbonates/hydroxides/oxides.

THICKENERS

Many existing filtration applications can be greatly enhanced if their present equipment, such as plate-and-frames and rotary vacuums, is used in conjunction with a thickening operation. Table 2.7 illustrates this point. Case 1 shows that if a feed slurry of 2 percent is concentrated in a filter to 50 percent (by volume), a total of 98 percent fractional removal of water is needed. If, however, a thickener is employed to concentrate from 10 percent, the fractional removal of water is 82 percent, thus leaving only 16 percent of the filter. This means that the present filter could be used about three times more effectively if supplemented with a thickener.

In Case 2, a 1 percent slurry is concentrated to 30 percent solids. A single filter would require 98 percent fractional removal of the water. By use of a thickener concentrating first from 1 percent to 7 percent, we fractionally remove 87 percent of the water. This leaves only 11 percent fractional removal of the present filter to go from 7 percent to the required 30 percent.

Many filter thickeners are simple settling tanks or decanters. Despite recent design improvements, thickeners of this type are generally large and bulky and have relatively slow rates. Centrifuges have a greater driving force but, in general, are expensive and can deliver cloudy overflow if fine particles are present.

High-velocity cross-flow thickeners are available, but operating experience often shows them to be highly dependent on the rheology of the slurry. Sometimes a slight increase in outlet concentration can result in filter blocking.

Dynamic Thickeners

Dynamic thickeners have appeared on the market in recent years. Special dynamic elements housed inside a thickening chamber keep the slurry continuously moving; hence, a concentrate of a paste-like consistency is possible

TABLE 2.7 EXAMPLES OF THICKENER OPERATION IMPROVEMENTS TO FILTRATION

Case	Solids (vol %)	Void Ratio, ϵ[a]	Fractional Removal of Water (F%)[b]
1	2	49	0
	10	9	82
	50	1	16
2	1	99	0
	7	13.29	87
	30	2.33	11

[a] ϵ = vol liquid/vol solids.

[b] F% = $(\epsilon_1 - \epsilon_2)/\epsilon_{feed}$.

without the danger of filter blocking. High flow rates per unit area, resulting from very thin cake formation, allow such units to be designed of relatively small size.

These systems are designed for little or no cake formation at lower levels of concentration. It can be shown that filtration rates increase with a reduction in cake thickness. However, some materials (especially gels such as aluminum hydroxides) are so compressible that 90 percent of the available pressure drop is absorbed by a "skin layer" formed on top of the filter media, while the remainder of the cake remains soupy and unconsolidated. Consequently, reducing the cake thickness (in examples such as unwashed "gels") in equipment that uses techniques of "thin cakes" would not result in any significant improvements. It is therefore advantageous to minimize the formation of a "skin layer." Bagdasarian [18] reports a fivefold increase in filtration rate on a rotary vacuum filter using a dynamic thickener.

Dynamic thickeners operate in a recycle mode of operation. The feed enters the thickener and the filtrate leaves the filtering plates while the steady-state-running, concentrated paste comes out of the modulating cake valve and reenters the feed tank. When the feed tank solids reach a predetermined concentration, e, the thickening operation is complete. To make the thickening operation continuous, one would install two feed tanks, so that after the first tank is completed the product can then be fed to continuous filtration equipment for further liquid removal as the second tank of feed solution is processed through the thickener. These two feed tanks can be set up with high-level and low-level audible signals and automatic switching three-way plug valves so that continuous operations are possible with any continuous filter. The operating scheme is illustrated in Figure 2.33.

Solids Washing

Washing of chemical solids in filtration is employed to enhance the purity of the product. High washing efficiency is the ultimate goal, along with minimum use of energy and wash liquid, clear filtrates, maximum flow rates, and a homogeneous washed product. Washing is usually accomplished in conventional cake-forming systems by forcing wash liquid through existing filter cakes. The initial efficiency is high as the mother liquor is being displaced; however, after breakthrough, the process is controlled by the diffusion rate of the solute, which explains why washing efficiency drops so rapidly with time.

Cake sagging in plate-and-frame filters reduces the effectiveness of the wash as most of the liquid flows through the area having the least solids buildup of cake. In rotary drum filters, an even more detrimental effect is cake cracking. Massive amounts of wash liquid short-circuit directly through the filter cloth, thus crippling the entire washing process. In contrast, with dynamic thickeners, the slurry or paste is washed instead of a cake. Solubles in the feed are dispersed into the wash liquid by strong agitation.

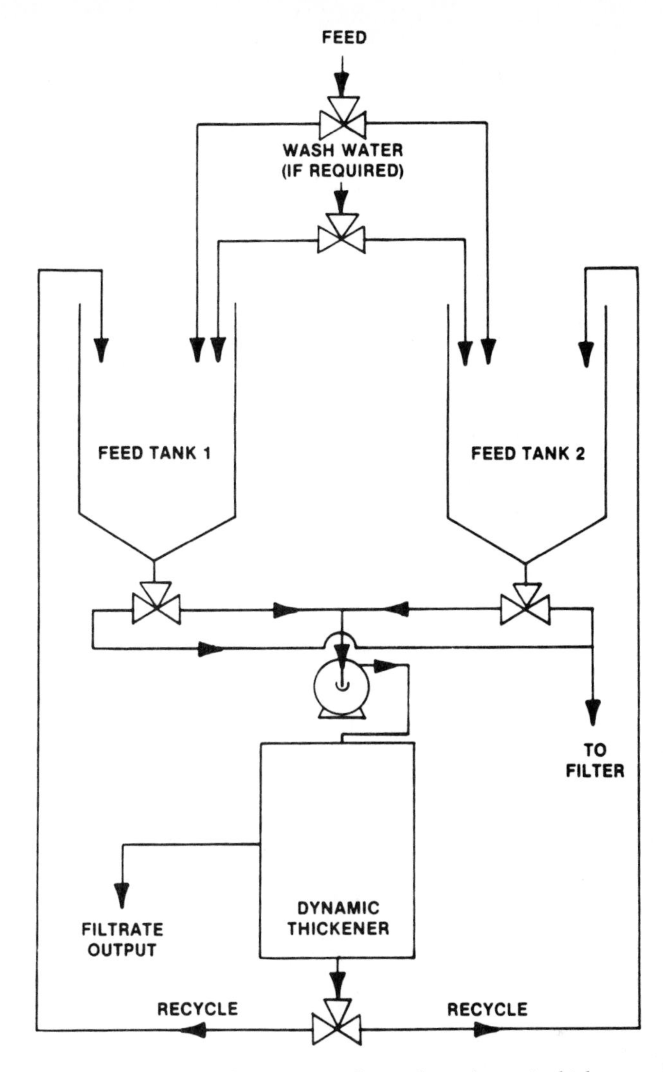

Figure 2.33 Typical operating scheme for a dynamic thickener.

The time required to reduce the solubles in a slurry to a desired level is a function of the feed solids concentration. Studies [18] on a single-stage bench-scale filter indicate that the optimum washing concentration can be determined from the slurry's filtration characteristics. For example, as shown in Figure 2.34, washing should begin at the feed concentration if the data produce a concave curve. On the other hand, a convex curve implies that washing at the higher solids concentration is best. Finally, the washing curve may contain an inflection point—an indication that the slurry should be thickened to a predetermined concentration before washing begins. Note that each curve shows its optimum slope line.

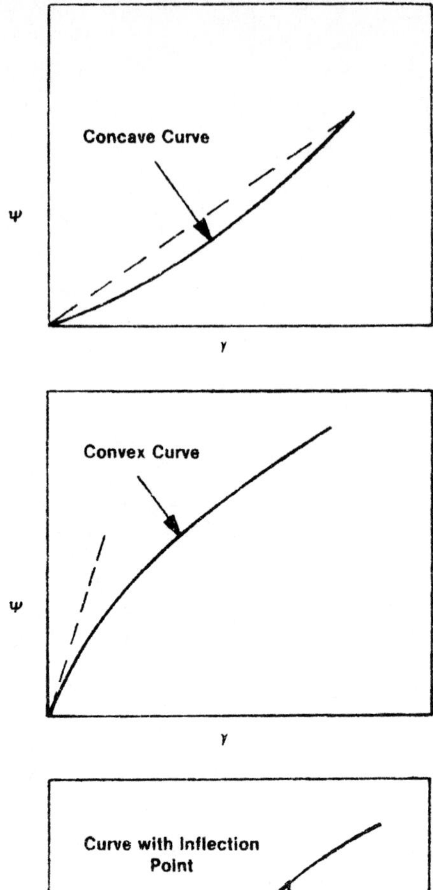

Figure 2.34 Optimum wash curves for a dynamic thickener. Wash at feed concentration (top); wash at thickened concentration (middle); wash at semithickened concentration A (bottom). ψ is a function of rate; γ is an inverse function of solids concentration.

As the first stage is run, information is automatically recorded (with a data logger) concerning temperature, torque, and filtrate weights. Thus, in one run filtrate rates can be obtained as a function of solids concentration.

This information may be fed to a computer that delivers a printout that graphs:

- filtrate rate q (gal/hr-ft^2)
- torque T (in.-lb)

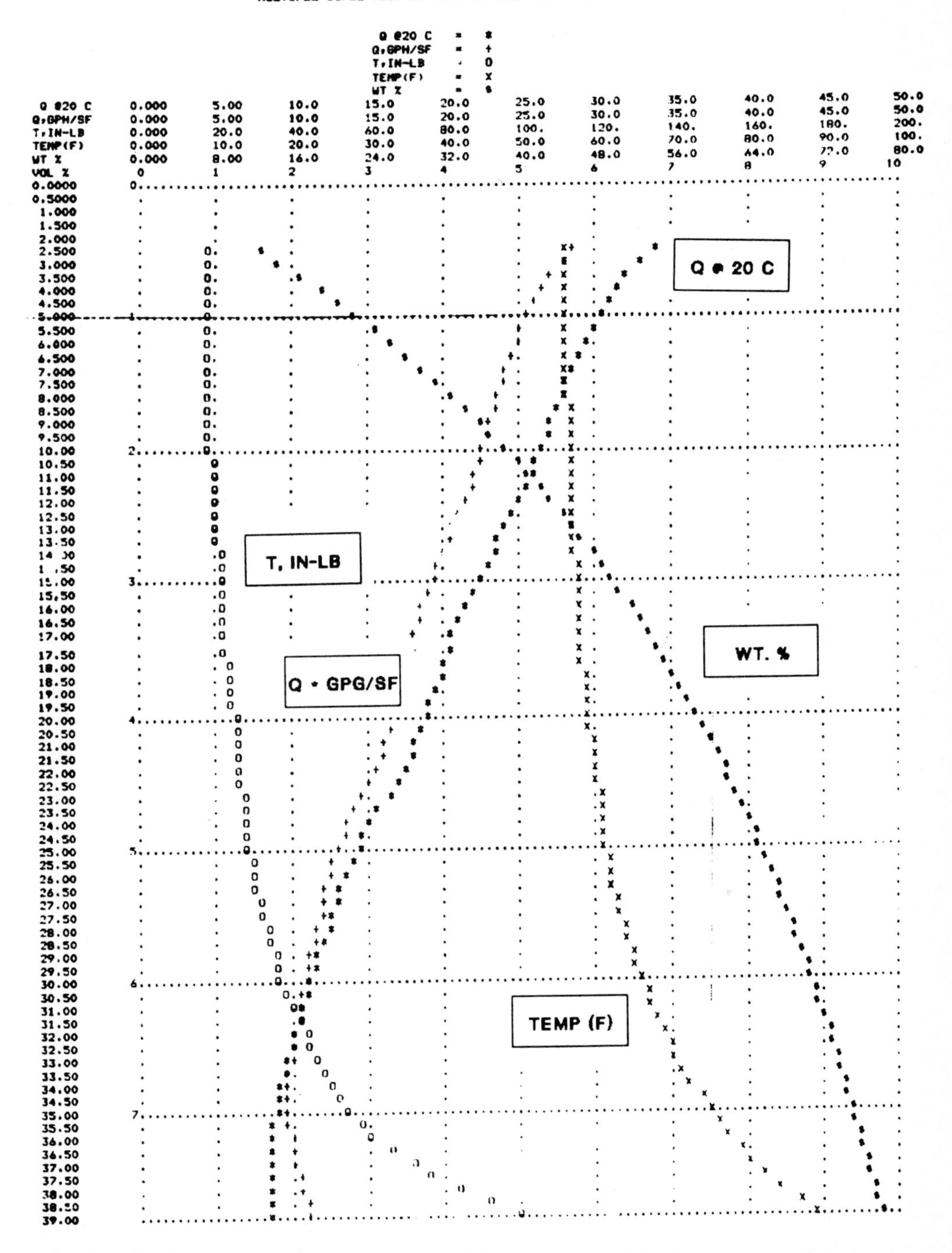

Figure 2.35 Computer printout plot of thickener data vs. solid volume concentration.

- temperature (°F)
- wt% solids in the cake

The preceding variables are shown plotted against the volume percent solids in Figure 2.35.

CENTRIFUGAL FILTRATION

Filtering centrifuges are distinguished from standard centrifugation by a filtering medium incorporated into the design. Slurry is fed to a rotating basket or bowl having a slotted or perforated wall covered with a filtering medium such as canvas or metal-reinforced cloth. The angular acceleration produces a pressure that transports the liquor through the filtering medium, leaving the solids deposited on the filter medium surface as a cake. When the feed stream is stopped and the cake spun for a short time, residual liquid retained by the solids drains off. This results in final solids that are considerably drier than those obtained from a filter press or vacuum filter.

Principal types of filtering centrifuges are suspended batch machines, automatic short-cycle batch machines, and continuous conveyor centrifuges. In suspended centrifuges, the filter medium is usually canvas or a similar fabric, or woven metal cloth. Automatic machines employ fine metal screens. The filter medium in conveyor centrifuges is usually the slotted wall of the bowl itself.

Figure 2.36 shows still another widely used design. The system combines the features of a centrifuge and a screen. Feed enters the unit at the top and is immediately brought up to speed and distributed outward to the screen surface by a set of vanes. Water or other liquid is forced by the sudden centrifugal action through the screen openings into an effluent housing. As solids accumulate, they are gently moved down the screen by the slightly faster rotating helix. With the increase in screen diameter, higher centrifugal gravities are encountered and solids are blown out the bottom of the rotor by a set of vanes into a conical collection hopper.

The theory of constant-pressure filtration may approximately be applied to filtration in a centrifuge. The following are assumed:

1. Effects of gravity and changes in the liquid kinetic energy are negligible.
2. The pressure drop developed from centrifugal action is equivalent to the drag of the liquid flowing through the cake.
3. Particle voids in the cake are completely filled with liquid.
4. The resistance of the filter medium is constant.
5. Liquid flow is laminar.
6. The cake is incompressible.

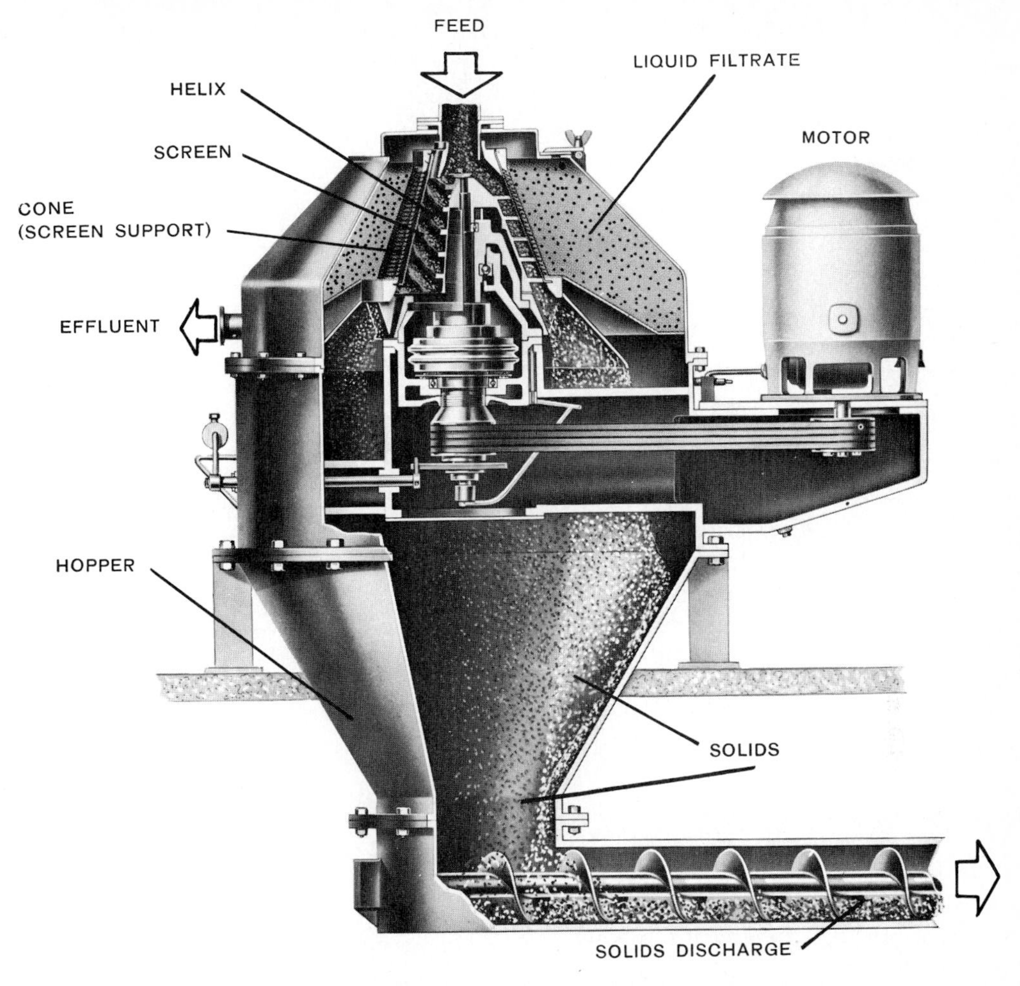

Figure 2.36 Cutaway view of one type of filter centrifuge. *(Courtesy of SciTech Publishers, Inc., Morganville, N.J.)*

Based on these assumptions, a simplified analysis can be developed for the cake after it has been deposited and during the flow of clear filtrate through it. The linear velocity of the liquid through the cake (u) is:

$$u = \frac{dV/dt}{A} = q/A \tag{2.3}$$

where

V = volume of filtrate
q = volumetric flow rate

The pressure drop from centrifugal action is:

$$-\Delta P = \frac{\rho\omega^2(r_2^2 - r_1^2)}{2g_c} \tag{2.4}$$

where

ω = angular velocity (rad/sec)
r_2, r_1 = radii of the inner surfaces of the liquid and face of the cake, respectively

Combining these two expressions yields

$$q = \frac{\rho\omega^2(r_2^2 - r_1^2)}{2\mu\,(\alpha' m_c/\overline{A}_L\,\overline{A}_a + R_m/A_2)} \tag{2.5}$$

where

A_2 = area of filter medium, i.e., inside area of centrifuge bowl (m^2)
$\overline{A}_a$ = arithmetic mean cake area (m^2)
$\overline{A}_L$ = logarithmic mean cake area (m^2)
α' = specific cake resistance (m/kg)
R_m = filter medium resistance (m^{-1})
M_c = mass of solids in filter cake (kg)

The average cake areas are defined as follows:

$$\overline{A}_a = (r_i + r_2)\,\pi b$$

$$\overline{A}_L = \frac{2\pi b\,(r_2 - r_i)}{ln\,(r_2/r_1)} \tag{2.6}$$

where

b = height of the bowl

Equation 2.6 is only applicable to a cake of definite mass (i.e., the expression is not integrated over an entire filtration starting from an empty centrifuge) and is accurate within the limitations of the derivation assumptions.

SCREW PRESSES

Dewatering is not only an important step in a filtration process—it is also one of the primary operations in processing materials. The necessary first step in the efficient drying or processing of many products is the extraction of excess moisture by screening and pressing. Sludge dewatering can be accomplished in several ways. However, in general, pressing tends to be a more energy-efficient operation than evaporation or other heat transfer methods.

Multistage screw presses can be used for dewatering chemical cellulose and for the removal of "black liquor" from kraft pulp, employing a recycling system with liquor flow countercurrent to the flow of stock, thus producing a much higher percentage of solids in the liquor fed to evaporators. Presses can also be employed in the continuous rendering industry, as well as in reconstitution processes, as for example, flax shive slurries, where four presses are used in conjunction with four slurry-blending tanks, operating as a four-stage countercurrent washing or leaching step for upgrading an otherwise waste material. On certain products, continuous four-stage presses can accomplish multistage counterflow washing in a single unit.

Screw presses may be used in the diffusion process for sugar cane, wherein the liquids for the diffusion of sugar solutions and fresh makeup water are extracted from the cut cane chips by the single or multistage unit.

Some other products that can be handled by continuous screw presses are:

- reclaimed and synthetic rubber,
- wood pulp
- waste paper pulp
- drugs
- miscellaneous chemicals
- brewer's spent grains and hops
- distiller's spent grain
- packing house cracklings
- paunch manure
- soybean and cereal by-products
- beet pulp
- tomato pulp
- citrus pulp and peels
- sweet and white potato pulp
- tobacco slurries
- cooked fish and fish cannery offal
- copra
- peat moss
- corn germ
- nitrocellulose
- castor seed or beans
- coffee grounds
- alpha-cellulose

Presses are available with many types of castings, designed to suit the characteristics of the material to be pressed, such as:

1. heavy 1-in.-thick carbon or stainless steel slatted casings
2. $\frac{3}{16}$-in.-thick naval alloy brass-drilled screens, tapered for self-cleaning
3. stainless steel perforated screens
4. stainless steel narrow bar-type super-drainage casing
5. fine-mesh Dutch-twilled filter cloth in stainless steel

The narrow bar-type drainage casing has approximately three times the drainage area of the steel-slatted casing, and twice the drainage area of the perforated or drilled-screen casings.

Low oil and moisture contents can be obtained with a continuous press, although output and final moisture content vary with the material being pressed, the speed at which the press is rotated, the uniformity of the feed

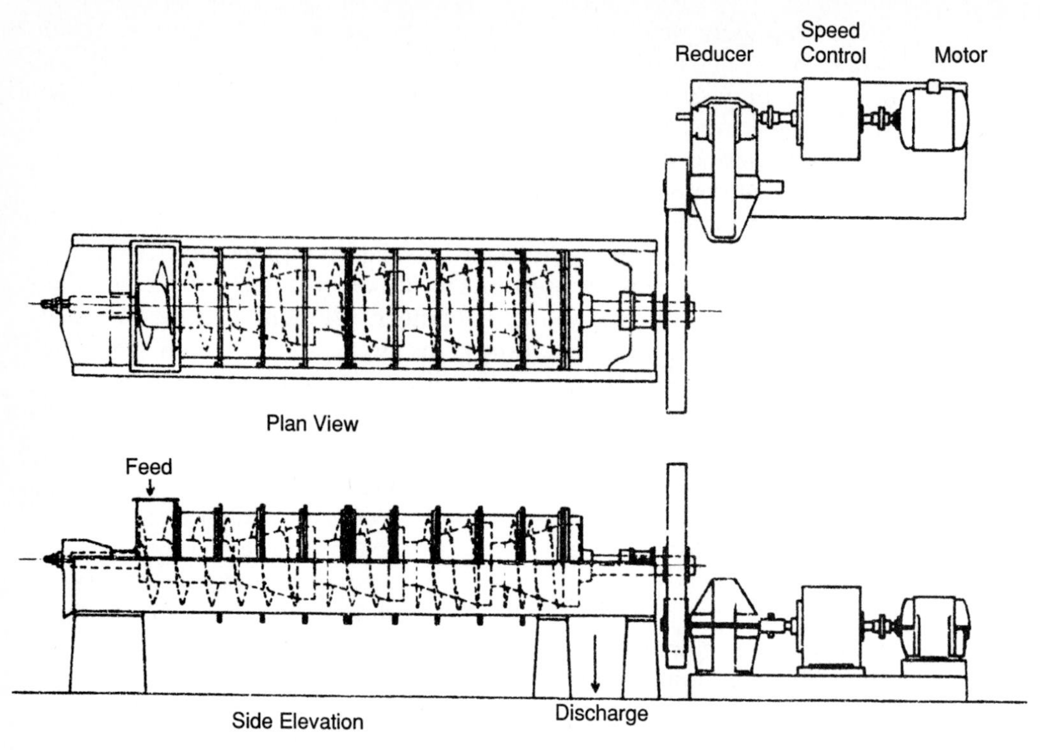

Figure 2.37 Plan and side elevation views of a four-stage press. Screws indicated by dotted lines. *(Courtesy of SciTech Publishers, Inc., Morganville, N.J.)*

to the press and the manufacturing process. A varispeed mechanical feeder with screw feed is available for forced feed when gravity feed is inadequate.

Presses are of extremely rugged construction. Various parts of the screw press, for example, may be chrome plated, of stainless steel or Monel, or furnished in other materials where corrosion and abrasion are severe.

Provisions can be made for steam, water, press liquor, or other liquids to be injected for cleaning or improved processing results.

The material enters the press through the intake hopper from a surge lank or conveyor and drops on the feed flights (with wide pitch) of the screw. The flights of the screw become progressively closer together and the cones of the various stages increase in diameter as they approach the discharge end. Each successive stage presses the material harder; the high pressure extracts the liquid, which passes through the perforated screens or other types of casings and leaves the press in all directions around the casing.

A single-stage continuous screw press has contact parts that are constructed of stainless steel, including the cover and catch pan. Such a unit is equipped with a variable-speed drive (but with the gear guard removed). This particular type of press is used to extract vitamins from alfalfa, citrus, and similar products.

Figure 2.37 shows a plan review of a 12-ft press equipped with a four-stage screw and stainless steel, self-recovering screws but with the covers, liquid catch pan, and gear guards removed. The press is driven by a 75-hp variable-speed drive to provide greater flexibility in processing a wide range of materials of varying consistencies, including citrus waste, precooked fish, and fish offal. With various design modifications, these presses are also used in the rendering industry and for reconstituting tobacco.

NOMENCLATURE

A = area (m^2)
b = bowl height (m)
F = temperature (°F)
G = mass of dry solids per time (kg/sec)
m_c = mass of solids in filter cake (kg)
q = filtration rate (m^3/m^2-sec)
Q = filtrate rate (gal/hr-ft^2)
R_m = filter medium resistance (m^{-1})
r = radius (m)
T = torque (in.-lb)
u = linear velocity (m/sec)
V = volume of filtrate (m^3)
Y = mass ratio of liquid to inert solids

Greek Symbols

α' = specified cake resistance (m/kg)
ρ = density (kgm^3)
ω = angular velocity (rad/sec)

REFERENCES

1. Kufferath, A., *Filtration und Filter*. Berlin: G. Bodenbender, 1954.
2. Daniells, K. J., *Filtration*, 1(4): (1964), 196.
3. Flood, J. E., H. F. Porter, and F. W. Rennie, *Chem. Eng.*, 73(13): (1966), 163.
4. Purchas, D. B., *Chem. Prod.*, 20(4): (1957), 149.
5. Purchas, D. B., *Filtration*, 1(6): (1964), 316.
6. Thomas, C. M., *Trans. Inst. Chem. Eng.*, 43(8): (1965), 233.
7. Trawinski, H., *Chem. Eng. Technol.*, 42(23): (1970), 1453.
8. Cheremisinoff, N. P., and P. N. Cheremisinoff, *Cooling Towers. Selection, Design and Practice*. Ann Arbor, MI: Ann Arbor Science Publishers, Inc., 1981.
9. Dickey, G. D., *Filtration*. New York: Reinhold Publishing Co., 1961.
10. Berline, R., *Cenie Chim.*, 73(5): (1955), 130.

11. Bulletin 315-181, Envirex—A Rexnord Co., Waukesha, WI (1982).
12. Michel, K., and V. Gruber, "Experience with Continuous Pressure Filtration in New Disk-Type Filter," *Chem. Ing. Technol.*, 34: (1970), pp. 773–779.
13. Malinovskaya, T. A., and V. F. Shevchenko, "Investigation fo Suspension Thickening and Washing Processes in a Filter with a Rotating Cylindrical Screen," *Teoreticheskie Osnouy Khimicheskoi Teknologii*, 1: (1973), pp. 592–598.
14. Zhevnovatyl, A. I., "Basic Principles of Filtration of Suspensions in Flow Without Formation of Deposits," *Zhurnal Prikladnoi Khimii*, 48: (1973), pp. 334–338.
15. Bagdasarian, A., and F. M. Tiller, "Operational Features of Staged, High-Pressure, Thin-Cake Filters," paper presented at Filtration Society Conference on Filtration, Productivity and Profits at Filtech/77, London, Septmber 20–22, 1977.
16. Jeffry, D. J., and A. Acrivos, "The Rheological Properties of Suspensions of Rigid Particles," *Am. Inst. Chem. Eng. AIChE J.*, 22: (1976), pp. 417–432.
17. Tiller, F. M., "Delayed Cake Filtration," *Filtrat.*, September 14, 1977, pp. 13–18.
18. Bagdasarian, A., Artisan Industries Inc., Metalfab Div., Jet-Vac Corp., Waltham, MA. Personal Communication (1982).

3

Ultrafiltration

PHYSICAL PARAMETERS IN ULTRAFILTRATION

Microporous Membranes, Filtration Ranges, and Limits

Figure 3.1 illustrates the three kinds of submicrometer semipermeable membranes. The type with the largest pores is used for microfiltration (MF). For the purposes of this chapter, MF has arbitrarily been assigned the range of 0.05 to 10 μm or perhaps 0.02 to 10 μm, depending on circumstances. MF separation generally involves removing particles from fluids based on size; osmotic pressure is negligible. Ultrafiltration (UF) generally involves separation of large molecules from smaller molecules, and overlaps somewhat with the porosity range of membranes used for reverse osmosis (RO). RO usually involves purification or concentration of small molecules or ionic constituents in a solvent. Thus, we have microfilters, ultrafilters (the subject of this chapter), and membranes used for RO.

Solubility and Pore Size Interactions

The overlap of the definitions for RO and UF membranes arises from the following considerations. Oversimplified, the "pores" in the skin of a membrane intended for removal of salt by RO are generally larger (e.g., 10 to 40

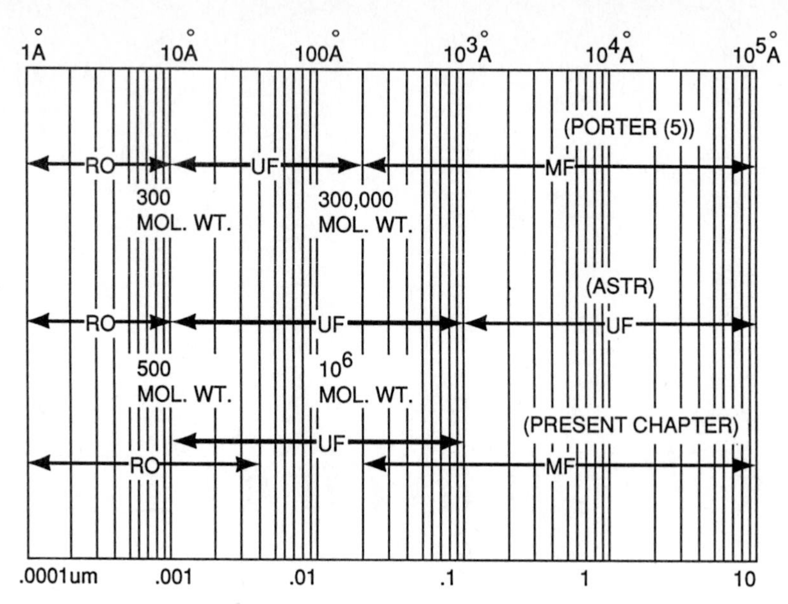

Figure 3.1 Microporous filtration ranges.

Å) than the hydrated ions (e.g., Na^+ Cl^-, Ca^+, SO_4^{2-}) they are intended to repulse. However, these pores are filled with water that is strongly influenced by the polymeric walls of the pores. Such water becomes *ordered water*, which, because of its ordering, has too low a dielectric constant to dissolve salt ions in contrast to the bulk water. Thus, salt rejection in a useful RO membrane (e.g., more than 85 percent salt rejection) is based on the lack of solubility of the hydrated ions in the ordered water within the pores, not on the size of the pores. It is not hard to imagine that the same membrane, or (at least) an inferior RO membrane (e.g., 5 percent to 20 percent salt rejection), would pass small molecules and reject larger molecules based primarily on size (UF) rather than on solubility (RO). Hence, we see the overlap of RO and UF ranges in Figure 3.1.

Effective Thickness of Various Membranes

RO, UF, and MF membranes are generally a few mils in thickness; however, the discriminatory layer may be either a tight skin supported by an open substructure (i.e., a very thin effective thickness and thus low frictional resistance to flow) or it may be the entire thickness of the membrane or gel involved in the pass/rejection mechanism (Table 3.1). In the latter case, the friction factors are much higher; that is, the entire thickness equals the effective thickness.

TABLE 3.1 TYPES OF MICROPOROUS MEMBRANE

	RO	UF	MF
Pressure Driven	Skinned[a]	Skinned[a]	Isotropic[b]
Concentration Driven		GEL[b]	

[a] Asymmetric.
[b] Homogeneous porosity; surface to surface.
[c] Graduated porosity; surface to surface.

Importance of Osmotic Pressure

Osmotic pressure across a semipermeable membrane arises from differences in concentration, which in turn arise from relative ratios in the numbers of impermeable individual ions or molecules on the two sides of the membrane. These osmotic pressures are dominant when salts are to be removed by RO. Osmotic pressures vary from 3.5 psi for good tapwater to 350 psi with average seawater, as the number of ions per unit volume is very high (35,000 ppm). At the other extreme (MF), there are essentially no dissolved species that cannot permeate through the membrane; it follows that the osmotic pressures are minimal. UF membranes lie in between, usually with very few impermeable species of very high molecular weight; and therefore, much lower osmotic pressures exist across the membrane. Exceptions can exist involving UF and may be circumvented, for example, the evaporation process.

Effective Thickness Coupled with Osmotic Pressure and Pressure-Driven and Concentration-Driven Processes

It follows that operations, such as RO, involving high osmotic pressures require higher pressures (i.e., consume more energy) than do low osmotic pressure operations. Therefore, very thin effective thicknesses are desirable for practical industrial or commercial RO installations to cut down on the frictional resistance to flow due to effective thickness (see Table 3.1). Concentrated brine is continually swept out of the RO elements and away from the membrane to avoid plugging and concentration polarization. At the other extreme, MF membranes involved in dead-end flow require low driving pressures; therefore, thicker membranes with higher dirt-holding capacities are generally found most useful. Skinned or pseudo-skinned varieties of MF membranes plug rapidly and account for some commercial failures of selected microfilters.

UF membranes lie in between RO and MF membranes and are of two kinds; both are useful. Of industrial importance, and the subject of this chapter, are the thin-skinned membranes, which allow enhanced flow rates (low friction factors) at given pressure differentials. Such UF membranes have larger pores in the thin skin than most RO membranes, and molecules of different molecular weights may be separated. Shape, size, and molecular weight are important. As the osmotic effect is less important with UF membranes than in the case of RO membranes, lower pressures (generally less than 100 psig) are sufficient to promote permeation, and molecules that differ by a factor of ten in their molecular weights may usually be separated. Fractionation of cheese whey into solutions of protein and lactose is one example.

Of medical and biotechnical importance are the thicker homogeneous gel membranes, such as Cuprophane™, which are used in the artificial kidney and/or concentration dialysis. With the Cuprophane membranes, diffusional migration, driven by concentration differences across the membrane, effects the transport of the various species across the membrane and little, if any, pressure differential is applied.

In kidney dialysis, toxic "middle molecules" diffuse across the Cuprophane membrane and out of the blood, while the larger desirable species are retained. Almost as much salt diffuses out of the blood as diffuses into the blood from the dialysate during this procedure. A small pressure is imposed that depletes the patient of a few pounds of accumulated water over a period of hours. Such processes are considered to be primarily concentration driven.

These thick gel membranes are biotechnically very important; however, the pressure-driven thin-skinned UF membranes, while perhaps somewhat less selective, produce product streams so much more rapidly that they are the materials of choice for industrial processes [1].

Molecular Morphology and Weight

The preceding paragraphs primarily consider the physical parameters of the various membranes and their porous properties. Particularly in the case of UF, serious consideration must be given to the species that penetrate or are rejected by the UF membrane.

Figure 3.1 shows that different sources attempt to relate molecular weight and pore size. Note that a 10 Å is presumably the cutoff point for either 300 or 500 mol-wt molecules, depending on the reference. Both sources could be correct, and the reasons that such uncertainty exists become even more important when larger molecular weights are involved.

As examples only, consider the behavior and properties of proteins that one might wish to separate. Proteins are said to have primary, secondary, tertiary, and sometimes quaternary structures. An oversimplified description of protein configurations in solution is useful. The various individual amino acids (about equal in number to the letters in the alphabet) may be strung together head to tail in an almost infinite number of sequences, just as randomly hitting the keys of a typewriter will give nonsensical words hundreds of letters long. Each of these random chains of amino acids (words) would correspond to the primary structure of a different protein. A particular sequence of amino acids depicted in two dimensions is considered the primary structure of a specific protein. There are, moreover, highly selective sites along these chains that are attracted to other specific sites along these same chains, forming loops held together by hydrogen bonds. A rendition of what sites are connected to what other specific sites and hence whether the resultant protein molecule would be forced to assume either helical or pleated sheet configurations reveals the secondary structure. As a result of these same hydrogen-bond interactions, the helical chains or pleated sheets become twisted, coiled chains, rods, or globular shapes. This morphology constitutes the three-dimensional or tertiary structure.

On occasion, two to four independent chains (based on primary structure) become intertwined via hydrogen bonds and van der Waals forces and

TABLE 3.2 INTRINSIC VISCOSITIES FOR MACROMOLECULES

	Molecular weight	Intrinsic viscosity (cm^3/g)
Compact Globular Particles		
Polystyrene Latex Particles	10^9	2.4
Ribonuclease	13,700	3.3
Lysozyme	14,400	3.0
Myoglobin	17,000	3.1
β-Lactoglobulin	35,000	3.4
Ovalbumin	44,000	4.0
Serum Albumin	65,000	3.7
Hemoglobin	67,000	3.6
Liver Alcohol Dehydrogenase	83,000	4.0
Hemerythrin	107,000	3.6
Aldolase	142,000	3.8
Ribosomes (yeast)	3.5×10^6	5.0
Bushy Stunt Virus	8.9×10^6	4.0
Randomly Coiled Chains		
Polystyrene in Toluene	45,000	28
	70,000	37
Reduced Ribonuclease	13,700	14.4
Oxidized Ribonuclease	14,100	11.6
Oxidized Ribonuclease in Urea	14,100	13.9
Ovalbumin in Urea	44,000	34
Serum Albumin in Urea	66,000	22
Reduced Serum Albumin in Urea	66,000	53
Myosin in Guanidine Hydrochloride	200,000	93
RNA	1.5×10^6	100
Heat-denatured DNA	5×10^6	150
Rod-like Particles		
Fibrinogen	330,000	27
Collagen	345,000	1,150
Myosin	620,000	230
DNA	5×10^6	5,000
TMV	4×10^7	29

TABLE 3.3 DIMENSIONS OF VARIOUS PARTICLES

Particle	Dimensions (μm)
Yeasts, Fungi	1–10
Bacteria	0.3–10
Viruses	0.03–0.3
Proteins (10^4–10^6 mol wt)	0.002–0.1
Enzymes	0.002–0.005
Antibiotics, Polypeptides	0.0006–0.0012
Sugars	0.0008–0.001
Water	0.0002

these also assume various three-dimensional morphologies. These multiple-strand agglomerates are said to have quaternary structure.

To complicate matters still further, these protein molecules may assume different morphologies in different environments or solutions. Table 3.2 [2] shows the intrinsic viscosities of various proteins where the intrinsic viscosity is defined as volume per mass of a given protein; it may be seen that the molecular weights of proteins bear little relationship to the intrinsic viscosity [3].

Note that ovalbumin (44,000 mol wt) is a compact globular particle that occupies 3.7 cm^3/g; if the sulfur-sulfur bonds are decoupled, it further opens to encumber 54 cm^3/g of protein.

Note also that the globular Bushy stunt virus (9×10^6 mol wt) occupies 4 cm^3/g, while the rod-like DNA (5×10^6 mol wt) occupies 5,000 cm^3/g. As an additional consideration, despite the greater volume per mass, might not coiled chains or rods "snake" their way through smaller pores than their globular counterparts?

Probable dimensions of variously sized particles are listed in Table 3.3 [4]. Further discussion exceeds the scope of this chapter. These examples illustrate that one should not jump to any filtrative conclusions based on molecular weight.

Therefore, while it is safe to say that a given UF membrane could separate the much smaller lactose from the much larger protein in whey, it is dangerous to assume that selected proteins could be separated from each other without experimental evidence.

FILTRATIVE OPERATIONAL MODES

Purification, Fractionation, Concentration, and Partition

All microporous filtration (MF, UF, and RO) deals with purification, fractionation, concentration, or partition. An example of purification is pressure-driven UF removal of particles and high-molecular-weight species from water subsequently to be used in hollow-fiber RO desalination. An example of pressure-driven fractionation is separation of protein and lactose from cheese whey for use as food additives (in the case of protein) and subsequent fermentation into alcohol (in the case of the lactose). Were the lactose merely defined as waste and dumped into a sewer, the process would be defined as pressure-driven UF protein concentration.

Another concentration-driven UF process is the concentration of protein solutions in the laboratory where an aqueous solution of protein is placed in a dialysis bag or tube and left for a period of hours in concentrated salt solution. Kidney dialysis also exemplifies concentration-driven partition filtration.

Dead-End versus Cross-Flow Filtration

Having distinguished between MR, UF, and RO, and identified the two prevalent kinds of UF membranes, we now discuss modes of operation: cross-flow (also tangential-flow and/or split-stream) filtration versus dead-end filtration. The numbers of particles per unit volume generally diminish in the order: RO (ions) > UF (molecules) > MF (bacteria, and so on). When the retained particles are comparatively small in number, as is usually the case in MF filtration, dead-end filtration is suitable (Table 3.4). At the other extreme, as is the case in RO, concentration buildup always demands cross-flow, tangential-flow, or split-stream treatment. The concentrate is continuously swept away, providing a relatively unchanged surface concentration. Pressure-driven UF also uses split-stream filtration to avoid membrane plugging or concentration polarization (also known as gel polarization). More recently, cross-flow filtration coupled with backflushing has also been implemented in MF filtration when the particulate load is particularly heavy or when long lifetimes of the MF membranes are desired.

Membrane in Element/Module Configurations

All three kinds of membranes (RO, UF, and MF) may be manufactured in either flat sheet, tube, or hollow tubular form. Generally, the hollow fiber (RO and smaller) or hollow tubular (UF and larger) configurations are less

TABLE 3.4 DEAD-END VERSUS CROSS-FLOW FILTRATION

	RO	UF	MF
Dead end	–	–	–
Cross flow			
Tangential Flow			
Split Stream	+	+	+ (energizing)

effective per unit area than are the flat sheet configurations, but this is offset by the greater effective area that can be packed into a volume of hollow tubules or fibers. The flat sheet configurations are usually plate-and-frame, spiral-wound, or pleated cartridges.

The third configuration, large tube (0.5 to 1.5 in.) is intermediate, possessing the performance characteristics of the flat sheet but lacking the surface-to-volume advantage of hollow fibers. Tube configurations can, however, cope with the most contaminated streams, primarily because they can be cleaned mechanically. At the two extremes, tiny hollow tubules and most flat sheet configurations can be cleaned by reverse flow, but certain clogging contaminants are difficult to remove.

MAINTENANCE OF UF ASSEMBLIES

Gel or Concentration Polarization and Fouling

The previous discussion brings us to one of the most important features of UF—gel polarization, which is important when the separation of macromolecules is involved in either flat sheet, tubular, or hollow fiber UF membrane configurations. As permeate containing the smaller molecules passes through the membrane, a layer of solution containing the larger rejected molecules accumulates adjacent to the membrane surface and may reduce the flow by plugging or fouling the membrane and/or forming a gelantinous filtration medium in series with the original membrane. This can increase frictional resistance and sometimes reduce its effective pore size, not allowing the passage of smaller molecules that were intended to pass through the unencumbered membrane.

In some cases the problem is so severe that UF is precluded. However, three approaches have been used successfully in restoring the utility of the fouled membranes and/or keeping them from becoming fouled. They are (in order of decreasing difficulty of application) periodic purging with cleaning solutions (e.g., chemicals or enzymes), introduction of turbulence (as follows) by one of a number of baffling arrangements, and periodic backflushing. Backflushing is most readily applied to hollow tubular devices and is responsible in no small part for their growing acceptance. Turbulence promoters, generally inapplicable in hollow tubule devices, are most commonly employed in flat sheet configurations. Here, for example, Vexar™, a coarse webbing, is placed next to the membrane surface to induce a sweeping action or eddy currents, which promote rapid mixing of the incipient boundary layer back into the bulk fluid. There is, of course, a maximum concentration of potential gel-forming material that can be tolerated, at which point further UF becomes ineffective. Such induced sweeping is employed in plate-and-frame, spiral-wound and pleated-membrane devices.

TABLE 3.5 CLEANING TECHNIQUES FOR UF

	Chemical	Reverse flow	Mechanical
Flat Sheet	+	±	±
Spiral Wound	+	±	±
Fluted	+	±	±
Tubular	+	±	±
Hollow Filter	+	+	−

Cleaning Techniques for UF Assemblies

Periodically, the cumulative effects of gel polarization, dirt accumulation, or biological growth render it necessary to renovate or clean the UF assemblies. These cleaning or antifouling techniques are of three kinds: chemical, reverse flow or mechanical. Combinations of these can be used. All are practicable, depending on element, module, or cartridge configuration (Table 3.5).

Chemical cleaning techniques are applicable to all configurations, although care must be taken to make certain that the membrane and other materials of construction are compatible with the chemical agents used. Reversing the flow is usually practicable, but with certain flat-sheet, spiral-wound, fluted, and tubular configurations, inadequate membrane support during reverse-flow operation may cause problems.

Because plate-and-frame and tubular configurations are used with the most contaminated fluids, mechanical cleaning techniques are used. In the case of plate-and-frame systems, the equipment may be disassembled and scrubbed, while in the tube configurations oversized soft foam plugs are driven through the tubes by pressure.

Preventive Cleaning Measures

The three methods of cleaning fouled UF membranes have been discussed previously and while induced mild turbulence (considered later) may be seen as a preventive measure, all of the procedures result from the necessity to counteract the effects of gel polarization.

At least five additional techniques are being investigated that fall into the preventive category:

1. The tube pinch effect no doubt takes place during rapid laminar flow in hollow tubules where hydrodynamic forces tend to cause particles to migrate toward the centers of the tubules and, hence, away from the walls.
2. Enzymes which decompose protein deposits have been incorporated into the UF membranes, either by postimmobilization or by inclusion

during the membrane's manufacture. Such membranes may be considered as self-cleaning to some extent.

3. Immobilized positive or negative charges have been attached to UF membranes. By repelling like-charged species, the tendency to foul is diminished (see section on electrodeposition of paints).
4. Electric fields have been imposed such that potentially fouling macromolecules or particles are attracted away (electrophoretically) from the UF membrane surface.
5. Emulsified surfactants are injected into the feed. The surfactants are selected depending on the specific surfactant's enhanced ability to attract specific foulants to the water-surfactant interface rather than to the membrane-water interface.

Turbulence (General) versus Turbulent Flow (Specific)

One of Webster's definitions of the broadly used term *turbulence* is "irregular (fluid) motion especially when characterized by up and down currents." Conceptually, then, the eddy currents and deflected and mixing flows described earlier could be considered turbulent. Other definitions of turbulence span the spectrum from mild ("any departure from smooth flow") to vigorous ("wild commotion, tumult and violence"). Therefore, *turbulence* is a broad term embodying any hydrodynamic departure from smooth flow. Turbulent flow in engineering parlance has a much more limiting definition. Webster's definition of turbulent flow is "a fluid flow in which the velocity at any given point varies erratically in magnitude and direction." This engineering term, *turbulent flow,* defines a subdivision of turbulence, in simple terms, and is at the "violent" end of the turbulence spectrum. In many cases, induced mild turbulence at the mild end of "mixing flows and eddy current" of the turbulence spectrum is much to be desired for eliminating concentration or gel polarization. But for biological fluids, violent engineering turbulent flow is to be avoided as biological cells may be destroyed, protein may be denatured and/or, in some cases, membranes may be damaged.

In scientific terms, the density and viscosity of the fluid, along with the geometry and velocity, form a dimensionless value called the Reynolds number. As a function of geometry, if this number is greater than 4,000 to 10,000, turbulent flow is assured. Laminar or smooth flow is assured if the Reynolds number is less than 2,000.

Providing that equipment can withstand the buffeting, turbulent flow may be useful when cleaning solutions are in use or fast flushing is involved, but generally not during the filtration process itself.

Special Hollow Tubule Maintenance Procedures

The hollow tubule configurations with lumens frequently on the order of 0.5 to 2.0 mm in diameter present a different set of constraints but also present opportunities. Consider a cartridge (Figure 3.2) composed of a large number

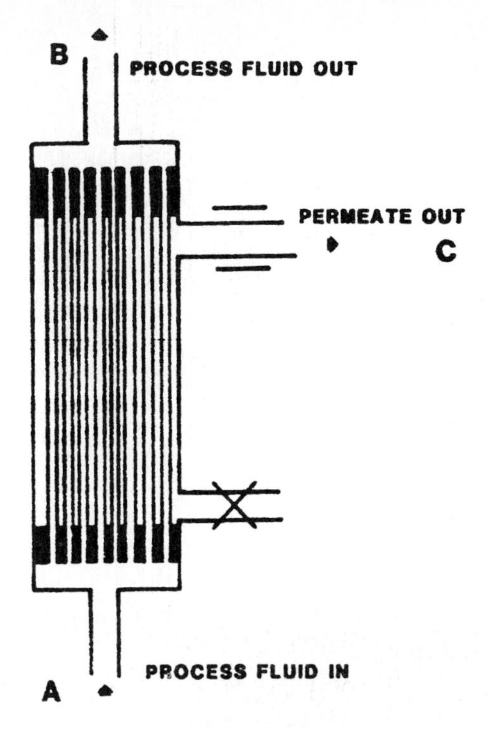

Figure 3.2 Hollow tubule ultrafiltration.

of hollow tubules potted at each end and encased in such a manner that the process stream can enter a plenum (*A*) at either end of the bundle of hollow tubules, proceed through the length of the tubules, losing fluid through the walls (UF), into the encasement and exit into either a drain or reticule tank (*B*). Provision is made for removing the permeate (C).

Figure 3.3 (left) illustrates a similar situation, where 90 percent of the material issues as permeate (C). When the permeate flow decreases below a certain point due to fouling, the device may be renovated (Figure 3.3, right) by closing transiently the permeate valve, reducing the average transmembrane pressure and concomitantly increasing the fluid through the tubules fourfold. In this fashion the fast flush may remove the accumulated debris.

A close look at fast flushing (Figure 3.3, right) reveals that backflushing is also taking place. The hollow tubular bundle has a substantial friction factor due to the small diameters; hence, there is a pressure drop between *A* and *B*. Assuming for convenience a 20-psi pressure drop down the tubules (from *A* to *B*) under fast-flow conditions, what would be the pressure of the encasement? Assuming symmetry, the pressure would be around the average at *A* and *B*. Thus, at the tubules near *A* there would be a 10-psi pressure drop between *A* and the encasement (encouraging ultrafiltration permeation), while the pressure would be reversed at the tubule endings near *B*, encouraging ultrafiltered fluid to backflush the tubules near *B*. Reversing the "fast-

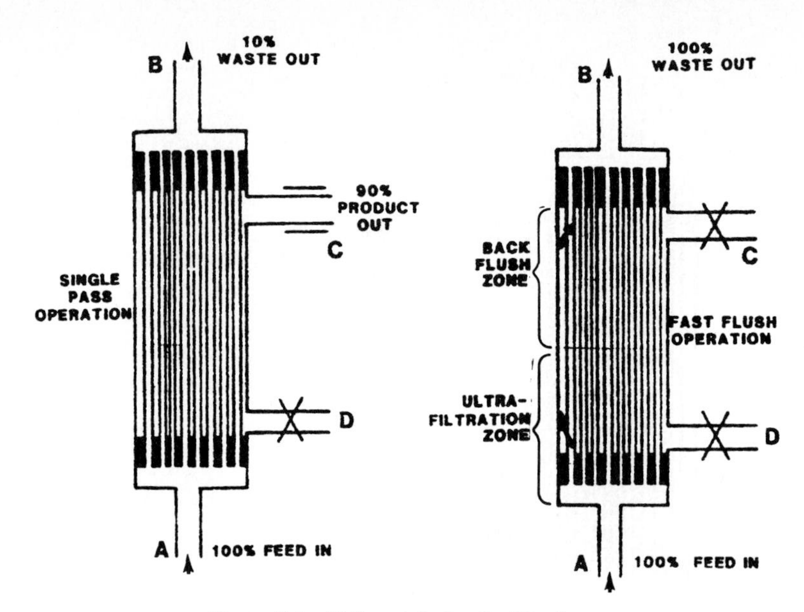

Figure 3.3 Hollow tubule ultrafiltration.

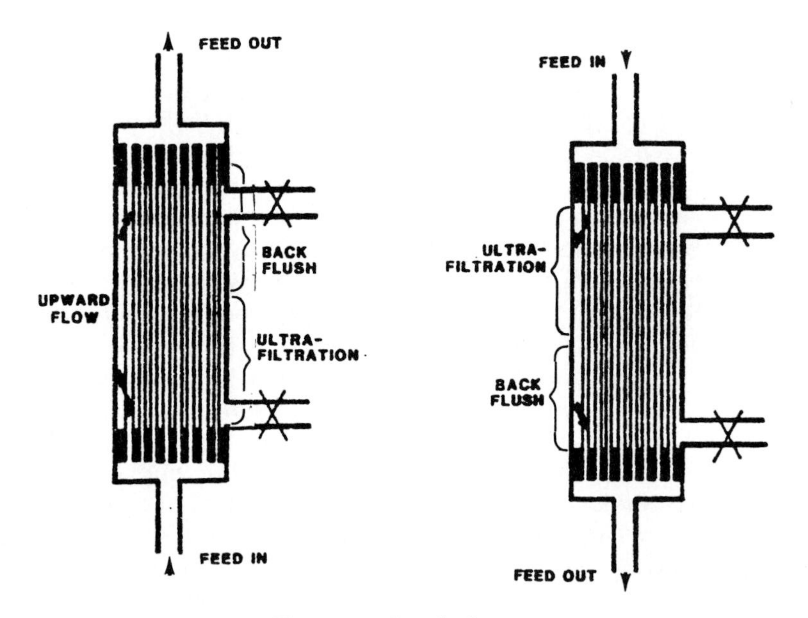

Figure 3.4 Fast flushing.

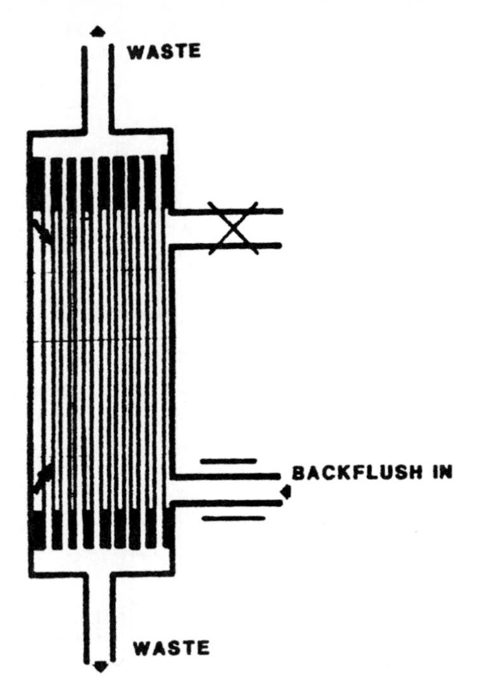

Figure 3.5 Backflushing.

flow" direction through the tubules would backflush in turn each end of the device (Figure 3.4).

As backflushing near the middle would be null from time to time, ultrafiltered fluid or other cleaning solutions could be injected through *D* and reclaimed or dumped through *A* and/or *B* (Figure 3.5).

UF APPLICATIONS

Although UF was first thought to be primarily applicable to the treatment of wastewaters such as treated sewage to remove particulate and macromolecular matter, it is now known to be useful industrially in producing high-grade waters, recycling electrocoat paint particles, separations involving whole and skim milk, vegetable protein isolates (especially soybean), fermentation products, fruit juices, biochemicals such as pyrogens, phages in general, and human chorionic gonadrotropin.

Environmental problems are also candidates for UF processes. Tanneries, for example, produce sulfides and protein in the dehairing process, which may be recovered. Polyphenols may also be removed selectively from must in the wine-making process.

In the dairy industry, UF has been used in making more than 30,000 ton/yr of Camembert and feta cheeses in France alone. New manufacturing

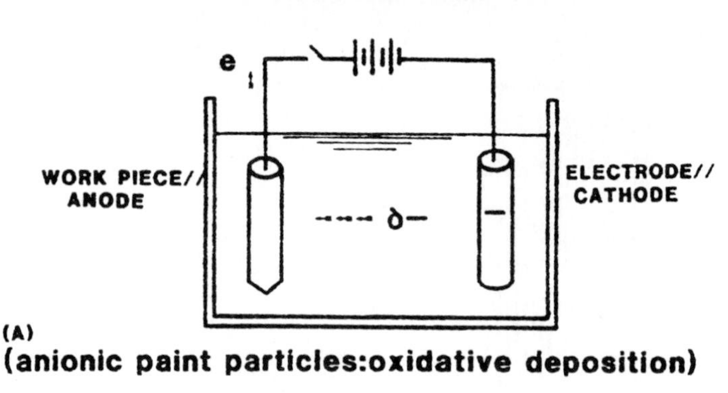

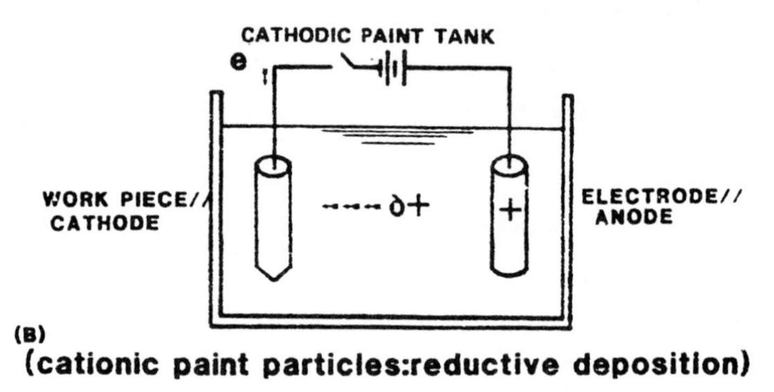

Figure 3.6 Electrodeposition of paint.

processes for the production of ricotta, cream, and St. Paulin cheeses also involve UF. A few specific examples are briefly reviewed later and the reader is referred to Cooper [5] for additional examples and references.

Electrodeposition of Paint

In electrodeposition of paint, the conductive unit to be coated is immersed in an aqueous solution of paint (usually primer) particles, and a voltage is impressed between the tank or an electrode and the work piece to be coated. If the piece to be coated is positively charged (i.e., the anode) and the paint particles are negatively charged, then the anionic particles migrate to the work piece and are deposited by an oxidative process (Figure 3.6A). During this oxidative deposition, electrons are removed from the paint particles, producing a neutral primer coating.

There is a troublesome side reaction wherein some neutral metal atoms from the work piece also give up electrons and dissolve in the anodic paint film, formation cations (e.g., $M^{\circ} \rightarrow M^{s+}$ + electrons). It has been observed that the inclusion of such metal cations reduces the corrosion resistance of the paint film. Nevertheless, anodically deposited paint, particularly in the automotive field, has become a large-scale business.

With the advent of the new cathodic paints, conversion to cathodic electrodeposition has been rapid (see Figure 3.6B). Reduction of the positively charged paint particles also causes a neutral primer film to be deposited, but the metal work piece has no tendency to dissolve under the reducing conditions.

After electrodeposition of the primary coat, the work piece is lifted from the tank and undeposited paint solution is washed off of the work piece. In principle, this wash water would be a much-diluted paint solution and thus would be an environmental hazard if simply dumped into a sewer.

UF of the aforementioned dilute waste stream into reclaimed paint and wash water was seen as a possible solution. However, a paint emulsion is in very delicate balance; too much dilution or too much concentration renders it unstable. An alternative was developed by generating wash solution on a continuous basis from the paint tank itself. The ultrafilter is operated at very low flux such that the paint solution remains within narrow concentration limits and the permeate (paint solution with additives but without particles) is used sparingly in a countercurrent fashion to wash the work piece, and the wash fluid and paint particles are returned to the paint tank. Under these conditions the composition in the paint tank remains relatively unaffected and amenable to periodic adjustments to compensate for particle loss and solvent evaporation.

Using Romicon hollow tubule configurations and XM-50 (vinyl copolymer) membranes at rapid paint velocity and with period backflushing, the membranes last for many months on anodic paints. The fact that the XM-50 membrane has excess immobilized negative charges and thus repel anodic paint particles accounts, in part, for the long lifetime of these UF membranes.

As the popularity of cathodic paint increased, wherein the paint particles are positively charged, the negatively charged XM-50 membranes fouled irreversibly 20 percent of the time due to the attraction between cationic paint particles and the anionic membrane.

CMX membranes, in which the membrane incorporates permanently immobilized positive centers, have been developed for use with cathodic paints. These CMX hollow tubule membranes are now usable for many months with periodic backflushing and cleaning, as are the XM-SQ membranes with anodic paints.

Cheese and Whey Processing

The earliest and still predominant use of UF in cheese making was primarily to minimize the environmental consequences of dumping millions of gallons of whey following the cheese coagulation process. Of secondary importance was reclamation of soluble proteins from the whey for use as food additives. Considerable progress has been made, but most whey still goes to waste. The permeate, following protein concentration by UF, is rich in lactose. Lactose is under consideration for fermentation into a wine of dubious distinction, but more probably may function as a source of industrial alcohol.

At the Dairy Research Laboratory in France, various UF schemes for removing the lactose from skim milk before cheese making, along with demineralization, promise to make cheeses more reproducible and of more homogenous composition than heretofore routinely possible.

The eleven-step process for making Camembert cheese could use UF three times, first to demineralize and remove lactose from the milk before adding the starters, thus producing liquid precheese, and later after adding the starters but before adding the Penicillium candidum and rennet to coagulate the proteins. Following syneresis, the residual fluid of protein-rich whey may also be concentrated by UF.

Similarly, schemes to produce feta and ricotta cheeses from standardized pasteurized milk have been developed that may employ UF at least twice.

Although the preceding procedures are the more existing new applications of UF to cheese whey, the predominant present use is to harvest whey protein concentrates (WPC). WPC are available commercially in the range of 30 to 80 percent to protein. The end uses of WPC vary dramatically. The properties of the WPC can vary as a function of pasteurization temperature, time, pH during filtration, levels of lactose, minerals present, and pH adjustment following UF as well as drying.

According to Cooper[5], a measure of the protein denaturation that reflects quality is the immunological activity of bovine serum albumin (BSA). Relatively undenaturated WPC registers a degradation of BSA activity of about 20 percent during the six steps involved (pH adjustment, pasteurization, ultrafiltration, pH adjustment, evaporation, and spray drying).

Cooper discriminates between three turbulent systems (large tubes, flat plate, and spiral wound) and the thin channel systems (plate and frame and hollow tubules), and observes that the trend is toward greater use of thin channel devices.

Oil/Water Emulsions

A distinction needs to be made between oil and water mixtures such as water in diesel fuel or oil in ship bilge waters on the one hand, and oil-water emulsions as are used in metal working and finishing. For the latter the desire is to reclaim the oil emulsion concentrate as the reject and semiclean water as the permeate.

Haulage costs, environmental impact, and improved performance are generally the initial driving forces behind considering oil-water separation via UF. If the oil-water mixture is not an emulsion, ME assemblies are most useful because of the more rapid permeation rates possible with MF. In the MF mode, the membrane is first saturated (wetted) with either oil or water, which determines which phase will be the permeate. Once wetted with oil, for example, the MF filter rejects the aqueous phase and oil permeates the MF filter. The reverse is true when the MF filter is first wetted with water.

If oil-in-water emulsions cannot be separated by water-wetted MF filters (where the excess water becomes the permeate), then UF becomes appropriate and the pore size is selected to reject the dimensions of the emulsified oil droplets. Care must be exercised not to exceed the pressure where the liquid oil droplets can be deformed sufficiently to displace water from pores in the UF membrane (much as gas displaces fluid in the "classical" bubble-point experiments) and thus plug or penetrate the filter. If the emulsion is not reusable and can be brought to 25–40 wt percent oil, the retentate may be burned to produce energy.

A partial list of UF equipment manufacturers is:

- Abcor
- Amicon Corp.
- Aqua-Chem, Inc.
- Asahi Chemical Company Ltd.
- A/S DeDanske Sukkerfabrikker
- Desalination Systems
- Dorr-Oliver Inc.
- Dynapol
- Envirogenics
- Gelman Sciences Inc.
- Harza Engineering Co.
- Millipore Corp.
- Osmonics Inc.
- Patterson Candy Int.
- Thone Polenc
- Romicon Inc.
- Torary Industries Inc.

The literature [6–10] gives more in-depth information.

REFERENCES

1. Parrett, T., *Membranes Technol.*, 2(2) (1982), 16.
2. Schaehman, H. K., *Cold Spring Harbor Symp. Quant. Biol.*, 28: (1963), p. 409.
3. Mahler, H. R., and E. H. Cordes, *Biological Chemistry*, p. 99. New York: Harper & Row Publishers, Inc., 1966.
4. Beeton, N. C., in *Ultrafiltration Membranes and Applications, Polymer Science and Technology*, vol 13, p. 375. New York: Plenum Publishing Corporation, 1980.

5. Cooper, A. R., ed., *Ultrafiltration Membranes and Applications, Polymer Science and Technology*, vol. 13. New York: Plenum Publishing Corporation, 1980.
6. Michaels, A. S., "Ultrafiltration: Adolescent Technology," *Chem. Technol.*, 11: (1981), p. 36.
7. Porter, M. C., "Membrane Filtration," in *Handbook of Separation Techniques for Chemical Engineers*, R. A. Schweitzer, ed. New York: McGraw-Hill Book Company, 1979.
8. Brewlan, B. R., and R. A. Cross, "An Introduction to Membrane Separation Technology," in *An Introduction to Separation Science* (2nd ed.). New York: John Wiley & Sons, Inc., 1982.
9. Flinn, J. E., ed., *Membrane Science and Technology*. New York: Plenum Publishing Corporation, 1970.
10. Kesting, R. E., *Synthetic Polymeric Membranes*. New York: McGraw-Hill Book Company, 1971.

4

Microporous Membrane Filtration

Filters are used most frequently to remove scattered solid particles from a fluid. In the case of filtrative sterilization, the solid particles of concern are microorganisms. Filtration can also be used to remove air bubbles from liquid streams, and may be employed to separate liquid droplets, such as water or oil mists, from air streams. Filters also separate oil droplets from water, and vice versa.

Technically stated, filters are used to separate, remove, and collect a discrete phase of matter from its dispersion within the matrix of another. Generally, process filtration involves removing contaminants and creating a clean fluid stream, where the effluent stream is of interest. Conversely, analytical filtration involves removing substances from the fluid stream, where the collected material is of prime importance.

PARTICLE REMOVAL BY SIEVE RETENTION

Filters can remove particles by several different mechanisms. One very important mechanism is sieve retention—the manner in which a sieve or a screen works. The particle is simply too big to pass through the pore, so it becomes arrested on the top surface of the sieve or screen.

This type of particle capture is absolute in its reliability. Its performance in particle removal is complete and certain as long as the smallest particle being filtered is larger than the largest pore of the filter.

For this reason, measurement of a filter's largest pore size is important (as by the bubble-point procedure to be discussed later). Therefore, where the absolute removal of particles is required, as in the removal of all organisms by filtrative sterilization of solutions, screen or sieve-type retention is desired.

PORE-SIZE AND PARTICLE-SIZE DISTRIBUTIONS

The scattered particles removed from a fluid are almost never all of one size. Instead, they have a certain particle-size spread or distribution; some are larger and some smaller than the mean or average size. As noted previously, filter pore size are also nonuniform. They show a pore-size distribution. To achieve the desired absoluteness of removal that derives from sieve retention, the largest pores of the filter selected must be smaller than the smallest size particles of the particle-size distribution.

In actual field settings, this is not always possible to do with assurance. The filter pore-size rating (its mean flow pore) is known, as also is its bubble-point value (related to the size of its largest pores) and, on occasion, even its total pore-size distribution. Generally, however, almost nothing is known about the particle-size distribution. The choice of filter pore size is thus usually based on experience and judgment. Most often the application is a familiar one, and a match is made of the membrane filter to the one already in use. Otherwise, one selects a membrane rated in accordance with its pore size relative to whatever is known about the particle size. However, of the many types of filters available, a membrane filter is chosen whenever security of particle removal is desired, hopefully based on absoluteness of sieve retention.

Membrane filters are selected because, as we shall see, they have a narrower pore-size distribution than other types of filters. This narrowness makes less likely an accidental overlap of particle-size and pore-size distributions, the kind of overlap that would rule out absoluteness of sieve retention. The broader the pore-size distribution, as with nonmembrane filters, the greater the likelihood of such overlap. For this reason, membrane filters are used whenever security of particle capture is the goal.

INFLUENCE OF TECHNOLOGY OF MANUFACTURE

To understand why membrane filters have relatively narrower pore-size distributions, we must consider the technology involved in their manufacture and compare its influences with those of the technology of manufacture of other types of filters.

Method of Microporous Membrane Manufacture

A solution is made of polymer, solvent, and usually of pore-forming agent. A thin coating of this casting solution is spread very evenly on a smooth surface. Usually, evaporation of the solvent forms the membrane into its wet gel state. However, other means such as moisture imbibation or temperature lowering may be employed. What is important in this process is that in any solution, whether polymer in solvent or sugar in coffee, the molecules of the dissolved solid or solute disperse evenly throughout the entire volume of solvent and therefore become spaced at equal distances from their neighbors; that is, they become separated by spaces of equal dimensions. The equidistant separation of the solute or polymer molecules from one another is not an accident. It is an invariant consequence of the thermodynamic laws that govern solutions. This is important because the pores of the microporous filters arise from these spaces, from the intersegmental spaces within the polymer solution (Figure 4.1).

In other words, because solution technology is used in the manufacture of microporous membranes, the spaces separating the polymer segments in the casting solution and the pores arising from these spaces are all rather equal to one another in size. To be sure, perfection does not prevail. There are some differences in the size of the pores, but these are small because the pore size is being directed toward uniformity by solution thermodynamics. Thus, because solution technology is used in their manufacture, microporous membranes have a narrow pore-size distribution.

Method of Depth-Type Filter Manufacture

This large and important class of filters is manufactured by positioning individual fibers, or bits and pieces of ceramic, metal, or plastic on a surface, and then matting, gluing, or sintering the fibers or particles into a mat or solid composition wherein the spaces among the fibers or particles constitute the filter pores.

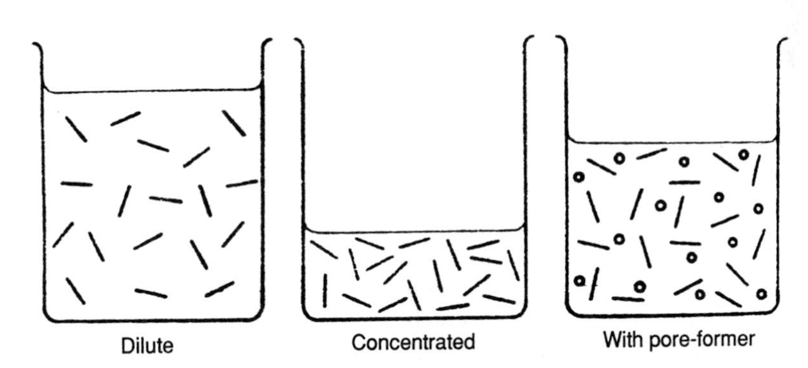

Figure 4.1 Polylmer solutions

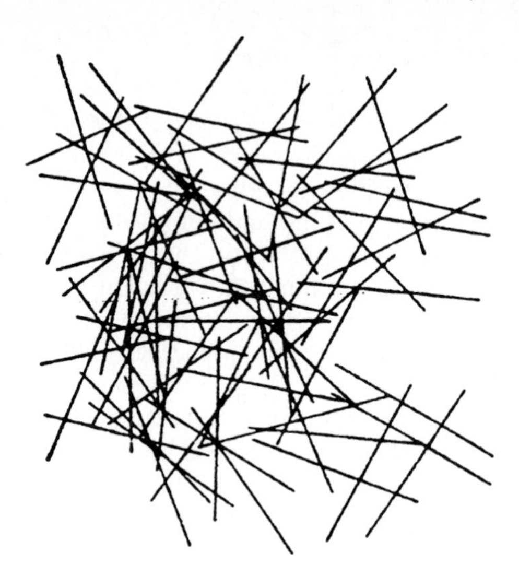

Figure 4.2 Modeling a fibrous filter by a system of lines drawn at random. N = 25.

In this method of manufacture, in principle, fibers are laid down one at a time until the desired filter mat is formed. As shown in Figure 4.2 this fiber deposition follows the laws of chance. The result is a randomness as to where the fibers become positioned and a consequent unevenness in their spacings. Therefore, the spaces among the fibers (i.e., the pores of these filters) vary very much in size. Thus, this technology of manufacture involving the randomness of the laws of chance produces filters that have broader pore-size distributions.

Depth-Type Filters

The breadth of the pore-size distribution is the one structural feature that differentiates among filters—specifically between membranes and depth-type filters. From this single structural difference, however, stems a long range of derivative properties.

Because of their broader pore-size distributions, depth-type filters have certain properties. They have a greater number of larger pores (as well as a larger number of smaller pores) compared to membrane filters of the same pore-size rating. Liquid flow during filtration preferentially takes place through larger pores. In fact, the rate of flow through a pore of micrometric or submicrometric dimensions varies as the fourth power of its radius. Therefore, liquid flow through larger pores is much greater than through pores of a smaller diameter. Thus, depth-type filters, being "more open," give greater rates of flow than would membrane filters of a similar pore-size rating.

When a particle contained in a liquid is subjected to depth-type filtration, it may encounter a pore smaller than itself in the surface of the filter and will, therefore, be retained by sieve capture. Because the depth-type filter has a broader pore-size distribution, however, the particle may also encounter one of the larger pores and will penetrate into the depth of the filter before it is retained. This is desirable because a large amount of particulate matter can be retained on the vast inner surfaces of the depth-type filter before it clogs. Thus, it is said that depth-type filters have a high particulate loading capability, or less elegantly, a large dirt-holding capacity.

This large dirt-holding capacity, coupled with the less expensive price of depth-type filters, means that depth-type filters should be used wherever possible for economic reasons. That is, one gets a lot of particulate removal at a relatively low cost with depth-type filters.

The same broad pore-size distribution that makes possible the penetration of particles into the body or depth of the filter can, however, provide pathways, on occasion, completely through the filter. Thus, depth-type filters are not totally reliable when it comes to particle retention. Such filters are said to have only nominal ratings.

NOMINAL AND ABSOLUTE RATINGS

Filters are often described as either absolute or nominal. Actually, this characterization does not describe an inherent quality of filters. Whether a given filter is nominal or absolute depends on the particle-size distribution that confronts its pore-size distribution. For instance, let us consider the wire mesh fence surrounding a tennis court. One of its functions is to keep tennis balls within the court. For this purpose it is an absolute filter. The ordinary cyclone fence used for this purpose will reliably retain all standard-size tennis balls. This type of fence is also an absolute filter for basketballs, and indeed for any size particle larger than its pores or holes. However, the same fence may or may not retain golf balls or marbles that may be driven at it. The retention in these cases is, therefore, said to be nominal. The same fence (the filter) is both absolute and nominal, depending on the size particle it encounters. Thus, a filter that is nominal in one context may be absolute in another.

Consider the encounter of particles with a filter of a given pore-size distribution. Let us begin with particles, the smallest of which are larger than the filter's largest pores. The filter, however broad its pore-size distribution, will be absolutely retentive to these particles. If we now consider, for the same filter, particles progressively smaller in size, then the point is reached at which the largest pores of the filter are larger than the smallest particles, and retention of these particles may or may not take place. Particle capture will depend on whether these smaller particles encounter the larger pores within the pore-size distribution. In this case, the filter, relative to this given particle-size distribution, will be said to be nominal.

As stated earlier, the broader the pore-size distribution of a filter, the more likely it is to be nominal in its retentivity, because the greater the likelihood of its larger pores overlapping the smaller particles of a given particle-size distribution. It is for this reason that in the filtrative removal of submicrometric and near-micrometric particles, microporous membrane filters with their narrower pore-size distributions are more likely to be absolute than are depth-type filters.

SURFACE RETENTION

When correctly sized for an application, membrane filters are absolute. That is, they retain all the particulate material present in the fluid. All of the particles, even the smallest, being too large to enter the filter pores, come to rest on the filter's surface. Depending on the nature of the particles, large loadings may block the filter's surface, possibly even to blind or seal it off, so that further filtration is interfered with, or perhaps made impossible altogether. This is the weak point of membrane filtration. In return for the absoluteness of retention (based on sieve-type capture), the filter may accommodate only small surface loadings of particulate matter before its useful life is compromised.

PREFILTER/FINAL FILTER COMBINATIONS

Because membrane filters can accommodate only limited amounts of particulate material, they are not used by themselves where large particle concentrations are involved. Instead, they are used in combination with prefilters, usually depth-type filters. The prefilter, upstream from the final membrane filter, removes most of the particulate material. Only the small portion of

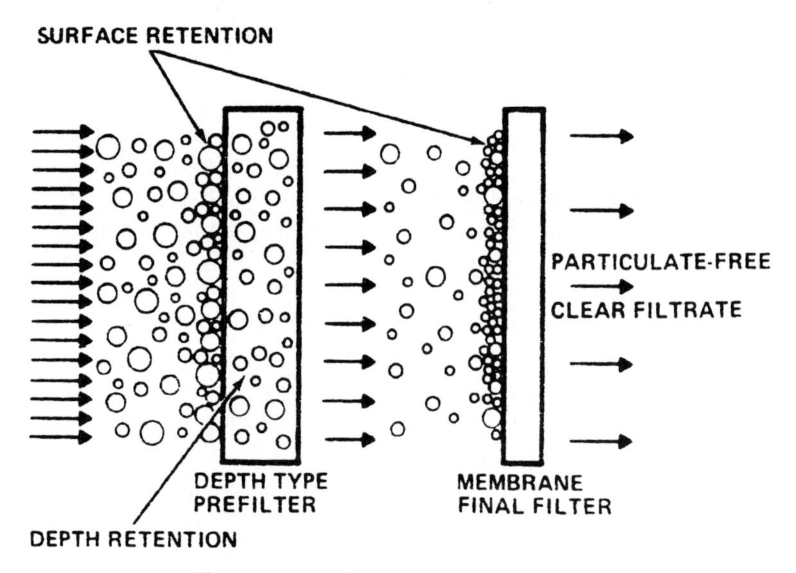

Figure 4.3 Prefilter/final filter combination.

particles that escapes capture by the prefilter because of its nominal retention emerges to confront the final filter. In correct prefilter/final filter arrangements, the particles will have been so reduced in concentration by the prefilters upstream that they will not prematurely block the final membrane. The final filter, correctly selected for its pore-size rating, will retain them absolutely while giving the filtered solution its ultimate characteristics of quality.

Even when a fluid has a high particulate count, it can be purified filtratively without prefilters by using more final filter area. Membrane filters, however, are more expensive than depth-type prefilters. Therefore, economical filtration requires that the bulk of the particulate matter be removed by prefilters upstream from the final filter. Particularly in processing contexts, one should not think in terms of a single filter, but rather in terms of prefilter/final filter arrangements (Figure 4.3).

DIFFERENTIAL PRESSURE LEVELS

A filter creates an impediment to flow by its presence in a system. At whatever pressure a fluid flows through piping, a higher pressure will be needed to continue the flow at the same rate if the blocking interference of a filter is present. The difference between the higher pressure upstream from a filter and the lower pressure downstream where the effluent is emerging from the filter is known as the applied differential pressure. This causes the liquid to overcome the resistance of the filter and to flow through it.

Filters work best at low ΔP. To be sure, at higher pressures, the rates of flow (the volumes of fluid flowing per unit time) are greater. However, the total throughput may or may not be as great because the life of the filter is compromised at higher ΔP, especially if the particulate material being removed is deformable and "blinds" the filter surface. Certainly the economics of filtration require the use of lower ΔP. Additionally, as we shall see, filter efficiency in terms of particle trapping is often enhanced at lower ΔP.

LIMITATIONS OF DEPTH FILTERS

Unloading

With membrane filters, when the system is properly sized, particle retention based on sieve capture will take place, regardless of whether ΔP is high or low unless, of course, the pressure is so high as to distort either the shape of the particle or of the pore in a way that lets the particle pass through the pore. This follows from the fact that the membrane is a continuous structure of permanent form. Such occurrences have not been reported at the modest pressures (below 90 psi) used in ordinary process filtration. (Actually, under higher ΔP, the membranes would compact, making particle passage more difficult.)

In the case of depth-type filters, this is not necessarily the case. On the basis of long experience it is known that certain depth filters may capture particles efficiently at low ΔP, but may "unload" at higher ΔP. That is, particulate matter already retained on the filter may be unloaded into the emerging clean filtrate stream when ΔP exceeds a certain value. The reason for this unloading is not always fully understood. In some cases it may be due to channeling (a dislocative movement of the fibers constituting the depth-type filter). Particles trapped by sieve retention might become dislodged when the fiber arrangement constituting the pore is disturbed. Also, retentions that can take place at lower ΔP may not occur as efficiently at higher ΔP. This can happen for all types of filters, but its occurrence is more noticeable in filters with broader pore-size distributions, i.e., depth filters.

Filter-Medium Migration

The U.S. Food and Drug Administration (FDA) enjoins against the presence of fibers in injectable solutions; a fiber is defined as any particle having an aspect ratio of greater than 3 to 1, that is, any particle whose long dimension is at least three times its short dimension. Because of their technology of manufacture, depth filters often have bits and pieces of fibers or other particles of their materials of construction imperfectly incorporated into the filter mass. When liquid flows through such a composition, the loose bits, pieces, and/or fibers are washed out of the filter composite and emerge in the filtered stream. This is known as filter-medium migration.

Depth-type filter manufacturers usually prewash such filters to eliminate unincorporated particles. However, according to the technical literature, when such a filter is exposed to a hydraulic shock, such as the water hammer caused when a filling machine valve closes, a new generation of particles or fibers is set loose. For this reason, wherever particles are strictly to be avoided in the filtrate, depth-type filters should not be used as a final filter.

Where depth-type filters are used for the advantages they offer, but where their fiber shedding is eschewed, they should be followed by final membrane filters. Because the pore passageways through membrane filters are very sinuous and convoluted, it is next to impossible for fibers to make their way through such filters. Therefore, membrane filters are excellent safeguards for the removal of fibers, regardless of origin.

OTHER MECHANISMS OF PARTICLE CAPTURE

Electrostatic Capture

Sieve retention is not the only means whereby particles are removed from fluids by filters. Membrane filters acquire electrostatic charges when gases such as air flow through them (or even over them) and, therefore, retain particles by this mechanism as well as by sieve retention. For this reason, filters are often ten times more efficient at particle removal in gas filtration than in liquid filtration. Thus, in coal mine atmosphere monitoring, it was found that coal-dust particles as small as 0.1 μm in diameter are completely retained on the walls of 5-μm-rated pores in polyvinyl chloride membranes.

Where electrostatic retention is involved, a particle may be small enough to enter a pore (escaping arrest through sieve retention), and yet be captured by electrostatic attraction. This mechanism does not operate to any significant degree where aqueous solutions are involved because water is a medium of high dielectric constant. Such media attenuate the interactions of the positive and negative electrostatic charges, thereby reducing their trapping efficiencies.

Electrokinetic Effects

A counterpart to the electrostatic attractions that operate in gaseous media are the electrokinetic attractions common to liquid media. In these, electrokinetic charges are generated on the filter's surfaces. These trap particles having opposite electrokinetic charges. Asbestos has the natural advantage of maintaining a positive zeta potential over a large pH range. It, therefore, serves as an excellent filter for most negatively charged particles.

Certain membrane manufacturers offer filters that can remove the mucopolysaccharides that are the seat of pyrogenic activity. This is done through use of the positive zeta charges that derive from grafting amine and certain other functional groups to nylon filter surfaces.

Positive Zeta Potential

According to Cohn's rule, when a solid particle is suspended in a liquid, the zeta potential charge characterizing its surface is a function of the dielectric constants of both materials. Consider an aqueous suspension of silica. Water has a dielectric constant of 70; silica, a dielectric constant of 2. Therefore, silica particles in aqueous systems have negative zeta charges. Asbestos, on the other hand, has a dielectric constant greater than that of water. Consequently, asbestos in water assumes a positive zeta potential.

Since most solid particles bear negative zeta potentials in water, their arrest by this particle capture mechanism requires filters with positive zeta charges. Asbestos is a superior filter material, among other reasons, simply because it naturally possesses this property. Regrettably, certain forms of asbestos are carcinogenic. The use of asbestos materials is, therefore, effectively prohibited. Positive zeta charges can be induced into a filter surface by introducing a density of amino groups onto its surface, as through chemical bonding.

Two manufacturers of microporous nylon membranes offer such products and advocate their use in pyrogen removal. By all accounts, within the restrictive limits of the useful pH ranges necessary to positive zeta potential development by these products, they do the job. There are, however, two objections to their use.

First, the quantity of negatively charged material arrested by the positive zeta charge depends on the magnitude of the latter. The relationship between the quantities of oppositely charged materials that become mutually attracted and fixed is stoichiometric, not limitless. Therefore, in actual filtration involving negatively charged pyrogen particles, these will be removed by the positively zeta-charged filter until its charges become satisfied and the charge density becomes reduced to zero. At that point, further pyrogenic material will no longer be retained by the filter. What is worse, however, is that the normal indicator of filter insufficiency (the onset of significant filter blockage) will never become evident. Its absence may cause the filter operator to conclude erroneously that pyrogen removal is still in progress. The situation does not make for reliable operations.

Second, the FDA requires that products be manufactured by processes whose steps have been validated to demonstrate that their functions yield product of the necessary high quality. To be sure, quality control operations are then used to spot-check, and to confirm the proper ongoing nature of these validated practices. This is far different from manufacturing a product, and then using quality control or other tests to sort out product of good quality from that of bad in a pick-and-choose type of operation.

Water for injection is supposed to be prepared pyrogen-free, as are the various chemical components eventually intended for compounding with it. It is not within the FDA guidelines for pyrogenic water to be prepared so as to necessitate additional cleansing to an acceptable level of quality. Indeed, the presence of pseudomonads in the prepared water is prima facie evidence that its mode of manufacture is unsuitable. In these circumstances, the use of pyrogen-removing filters would not seem appropriate. Their use would make acceptable a product resulting from an unacceptable manufacturing procedure. Nevertheless, there undoubtedly are situations in which filtrative removal of pyrogens will be of great value, and where filters with positive zeta potentials will be useful.

Absorptive Sequestration

Only retention by sieve-type capture requires that particles be too large to enter the filter pore. There are, however, many types of capture mechanisms, two of which have already been discussed, that operate to remove particles from fluid streams even when these particles are small enough to enter a filter pore.

It is known from air filtration experiences that when a particle comes within a very small distance (the attractive forces acting in accordance with the inverse-square law), capture will result through some type of contact or adsorption force. Thus in air filtration, particle arrests arise from inertial impaction of larger particles with the filter surface, and from Brownian motion that adventitiously brings smaller particles into contact with the filter.

It has been shown that influenza vaccine purification by filtration takes place through adsorptive sequestration of proteinaceous impurities, presumably through hydrogen bonding. Certain membrane materials, especially the mixed esters of cellulose, which can remove proteins from solution are capable of removing many soluble dyes from solution, and can capture and remove viruses and organisms from liquid suspensions. Membrane materials differ in their adsorptive properties. Thus, the mixed esters of cellulose are used when adsorptive arrest of particles or molecules onto the membrane surface is sought. Other polymers, such as polysulfone, are employed when protein adsorption is to be avoided, as in serum filtration.

Organisms are also retained by filters through contact or adsorptive capture. A 12-μm-rated pore-size Nuclepore® filter can effectively capture Staphylococcus aureus organisms by adsorption onto its polycarbonate surface. Adsorptive forces are strong enough to require a very definite water pressure to break loose the microbial plaque; this is an index of the strength of the adsorptive bond.

In sieve retention, a particle is sterically arrested by a filter only if it is too large to enter the pores. In adsorptive sequestration, a particle may be small enough to enter a filter's pores, but nevertheless may be retained. In making its way along the pore passageway, the particle may come close enough to the pore wall to become subject to such attractive forces as hydrogen bonding, electrokinetic attractions, or other electrical charge-induced phenomena, such as van der Waals or secondary valence forces, or even hydronobic interactions. In such cases, particle capture will result.

Adsorptive sequestration is therefore a redundant mechanism of particle capture, reinforcing sieve retention in instances where the pore-size and particle-size distributions render the latter imperfect. An organism may well escape sieve retention of a filter, yet undergo reliable capture through adsorptive sequestration. It is, perhaps precisely because of such reinforcing capture mechanisms that membrane filters far exceed performance predictions based solely on their pore-size measurements.

RETENTION BY 0.45 μm-RATED MEMBRANES

Evidence of adsorptive sequestration effects on organisms is forthcoming from the retention of Pseudomonas diminuta by 0.45 μm-rated membranes.

Interesting evidence is derived from flow-decay studies. As a filter progressively accumulates the particulate matter its purpose is to retain, interdiction of the filter pores increasingly occurs. Flow decay results and the rate of filtration declines; indeed, such flow decay is an index of the extent of particle removal by the filter.

Where sieve retention is the particle capture mode, and where the particles are nondeformable (or are not deformed at the applied differential pressures used), a filter cake builds on the surface of the filter. Liquid can permeate such a cake but at an impeded rate. At a first approximation, the flow decay of such systems is described by the equation:

$$\frac{t}{V_{(t)}} = \frac{k}{2}(V_{(t)} + 2V_f)$$

where V_f = volume of filtrate required to produce a change in total resistance equal to that of the filter
k = filtration constant

If the flow-decay data are plotted in conformity with t/V versus V and a straight line results, it is presumed that particle capture and pole blockage are the consequences of surface retention, i.e., sieve-type capture.

Where the particles are small enough to enter the pores before becoming adsorptively fixed to the pore walls, flux declines due to pore clogging results. In these instances, flow decay is expressed by the equation:

$$\frac{t}{V_{(t)}} = \frac{kt}{2} \frac{1}{J_u(O)}$$

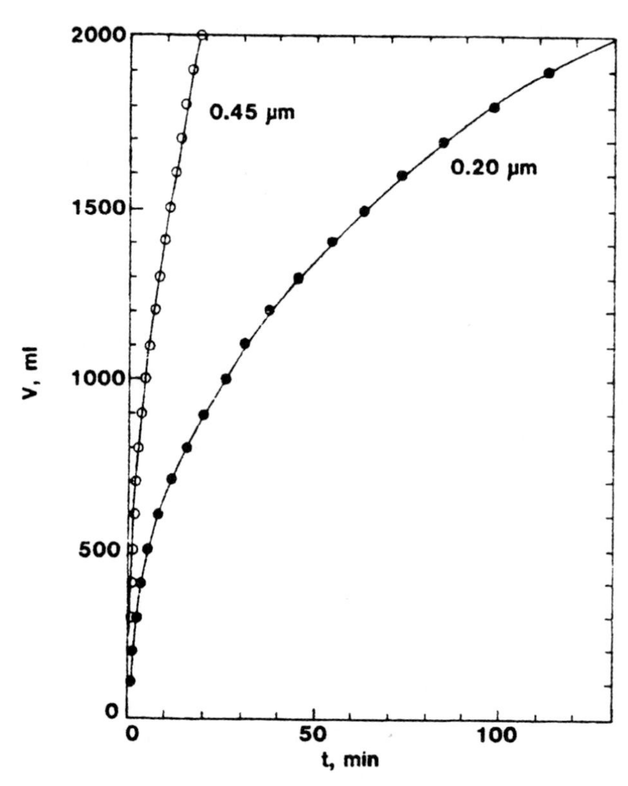

Figure 4.4 Plotting flow decay data. Volume throughputs using 0.2- and 0.45-μm-rated cellulose triacetate membranes.

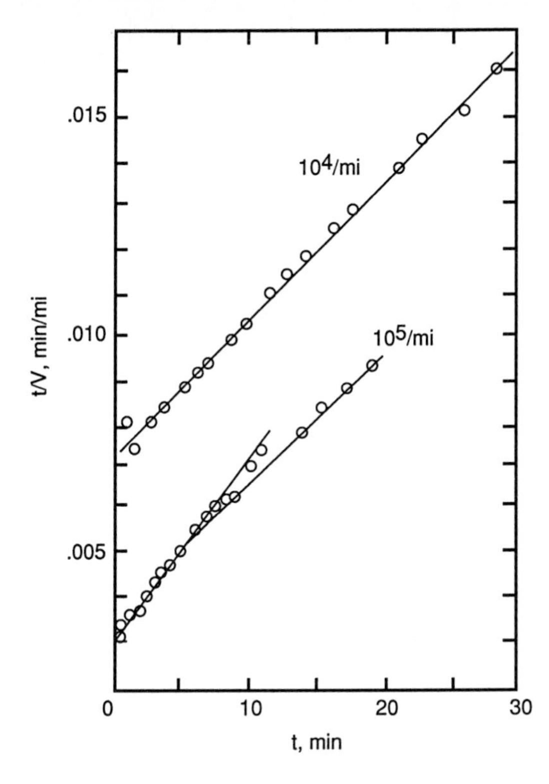

Figure 4.5 Adsorptive retention plots of P. diminuta at two concentrations using 0.45-μm-rated cellulose triacetate membranes.

Plotting the flow-decay data in accordance with $t/V(t)$ versus t will yield a straight line if pore clogging is the result of particle retention, presumably through adsorptive arrests.

Therefore, it is possible to carry out flow-decay measurements using organism suspensions and membrane filters, and to ascertain the prevalent capture mechanisms leading to the filter's impairment of flow.

Tanny and Meltzer [5] carried out such experiments using 0.45 μm-rated and 0.2-μm-rated membranes. The flux decline shown in Figure 4.4 for the 0.45-μm-rated cellulose triacetate membrane shows an almost straight line decrease, a consequence of its rather open porosity. In contrast, the flow decay of the corresponding 0.2-μm-rated filter shows a more precipitous decrease, as the more finely sized pores are more immediately blocked.

Plotting the data in Figure 4.5 in accordance with the adsorption mechanism equation $(t/V)/t$ yields a straight line for the bacterial feed concentration of 10^4/ml retained by the 0.45-μm-rated filter. This straight line indicates that the bacterial arrest is the result of adsorptive sequestration.

The line in Figure 4.5 derived for the 0.45-μm-rated membrane using a bacterial feed challenge of 105/ml shows an initial straight line followed by the onset of a curve. This shape shows adsorptive retention leading to pore clogging (more rapidly realized as a result of the higher bacterial concentration), followed by sieve retention of the bacteria subsequently filtered out by

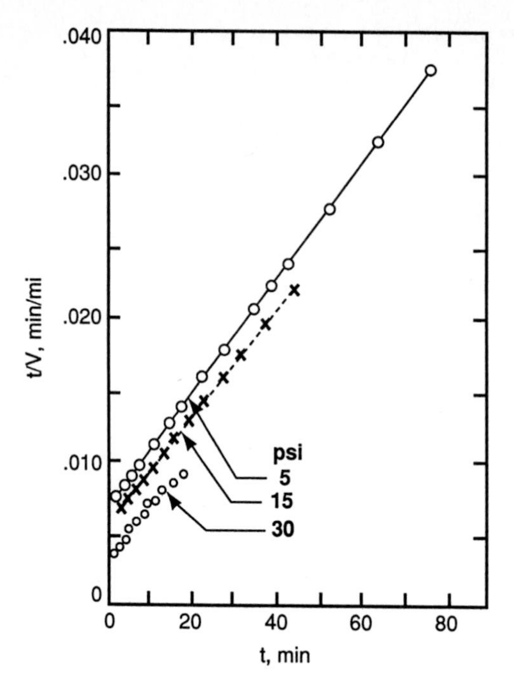

Figure 4.6 Adsorptive retention plots of P. diminuta at 10^5/ml concentration filtered on 0.45-μm-rated cellulose triacetate membranes at three different pressures.

the now clogged, and, hence, smaller-diameter pores. The total organism challenge level + $2 \times 10^7/cm^2$ of membrane surface.

Tanny and Meltzer demonstrated an inverse relationship between the extent of *Pseudomonas diminuta* retention by the 0.45-μm-rated membranes of the ΔP of the filtration.

Figure 4.6 shows that the adsorptive retention mechanisms still describes the filtration, with the higher pressures making for a more rapid clogging (abetted by the compressive deformation of the bacterial plaque), and showing evidences of sieve retention at 30 psi, the highest ΔP used.

If instead of the 105/ml challenges of Figure 4.5 that translate to 2×10^7 organisms/cm^2 membrane surface, substantially lower bacterial levels of 10^3 or 10^2/ml are used, then even at applied differential pressures of 30 psi, bacterial retention by adsorptive sequestration is apparently absolute. This work offers confirmation to the findings of Bowman et al. [12] that *Pseudomonas diminuta* are retained by 0.45-μm-rated membranes at concentrations up to 105 organisms/cm^2 of filter surface. The appropriate flux-decline data plots indicate that the 0.45-μm-rated filters are involved in adsorptive sequestration of the organisms. These conclusions seem confirmed by the retentivity response to the ΔP. Sieve retention would have been independent of the filtration pressures used.

PROBABILISTICS OF SIEVE RETENTION

Only in the circumstances where all the particles involved are larger than all the pores present is filtration (i.e., sieve retention) absolute in its nature; this condition is axiomatic in solid geometry.

There are, by some reckonings, approximately 10^{10} pores/cm^2 of 0.2-μm-rated microporlous membrane having some 80 percent total porosity. As stated, the arrangement of these pores reflects a pore-size distribution. The particles whose removal by these filters is being sought exhibit, in their turn, a particle-size distribution. Depending on the number of particles present relative to the number of appropriately sized pores, it is possible that a particle small enough to escape sieve retention by a larger pore may not encounter such a pore during an actual filtration. Retention of such a particle would thus occur, despite the existence of conditions making it possible to elude capture (i.e., pores of larger sizes). The larger the number of such particles and such pores, the greater the probabilities of their encounter. There is, thus, a probabilistic nature to sieve retention. It depends on the particle-size and pore-size distribution relationship.

The encounter of the rare smaller particles with the occasional largest pores does not depend on randomness alone. The flow of liquid through a pore is a function of the fourth power of its radius. Hence, liquid flow becomes hydrodynamically directed to the largest pores regardless of their paucity. In consequence, the particles suspended in the liquid, including the smallest, are also preferentially guided to the larger pores. Nevertheless, except for the condition where all particles, even the smallest, are larger than the smallest pore, particle retention by the sieve mechanism is probabilistic in character.

PROBABILISTICS OF ADSORPTIVE SEQUESTRATION

When a particle is small enough to enter a filter pore it can meet one of two destinies. It can negotiate the entire pore pathway and escape with the convective stream, or during its progression through the pore passageway, it may encounter the pore wall and become arrested by contact capture. Two considerations govern the fate of the particle: ΔP and the number of particles present. These two factors imbue adsorptive captures with a probabilistic stamp.

The larger the number of attempts by particles to run the gauntlet of the pore pathway, the greater the probability that some will make their way unimpeded. Therefore, the surety of particle retention by adsorptive sequestration is sensitive to the particle challenge level.

There is a practical aspect to the enhancement of particle retention by adsorptive capture. As stated, such particle arrests reflect the probability of pore wall encounter, an occurrence dependent largely on the residence time of the particle within the filter pore. The longer the duration of the particle within the pore, the greater the probability of its encounter with the pore wall, leading to its capture. What determines the length of time the particle is resident in the pore? Essentially, the velocity of the liquid stream going through the pore does. This, in turn, is largely dependent on ΔP. It follows, therefore, that the use of lower ΔP gives a lower velocity to the stream going through the filter, and enhances the probability of particles carried by that stream being captured by adsorption.

Use of low ΔP is impractical in processing large volumes of solution. Longer processing times will be needed. What is required, however, is not higher ΔP but more extensive membrane surface area coupled with lower ΔP. This results in lower velocities of the liquid through each of the many more pore paths of the more extensive filters. The lower face velocity, flux, or rate of flow per given area that results from the use of lower pressures is compensated by the added flow through the increased filter area. Increased efficiency in particle capture by adsorption is thus favored.

EARLY WORK

The concern of the Health Industries Manufacturing Association (HIMA) committee dealing with sterilizing filters was focused on organism challenge levels and on ΔP as a consequence of the work of Wallhausser and Tanny et al. [9] who demonstrated, as had Bowman et al. that the retention of P. diminuta by membrane filters was a function of these two operating factors. Wallhausser had reported the passage of pseudomonads through 0.2-μm-rated sterilizing filters, although these conceivably could have been natural water varieties smaller than P. diminuta. The work of Tanny et al. is uncertain concerning the absoluteness of P. diminuta retention by 0.2-μm-rated membranes at the initial stages of filtration, when the largest pores would have been sought out by the largest quantity of liquid flow.

Literature searches revealed that Elford [15] had elucidated a curve of organism challenge level versus pore size. The organism involved was Bacillus prodigiousus. Elford found that the regions that gave "active" filtrates (those that were unsterile) were most extensive when the operating factors of organism challenge level and pore size were highest. As the challenge level decreased, even for large pore sizes, the region of sterile filtrate increased. Similarly, as the pore size decreased, the region of sterile solution production increased even when the organism challenge level became greater. Clearly, both the organism challenge level and the pore-size influences asserted themselves in Elford's earlier work in accordance with what is now rationalized on the basis of adsorptive sequestration. Elford found that below a certain pore size, sterile filtrate was always produced, regardless of the organism challenge level, at least within his capacity of experimental measurement.

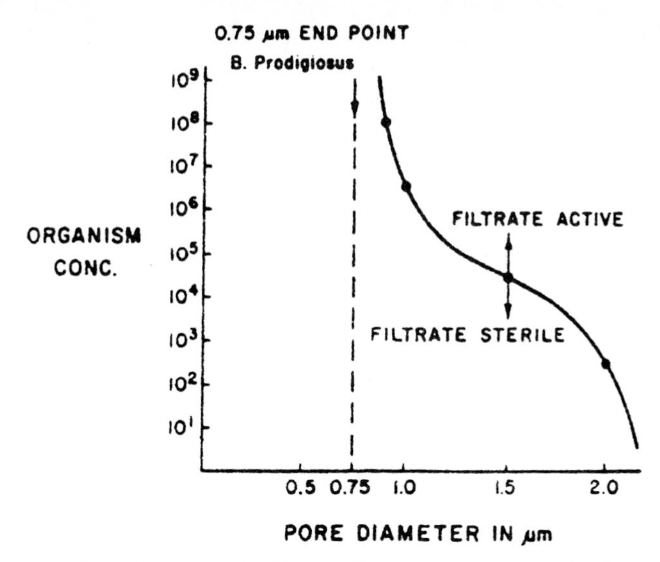

Figure 4.7 Pore diameter vs Bacillus prodigiosus concentration. Sterile or active effluents are a result of these parameters.

This region, it is now rationalized, reflects the absoluteness of sieve retention, the consequence of the smallest B. prodigious (now commonly called Serratia marescens) exceeding in size the pore dimensions of Elford's "cutoff" pore size (Figure 4.7).

ADSORPTIVE SEQUESTRATION HYPOTHESIS

The concept of adsorptive capture of organisms, specifically P. diminuta, that are small enough to enter a filter's pores has been discussed. Not all investigators are in agreement concerning this hypothesis. Sieve retention would be responsive to challenge levels on the basis of large-size pore/small-size particle encounter probabilities, however moderated by predominant flow through larger pores. The effects of higher ΔP on the retention of diminuta from aqueous suspensions can also be rationalized on the basis of sieve retention by theorizing that at higher pressures the rod-shaped organisms may assume an alignment within the rapidly flowing solution, which might increasingly preclude their long dimensions encountering the blocking pore orifices.

The opponents of adsorptive sequestration have demonstrated that the efficiency of P. diminuta trapping by 0.45-µm-rated membranes are uninfluenced by ionic strength, pH variations, wetting agent, or protein concentrations; these latter two conditions are seen as supplying modifiers to the membrane surface. However, adsorption phenomena are complex and their influences on organism retention are too little understood. In any case, the preceding treatments would not necessarily affect hydrophobic interactions, however influential on charge-dependent phenomena. As regards sieve retention, explanations are required for Wallhausser's findings that 0.35-μm-rated latex beads of polystyrene can negotiate 0.2-μm-rated membranes while 1- × 0.25-μm rod-shaped pseudomonads do not.

Of greater significance is that whether by the workings of sieve retention or adsorptive sequestration, the response of organism capture efficiencies to organism challenge levels and to ΔP gives practical importance to the control of these operational factors in the filtration processes.

EXPERIMENTAL ELUCIDATION OF CAPTURE MECHANISMS

Where knowledge of the particle-size distribution is known, filters of suitable pore-size distributions can be selected to make sure that absolute sieve retention is achieved. When the particle-size distribution is unknown, absolute sieve retention may not prudently be assumed. In such cases, the completeness of particle retention must be expressed experimentally as a function of a defined particle challenge level and ΔP. It is possible, in such cases, that adsorptive sequestration is involved and that complete retention at one condition of particle level and/or ΔP may not be as reliable at another higher level. Thus, Bowman et al. found that 0.45-μm-rated membranes retained some 10^5 P. diminuta/cm^{-2} filter surface at 15 psi, but failed at such complete retention at higher challenge levels. Clearly complete organism retention at the 105/cm^2 level was not a consequence of absolute sieve retention. Rather it reflected an experimental matching of that organism challenge level at the 15-psi ΔP to the adsorptive capture properties of the filter sufficient for complete organism retention to occur.

Were complete organism capture to take place at one level of challenge and ΔP, and remain uncompromised at successively higher levels of both challenge and pressure, it could be concluded that the complete retention was independent of these factors, an indication that the particle-size and pore-size distributions for that filtration experience reflected absolute total sieve retention. There is evidence that the retention of P. diminuta by sterilizing membranes is a case of absolute retention.

ABSOLUTE RETENTION OF P. diminuta BY 0.2-μm-RATED MEMBRANES

Work at the Gelman Sciences laboratories showed that P. diminuta were retained completely by 0.2-μm-rated membranes at challenge levels of 2 × 10^7 organisms/cm^2 filter surface at ΔP up to 45 psi (3 bars).

Subsequently, work at the Pall and Millipore laboratories showed variously that the same completeness of retention of P. diminuta by 0.2-μm-rated membranes eventuated even when ΔP was as high as 90 psi (6 bars).

TABLE 4.1 IMPACT OF PRESSURE ON PASSAGE (RATIO)

Filter type	Pore size (μm)	0.5 psid	Ratio 5 psid	50 psid
GS	0.22	$>10^{10}$	$>10^{10}$	$>10^{10}$
HA	0.45	10^8	10^7	10^6
DA	0.65	10^4	10^4	10^3
AA	0.80	10^2	10^1	10^0

Moreover, sterilizing-grade filters retained challenge levels of 10^{10} organisms/cm^2 filter surface.

Work at the Millipore laboratories (Table 4.1) confirms work earlier reported. The smaller the pore size, the greater the organism challenge level it can withstand. Concomitantly, the lower the ΔP for a given pore-size rating, the higher the organism challenge level the filter can endure. All these results are in keeping with organism capture by adsorptive sequestration. In the case of the 0.2-μm-rated membrane, however, P. diminuta retention is complete even at 101° organisms/cm^2 filter surface and at ΔP as high as 60 psi—a manifestation of sieve-type arrest.

It is evident that 0.2-μm-rated membranes being produced by various filter manufacturers for sterilizing purposes are complete in their retentivity of P. diminuta regardless of the challenge levels and the applied differential pressures. From this it can be concluded that the sterilizing membrane filters are absolute in this retention of P. diminuta.

MICROBES SMALLER THAN P. diminuta

The 0.2-μm-rated membranes defined as sterilizing against P. diminuta need not necessarily be absolute in their retention of smaller organisms. Thus, Wallhausser's natural waterborne pseudomonads could conceivably penetrate such filters at lower ΔP and even at lower organism challenge levels. The same holds true for the new microbes discovered in the Pall laboratories, even for the mycoplasmas periodically encountered in sera originating at Midwest packinghouses.

To be absolute against such organisms, it is conceivable that membranes with smaller pore-size ratings would be required. These tighter filters, however, have some disadvantages. Membranes with smaller-diameter pores will exhibit sharply reduced flow rates. This will result in appreciably higher filtration costs.

The designation of P. diminuta as the qualifying organism to be used by filter membrane manufacturers in their definition of *sterilizing filters* already bespeaks a worst-case situation. Pseudomonads are waterborne organisms.

Their presence in water supplies, particularly in pharmaceutical manufacturing contexts, is prima facie evidence that the water for injection is being manufactured improperly.

Whether a sterilizing filter is indeed suitable for the total organism removal required in a particular manufacturing process must be ascertained in the validation exercise by the filter user. That is the purpose of the validation. If a particular process does involve organisms smaller than P. diminuta, and these are normal to that manufacturing practice, then filters of smaller pore-size ratings may indeed by indicated.

PORE-SIZE RATINGS

The emphasis on particle retention brings into focus considerations of pore sizes and their measurement. There is, perhaps, no completely reliable way of directly measuring the pore sizes of microporous filters. Consequently, pore-size ratings are numbers that result from indirect measurements. These measurements involve assumptions regarding the pore structures that are oversimplifications of the facts. As a result, the numbers that result are only approximations. They do not represent real values of actual pore sizes.

Early on, pore sizes were characterized by the mercury intrusion method. In principle, mercury under increasingly higher pressures is progressively forced into smaller and smaller pores of a membrane. At each stage of pressure, the mercury uptake by the filter pores is measured. In this way, a histogram of pore size is made. This is not a very accurate method. In any case, claims that on its basis membrane pore-size distributions have been shown to possess a spread of 0.02 about the pore-size rating were shown to be untrue. One manufacturer states that a filter membrane of a given pore size will retain all "rigid" particles 0.2 μm larger than the pore-size rating. This too has been shown to be untrue by the work of Wallhausser, who demonstrated that 0.342-μm incompressible polystyrene latex particles can penetrate 0.2-μm-rated filters. Wallhausser's work, however, is largely unknown to the public, and the erroneous teaching is still largely accepted.

Numerous efforts have been made to size the largest membrane pores by elucidating experimentally what particle sizes the filters can absolutely retain. Adsorption phenomena cloud the results. Thus, Wallhausser found that 0.342-μm polystyrene latex beads can pass through 0.2-μm-rated filters under conditions that result in the total retention of organisms smaller in dimension (P. diminuta). Also, Johnston [23] showed that when aqueous slurries of two inorganic materials of the same particle-size distribution (silica and black iron oxide) were separately filtered through a given membrane, different particle retention efficiencies were manifested. Efforts at rating filters by bacterial sizing have been similarly ineffective [20].

There is an American Society for Testing and Materials (ASTM) method for membrane filter rating based on the mean flow-pore measurement. Airflow through a dry membrane is plotted as a function of pressure. The filter is then placed onto a pool of water, whereby water replaces the air in the pores. A plot is then made of airflow through the wetted filter as a function of pressure. A sigmoidal curve results. At half the angle of the dry airflow a line is now drawn to where it intersects the wet flow. This point is intensified, by definition, as the mean flow pore. A line is then extended from this point on the wet-filter flow to the pore-size scale and the value is read (Figure 4.8). This is not a direct measurement of the pore size. It is a hydrodynamic measurement. What it means is that half the flow through a wet filter is carried by pores larger than the mean pore size, and half the flow is carried by pores smaller than the mean pore size. It does not signify that half the pores are larger than the mean pore size and half are smaller; only that half of the volume flow is carried by pores smaller and larger than the mean pore size.

In the measurement of pore size by the mean flow-pore method, one is on much firmer ground than in using mercury porosimetry. Membrane filters are not used in applications involving the insinuation into their pores by mercury under very high pressures. They are, however, used in contexts wherein fluids such as air or water flow through them at moderate pressures, seldom exceeding 60 psi. Any mode of measurement is apt to have a greater pertinence when its conditions are similar to those of use.

The mean flow-pore method discloses membrane filters as having a pore-size distribution of some four times the mean-pore rating. Thus, if the

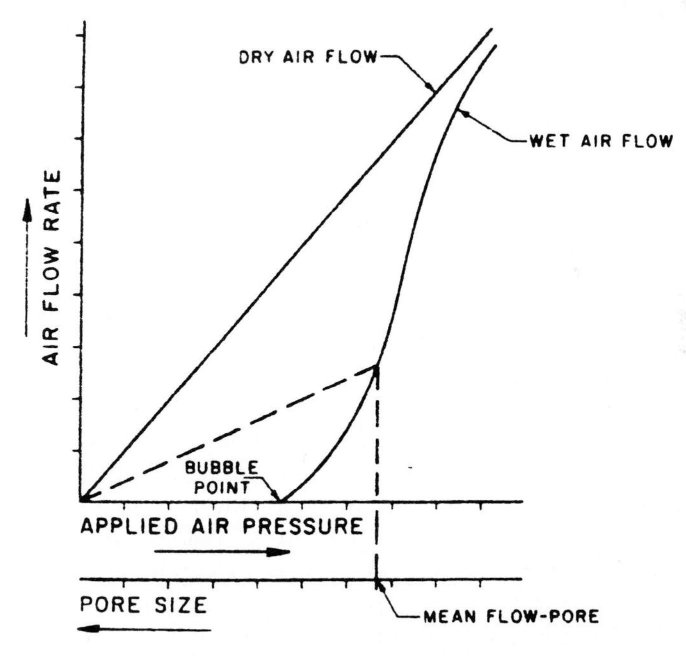

Figure 4.8 Pore size ratings: mean flow-pore measurement.

mean flow-pore has the value X, the largest pore size would be about X. This pore-size distribution is larger than had been suspected, but still much narrower than that of depth-type filters.

PORE SIZE AND RETENTION

On the basis of experience, filter engineers often learn when to use, say, 0.45-μm-rated or other "size" filters. The dimensional numbers for the filters should, however, not be taken too literally. The values are only approximations of the real but unknown dimensions. For this reason, indications that a membrane's pore-size distribution is sufficiently wide so that the largest pore is double the size of the mean pore leads to unduly pessimistic forecasts regarding particle retention. In actual operations, membrane filters far exceed in their actual retentive performances the prognostic implications of their pore-size ratings.

The pore size of the filter is critical where sieve retention is involved. As stated, if the smallest particle is not larger than the largest pore, sieve retention will not be absolute. Where adsorptive arrest is involved, the pore diameter is less critical but is not without influence. Given the situation where the particle is small enough to enter the pore, the narrower the pore diameter, the more likely the particle is to encounter the pore wall and to become captured before it exits with the convective stream. For this reason, greater filter thicknesses also make for increased particle trapping efficiencies. Thicker filters mean longer pore paths and, hence, longer residence times for the particle negotiating these pore passageways. This equates to greater probabilities of pore wall encounter and enhanced particle captures.

CAPILLARY RISE

Mean flow-pore measurement involves the evacuation of a water-filled pore by air under a suitable pressure. The water is held within the pore as a consequence of the capillary-rise phenomenon. Water will rise in a capillary because of the mutual bonding attraction between it and the pore surfaces. Thus, where no bonding attraction exists, such as in the case of mercury in a glass capillary, no liquid rise is evident. In such cases, the shape of the meniscus will be convex. Where attractive forces are at work, the meniscus is concave in shape, a result of the liquid lifting itself along the capillary walls. This rise occurs until the weight of the liquid column is just equal to the balancing pull of gravity. The forces involved in this phenomenon are set forth in the classical capillary-rise equation of Laplace. The liquid rise is expressed as $2\ r\ 2\pi r \times \gamma \times \cos\theta$ where $2\pi r$ describes the locus where the attractive forces exert themselves, at the perimeter of the circular capillary.

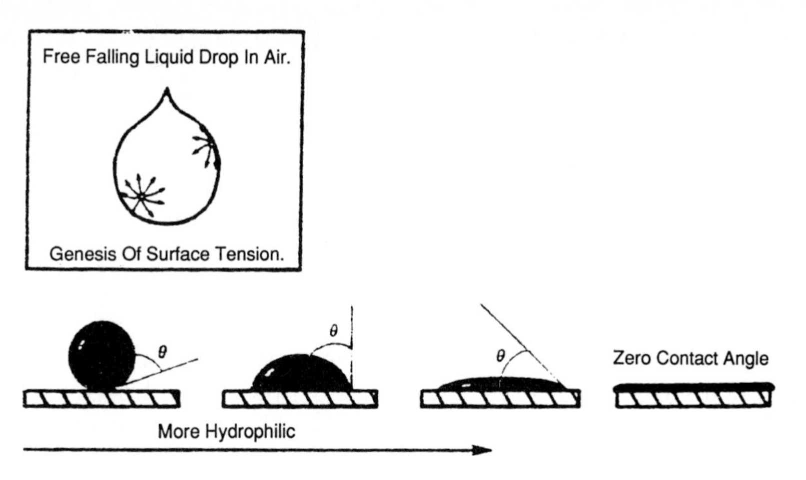

Figure 4.9 Wetting of solids surfaces: contact angle.

The surface tension of the liquid is represented by γ. Where perfect wetting is involved, the angle of wetting $\theta = 0$ (Figure 4.9). In this event, $\cos \theta = 1$. Balancing this force is the weight of the column of liquid, expressed as πr^2 for the area of the capillary occupied by the liquid, multiplied by the height of the column expressed as P, the pressure it exerts. The equation reads:

$$D = \frac{4 \times \gamma \times \cos \theta}{P}$$

The constant 4 relates to D, the diameter of the capillary (2 is the appropriate constant when the radius of the pore is being considered). From this equation calculations are made regarding the pore sizes being measured by this method. The numerator of the equation, where perfect wetting is involved, simplifies to K, and the equation to $D = K/P$. Different liquids confer different values to the numerator constant. Thus, water yields a numerator value of 30; kerosene, of 12.5.

The reason that this method does not accurately reflect the pore dimensions of membranes is that these filters are not characterized by circular capillaries. Hence, the $2\pi r$ factor, the perimeter of the circle, and πr^2, the area of the circle, do not apply. Assuming the circularity of pores where they are actually irregularly convoluted yields false values. Actually, a circle represents the smaller perimeter to enclose a largest area. Irregular perimeters will enclose smaller areas. For this reason, the calculated pore dimensions are too large [24]. Membrane filters exhibit a greater degree of retentivity than would be predicted from measurements based on the capillary-rise phenomenon.

A shape factor *L*/*A*, describing the actual length of the perimeter over the area it encloses, is required if true "pore diameters" are to be thus calculated. It has been found possible to elucidate such a shape factor for Dutch-twill cloth, a filter whose weave yields pores of equilateral shapes. It has not been found possible to describe such a shape factor for the irregular pores of microporous membranes.

FILTRATE CONTAMINATION BY WETTING AGENTS

Until the advent of the nylon microporous filter, wetting agent was added to all microporous filters to assure complete wettability by aqueous media. This assured prompt realization of the maximum flow rates of which these filters were capable. Additionally, it made possible the complete wetting necessary to measurements based on the capillary-rise phenomenon that also underlies the bubble-point determination. This latter consideration will be dealt with shortly. In any case, the use of wetting agents gave rise to extractables.

In one instance, polyhydrols (substances containing multiple hydroxy groups, which have a natural tendency to hydrogen bond and therefore to wet with water) were chemically grafted to the surface of an otherwise hydrophobic filter. The intent was to endow the otherwise hydrophobic polymer with a permanent film of wettable surface. This type of chemical grafting can undoubtedly be brought to an adequate state of development. At least in the early stages of manufacture, however, the resulting filter did yield extractables, presumably of the polyhydrol substance or its homopolymers.

Extractables issuing from the microporous membranes are undesirable as they contaminate the filtrate. Such extractables can particularly be offensive in the manufacture of small batch preparations where their concentrations would be greater.

The nylon filter is made of a naturally hydrophilic polymer. It requires no addition of wetting agent to assure the adequacy of its wetting. Therefore, such membrane filters can have admirably reduced levels of extractables. Much is made of this quality by nylon membrane manufacturers. However, all filters, no matter how clean the process of their preparation, must be presumed to have dust particles adhering to their surfaces as a result of the electrostatic charges generated by the passage of the air over their surfaces. Therefore, all filters should be rinsed, including nylon membranes, before flowing product solutions through them, to overcome the attracting electrostatic forces, enabling removal of the dust from the downstream side of the membrane. Rinsing the membrane filter is also advisable to ensure the removal of adventitious pyrogenicity, whatever the composition of the filter.

One nylon membrane manufacturer recommends a 10-liter flush per 10-in. cartridge to remove extractables, and a 20-liter flush to remove downstream residual particulates.

In the case of the more conventional membrane filters, those containing wetting agents, glycerol and hydroxypropylcellulose (a substance whose use for this purpose is accepted in pharmaceutical manufacture), a prior aqueous flush (approximately 20 to 30 liter/10-in. cartridge containing 5 ft^2 of membrane) is required to ensure the complete wetting necessary to integrity testing. A 10-liter preflush at 5 liter/min also removes the extractable wetting agent to 0.01 mg/l, the minimum level of measurement, so that its contaminative presence is avoided.

It would seem, therefore, that the advantages in this regard, of the nylon membranes are more apparent than real. What is required for any filter is that it be manufactured as cleanly as possible to avoid downstream particulates. Fabrication under good manufacturing practices (GMP) is highly desirable.

INTEGRITY TESTING OF MEMBRANE FILTERS

As stated, the capillary-rise equation can be used to calculate, however approximately, the pore-size rating of a membrane filter. When a membrane characterized by a pore-size distribution is dropped onto a pool of water, the liquid replaces the air in the pores. If such a filter is now mounted in an appropriate holder between two screens, and air pressure is applied from below, then the growing air pressure will expel the water first from the pore systems that have the largest diameters. This is so because these pores will have the minimum "wall effect" on the liquid they contain. The narrower the pore diameter, the greater the proportion of water contained in the capillary that will be involved in bonding attractions to the pore walls. For this reason, regardless of the number of pores a filter contains, the largest pore (or, more correctly, the largest pores, for there may be many of that size) will be evacuated first. This method, then, provides a means of measuring the "largest" pores of a membrane filter.

In the bubble-point integrity test [25], a completely wetted membrane is subjected progressively to incrementally increasing air pressure. (The imperatives of this thorough wetting dictate the use of wetting agents in membrane fabrication.) Overlying the wetted filter is a small depth of water. When the air pressure has reached the stage where the largest pores are evacuated of their liquid content, the air begins its passage through these channels and its egress is manifested in the form of bubbles rising through the overlying pool. The bubble point of a filter thus becomes characterized by the air pressure necessary to create its occurrence. It will be recalled that the bubble-point pressure is inversely proportional to the pore size of the membrane; the lower the bubble point the larger the pore and, conversely, the higher the bubble-point value the smaller the pore.

A bubble-point determination is carried out on a membrane filter at the beginning of filtration. This procedure may be repeated at intermediate stages of the operation, but in any case is performed at the end of the filtration. What is sought is invariance in the bubble-point value. This is taken to indicate that the filter did not undergo pore enlargement, whether as a consequence of damage or of some incompatibility with the solution being filtered.

There are other integrity tests as well, such as diffusive air passage measurement (also called the toward flow test), and the pressure-hold test. These are not covered in this book. Whatever the test, it serves only as a surrogate for the destructive but (ultimately) sole meaningful measurement of bacterial retention. To make the more convenient proxy test significant, an established correlation between its values and the organism retention of the filter must be demonstrated. Because the experimentally established correlation may depend on the precise procedures whereby the bubble-point or other integrity test was performed, it is essential that the filter manufacturer's procedure be adhered to precisely. To be sure, there are complexities even to these integrity tests. They are, however, well established and their practice has proved adequately dependable.

SUMMARY

Filtration is a complex operation involving the intricate interaction of many factors, some of which may as yet be unrecognized. In numerous applications, however, over long periods of experience, membrane filters have proved reliable. It may well be that our understanding of the total processes involved in filtration does not yet account for that high degree of dependability. Particle removal by membrane filtration is, however, a well established practice, successful when properly performed. This success augurs a growing future for the filtration technique and its expanding applications.

REFERENCES

1. Marshall, J. C., and T. H. Meltzer, "Certain Porosity Aspects of Membrane Filters: Their Pore-Distributions and Anisotropy," *Bull. Parenteral Drug Assoc.*, 30(5), 214–225.
2. Nickolaus, N., "What, When and Why of Cartridge Filters," *Filtrate*. September 12, 1975, pp. 155–163.
3. Oulman, C. S., and E. R. Bauman, "Streaming Potential in Diatomite Filtration," *J. Am. Water Works Assoc.*, 56. (1970), p. 915.
4. Oulman, C. S., and E. R. Bauman, *Filtrate*, September 7, 1970, p. 687.
5. Tanny, G. B., and T. H. Meltzer, "The Dominance of Adsorptive Effects in the Filtrative Sterilization of a Flu-Vaccine," *J. Parenteral Drug Assoc.*, 32(6) (1978), 258–267.
6. Zierdt, C. H., "Unexpected Adherence of Bacteria, Blood Cells, and other Particles

to Large Porosity Membrane Filters," paper presented at the American Society of Microbiology, 78th Annual Convention, Las Vegas, NV, May 1978.

7. Ziert, C. H., R. L. Kagan, and J. D. MacLowry, "Development of a Lysis-Filtration Blood Culture Technique," *J. Clinical Microbiol.*, 5(1) (1977), 46–50.
8. Leahy, T. J., Millipore Corporation, Microbiological Laboratories. Personal communication (1980).
9. Tanny, G. B., D. K. Strong, W. G. Presswood, and T. H. Meltzer, "The Adsorptive Retention of Pseudomonas diminuta by Membrane Filters," *J. Parenteral Drug Assoc.*, 33(1) (1979), 40–51.
10. Ruth, B. F., G. H. Montillon, and R. F. Montanna, "Studies in Filtration: Part I. Critical Analysis of Filtration Theory," *Ind. Eng. Chem.*, 25 (1933), pp. 76–82.
11. Hermans, P. H., and H. L. Bredee, "Zur Kenntnis der Filtrationsgesetze," *Rec. Tav Chim.*, 54 (1935), pp. 680–700.
12. Bowman, F. W., M. T. Calhoun, and M. White, "Microbiological Methods for the Quality Control of Membranes," *J. Pharm. Sci.*, 56(2) (1967), 222–225.
13. "Microbiological Evaluation of Filters for Sterilizing Liquids," HIMA Document no. 3, vol. 4, Health Industries Manufacturing Association, Washington, DC (1982).
14. Wallhausser, K. H., "Bacterial Filtration in Practice," *Drugs Made in Germany*, 19 (1976), pp. 85–98.
15. Elford, W. J. "The Principles of Ultrafiltration as Applied in Biological Studies," *Proc. Royal Soc. London*, 112B (1933), pp. 384–406.
16. Leahy, T. J., and M. J. Sullivan, "Validation of Bacterial-Retention Capabilities of Membrane Filters," *Pharm. Technol.*, 2(11) (1978), 65–75.
17. Wallhausser, K. H., "Recent Studies on Sterile Filtration," *Pharm. Ind.*, 41 (1979), pp. 475–481.
18. Johnston, P. R., "Submicron Filtration," *Chem. Eng. Process.*, 71 (1975), pp. 70–73.
19. Johnston, P. R., "Submicron Filtration with Cartridges," *Filtrate*, September 12, 1975, pp. 352–353.
20. Rogers, B. G., and H. W. Rossmore, "Determination of Membrane Filter Porosity by Microbiological Methods," *Devel Ind. Mierobiol.*, 11 (1976), pp. 453–459.
21. "Pore-Size Characteristics of Membrane Filters for Use with Aerospace Fluids," *ANSI/ASTM*, (1976), p. F316–70.
22. Pall, D. B., and E. A. Kirnbauer, "Bacterial Removal Prediction in Membrane Filters," paper presented at the 52nd Colloid and Surface Symposium, University of Tennessee, Knoxville, TN, June 12, 1978.
23. Johnston, R. R., and T. H. Meltzer, "Suggested Integrity Testing of Membrane Filters as a Robust Flow of Air," *Pharm. Technol.*, 4 (1980), pp. 49–59.
24. Lukaszewicz, R. C., G. B. Tanny, and T. H. Meltzer, "Membrane Filter Characterizations and Their Implications for Particle Retention," *Pharm. Technol.*, 2(11) (1978), 77–83.
25. Johnston, P. R., R. C. Lukaszewicz, and T. H. Meltzer, "Certain Imprecisions in the Bubble Point Measurement," *J. Perenteral Sci. Technol.*, 35 (1981), pp. 36–39.

5

Reverse Osmosis

Reverse osmosis (RO) for water and wastewater treatment and for reuse at electricity-generating power plants has increased in recent years. Uses of this unit operation include: recirculating condenser water, ash sluice water, boiler blowdown, boiler makeup, and wet sulfur dioxide scrubber waste. This chapter examines five cases of boiler makeup water in the utility industry. The different water supplies range from 100 ppm to as high as several thousand ppm.

Use of RO for desalination of seawater for boiler makeup is reviewed in one installation. The availability of this system has opened up the use of heretofore unavailable water supplies. It has been used by the industry as a pretreatment to ion-exchange demineralization. RO acts as an economical roughing demineralizer, bringing down the overall cost and improving the life of resins and operation of ion-exchange equipment.

GENERAL PRINCIPLES

Osmosis is the spontaneous passage of a liquid from a dilute to a more concentrated solution across an ideal semipermeable membrane that allows passage of the solvent (water) but not the dissolved solids (solutes) as shown in Figure 5.1. If an external force is executed on the more concentrated solution, the equilibrium is disturbed and the flow of solvent is reversed. This phenomenon (RO) is depicted in Figure 5.2.

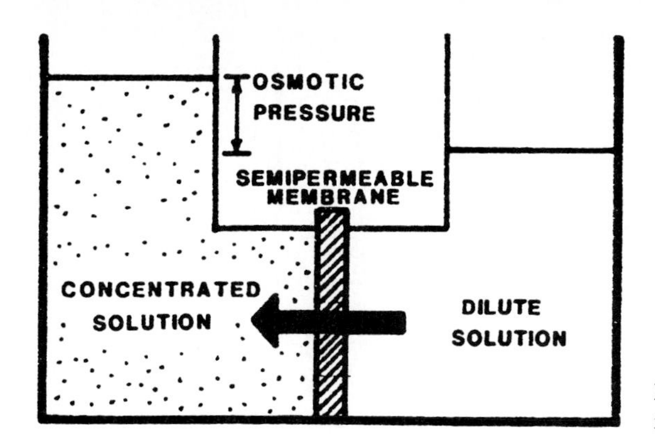

Figure 5.1 Osmosis: normal flow from low to high concentration.

A basic RO treatment system consists of components illustrated in Figure 5.3. Feed water to the RO system is pumped first through a micrometer filter. This is a replaceable-cartridge element filter nominally rated at 10 μm. The purpose of this filter is to remove any turbidity and particulate matter from the feed water before it enters the RO system.

The filtered raw water then flows to a high-pressure pump, which feeds the raw water at a pressure of 400 psi through the RO membrane system. Valves and pressure gauges between the micrometer filter, the high-pressure pump, and membrane modules control the flow of water through the system and monitor its operation.

The RO system consists of two stages, The raw water is pumped through the first stage, which contains twice the number of membrane modules as the second stage. The first stage purifies 50 percent of the water fed to the system and requires the remaining 50 percent which contains all of the contaminants. This reject water from the first stage is then passed through the second stage, which purifies 50 percent of the water fed to it and rejects the remaining 50 percent to waste. This second stage reject now contains all of

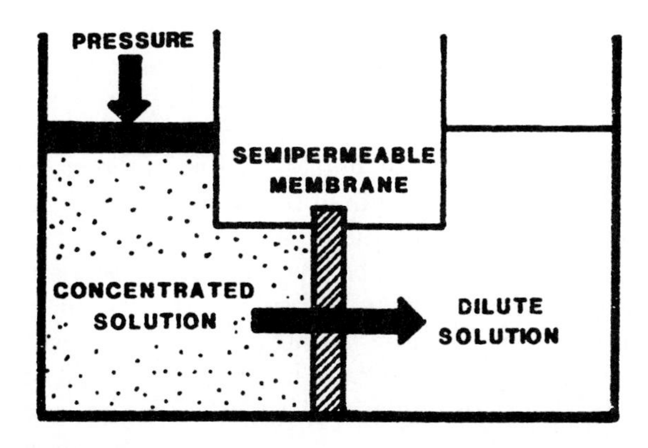

Figure 5.2 Reverse osmosis: flow reversed by application of pressure to high-concentration solution.

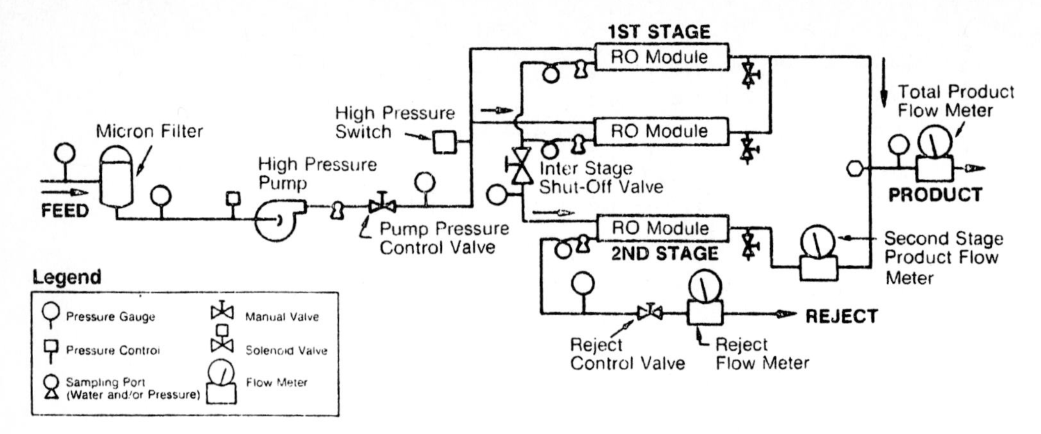

Figure 5.3 Permutit Corporation RO system.

the contaminants removed by both stages. Thus, the total flow through the system is 75 percent purified product water and 25 percent reject water.

The RO system removes 90 percent to 95 percent of the dissolved solids in the raw water, together with suspended matter (including colloidal and organic materials). The exact percent of product purity, product recovery, and reject water depends on the amount of dissolved solids in the feed water and the temperature at which the system operates.

RO membrane performance in the utility industry is a function of two major factors: the membrane material and the configuration of the membrane module. Of the four RO membrane module types, most utility applications use either spiral-wound or hollow-fiber elements. Hollow-fiber elements are particularly prone to fouling and, once fouled, are hard to clean. Thus, applications that employ these fibers require a great deal of pretreatment to remove all suspended and colloidal material in the feed stream. Spiral-wound modules, due to their relative resistance to fouling, have a broader range of applications. A major advantage of the hollow-fiber modules, however, is the fact that they can pack 5,000 ft^2 of surface area in a 1-ft^3 volume, while a spiral-wound module can only contain 300 ft^2/ft^3.

The hollow-line fiber configuration consists of a bundle of porous hollow-line fibers. These fibers are externally coated with the actual membrane and form the support structure for it. Both ends of each fiber are set in a single epoxy tube sheet, which includes an O-ring seal to match the inside diameter of the pressure vessel. Influent water enters one end of the pressure vessel and is evenly distributed along the length of the vessel by a concentric distributor tube. As the water migrates out radially, some of it permeates the fibers and exits the pressure vessel via the tube sheet on the opposite end. The direction of permeate flow is from outside to inside the fibers. The concentrated solution, or reject, completes its radial flow path and leaves the vessel at the same end at which it entered. Figure 5.4 is a representation of this configuration. For clarity, the vessel and inlet distributor have been omit-

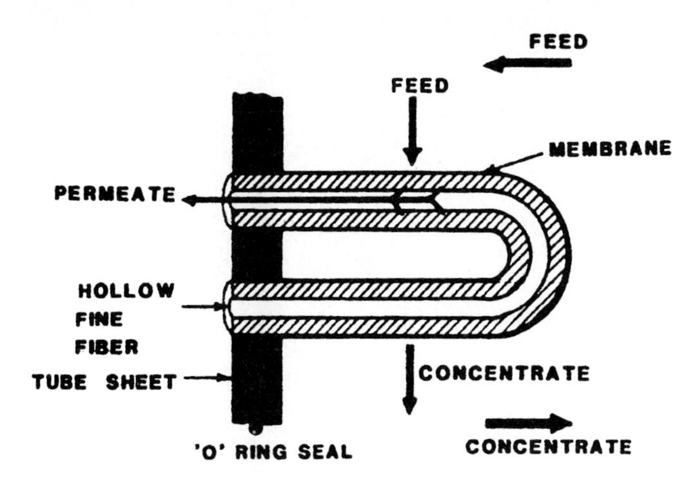

Figure 5.4 Hollow-fiber membrane.

ted. The actual outside diameters of individual fibers range from 3 mils to 10 mils, depending on manufacturer. Figure 5.5 depicts a complete module.

The spiral-wound configuration consists of a jelly roll-like arrangement of feed transport material, permeate transport material, and membrane material. At the heart of the wall is a perforated permeate collector tube. Several rolls are usually placed end to end in a long pressure vessel. Influent water enters one end of the pressure vessel and travels longitudinally down the length of the vessel in the feed transport layer. Direct entry into the permeate transport layer is precluded by sealing this layer at each end of the roll. As the water travels in a longitudinal direction, some of it passes in radially through the membrane into the permeate transport layer. Once in the transport layer, the purified water flows spirally into the center collection tube and exits the vessel at each end. The concentrated feed continues along the feed transport material and exits the vessel on the opposite end from which

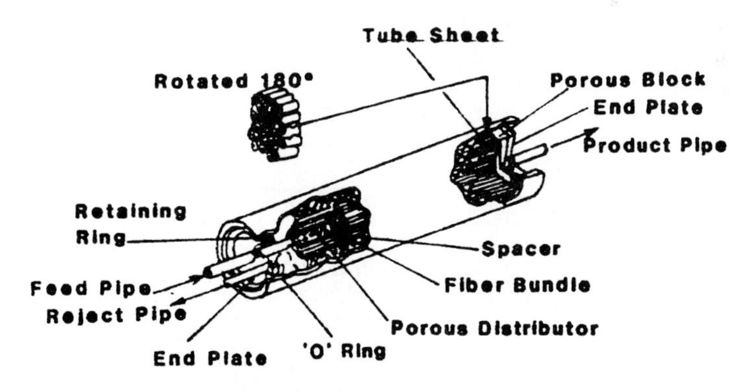

Figure 5.5 Hollow-fiber module.

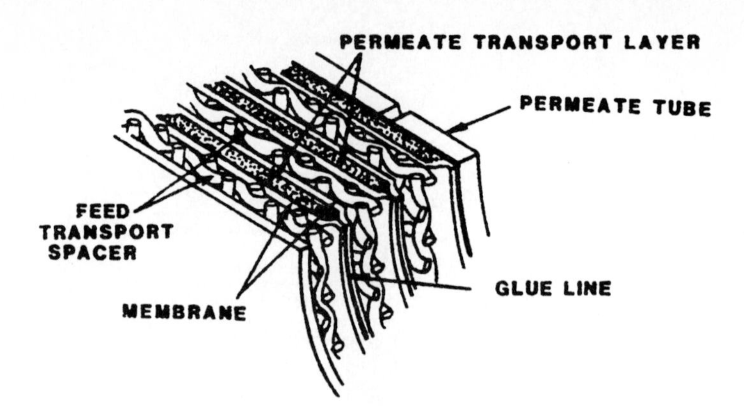

Figure 5.6 Spiral-wound membrane.

it entered. A cross section of the spiral configuration is depicted in Figures 5.6 and 5.7 and a typical module assembly is shown in Figure 5.8.

The two types of membrane materials used are cellulose acetate and aromatic polyamide membranes. Cellulose acetate membrane performance is particularly susceptible to annealing temperature, with lower flux and higher rejection rates at higher temperatures. Such membranes are prone to hydrolysis at extreme pH, are subject to compaction at operating pressures, and are sensitive to free chlorine above 1.0 ppm. These membranes generally have a useful life of two to three years. Aromatic polyamide membranes are prone to compaction. These fibers are more resistant to hydrolysis than are cellulose acetate membranes, but they are more sensitive to free chlorine. Some organizations use RO to reduce operating costs in makeup demineralizers in the utility industry, as shown in Figure 5.9.

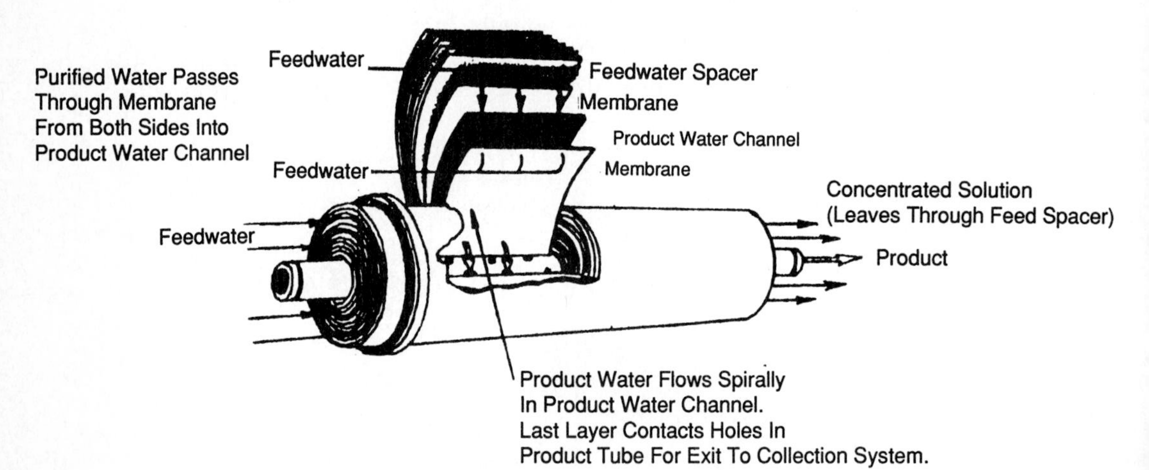

Figure 5.7 Spiral-wound membrane.

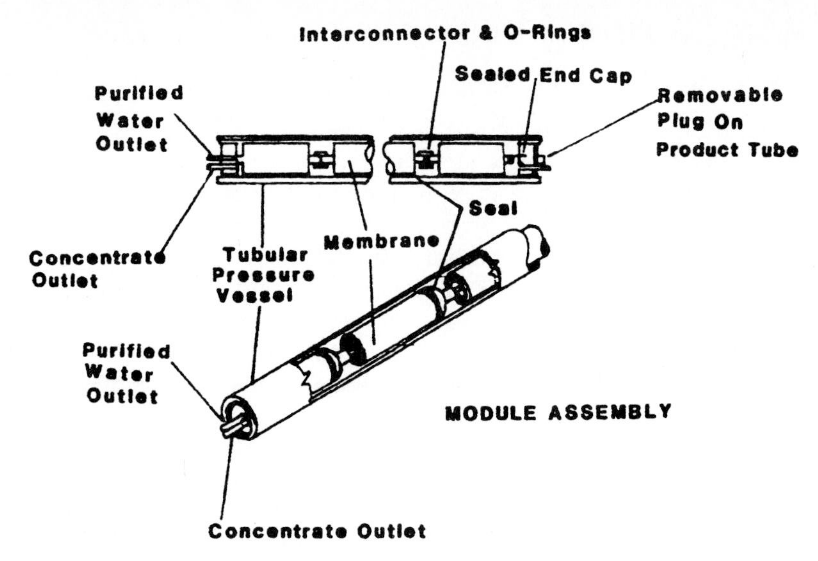

Figure 5.8 Spiral-wound module.

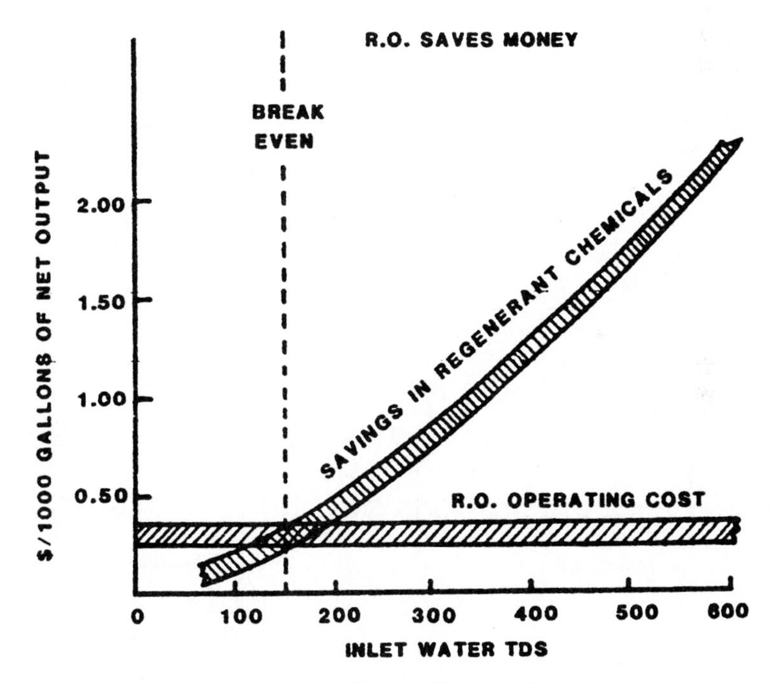

Figure 5.9 Chemical savings vs. RO operating costs.

INSTALLATIONS AND CASE HISTORIES

Case 1

A man-made lake supplies cooling water to a Midwest utility. The inlet total organic solids (TOS) was in the range of 600 ppm. The process flow diagram is shown in Figure 5.10. High hardness, alkalinity, and total suspended solids (TSS) are reduced by cold lime softening. The present RO system consists of two-stage trains using 8-in.-diameter spiral-wound cellulose acetate membranes. Following this system, a two-train demineralizer consisting of four beds of alternating cation and anion resins polishes the permeate flow.

The plant consisted of two 3-stage trains, rated at 237,500 gal/day with a 75 percent recovery. A retrofitted plant was placed in service three years after startup. The new system employed two two-stage trains operating at 75 percent recovery. The two trains use membranes from different manufacturers having different flux ratings. The permeate output is 216,000 gal/day from train A and 250,000 gal/day from train B. The performance data can be seen in Figure 5.11 and Table 5.1.

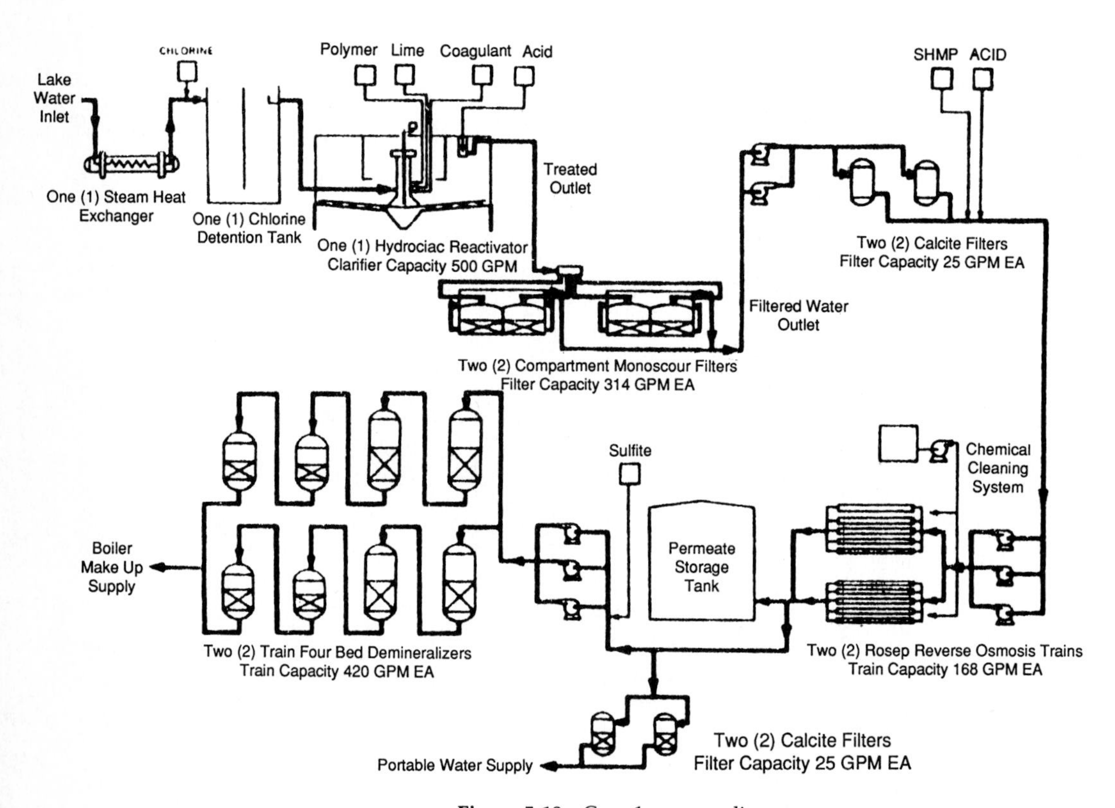

Figure 5.10 Case 1 process diagram.

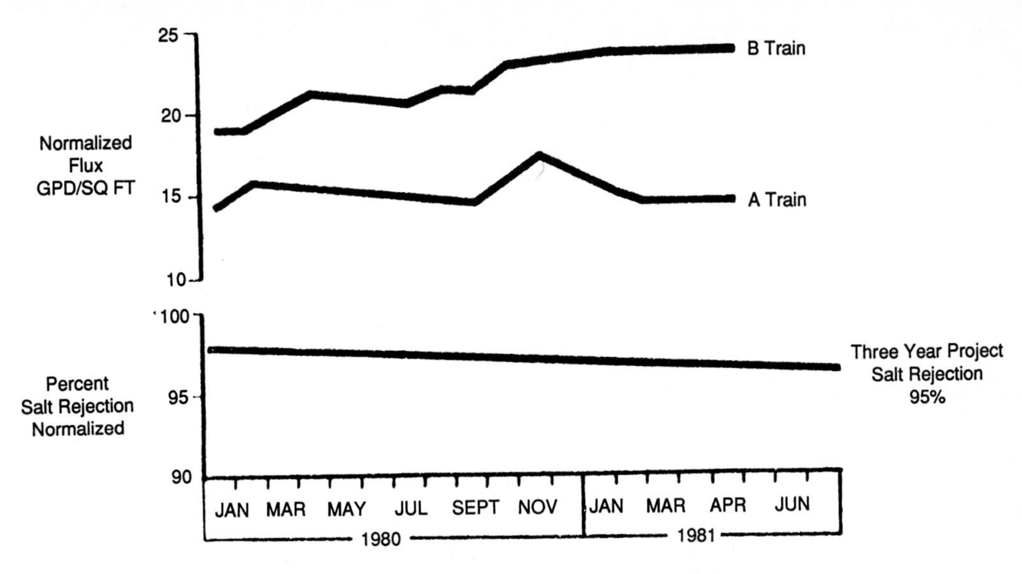

Figure 5.11 Case 1 performance history.

Cases 2 and 3

A West Coast utility began to use makeup water from a tidal river source. This utility installed two similar plants separated by a few miles along the same river. Both systems were virtually identical, producing 700,000 gal/day on an inlet feed of 5,000 ppm total dissolved solids (TDS). The process flow diagram is shown in Figure 5.12. The pretreated water is fed to two-stage RO trains using cellulose acetate spiral-wound elements operating at 75 percent recovery. From the RO system the water is pumped to two parallel trains of cation, anion, and mixed-bed ion-exchange vessels.

TABLE 5.1 CASE 1 RO PERFORMANCE ANALYSIS

	Average ion rejection (%)		
	Train A	Train B	Plant average (%)
Total Hardness	97.2	98.3	97.8
Calcium	99.3	97.9	98.6
Chloride	97.9	98.5	98.2
Sulfate	99.9	99.2	99.6
Silica	79.1	80.8	80.0
Salt	97.3	95.8	96.4
Average Water Recovery	71.7	73.1	72.4
Flux	13.7	25.9	19.8

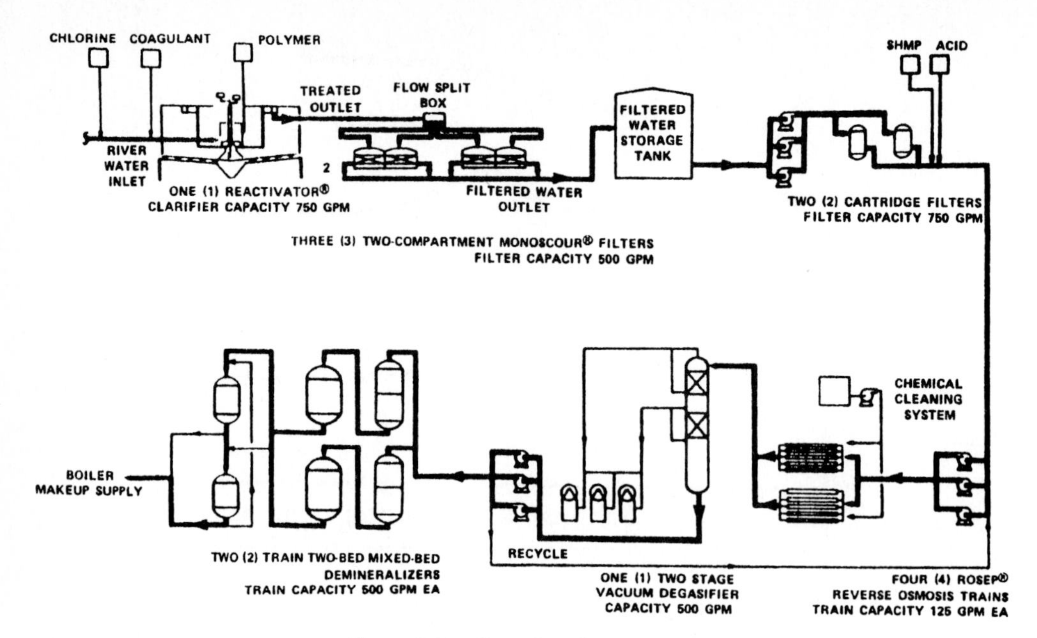

Figure 5.12 Cases 2 and 3 process diagram.

The Case 2 system started operation in September 1978; and the Case 3 system in January 1979. In July 1979 the Case 2 RO system experienced a rapid decline in salt rejection. The cause of the problem was due to an overload of the units with silt. The performance data can be seen in Figures 5.13 and 5.14 and Table 5.2. Case 3 has had a steady but reasonable decline in salt rejection with excellent operating experience. The sludge volume index (SVI) values for plant 3 averaged 2 to 5. Table 5.3 shows the performance history for this plant.

TABLE 5.2 CASE 2 RO PERFORMANCE ANALYSIS

	Average ion rejection (%)			Plant average (%)
	Train A	Train B	Train C	
Sodium	80.3	81.1	81.1	80.8
Calcium	90.9	88.3	88.7	89.3
Chloride	83.2	86.2	86.5	85.3
Sulfate	93.8	93.5	93.9	93.7
Silica	70.1	70.9	71.7	70.9
Salt	84.4	86.5	85.2	85.4
Average Water Recovery	73.2	72.7	74.1	73.3
Flux	10.5	10.0	11.2	10.6

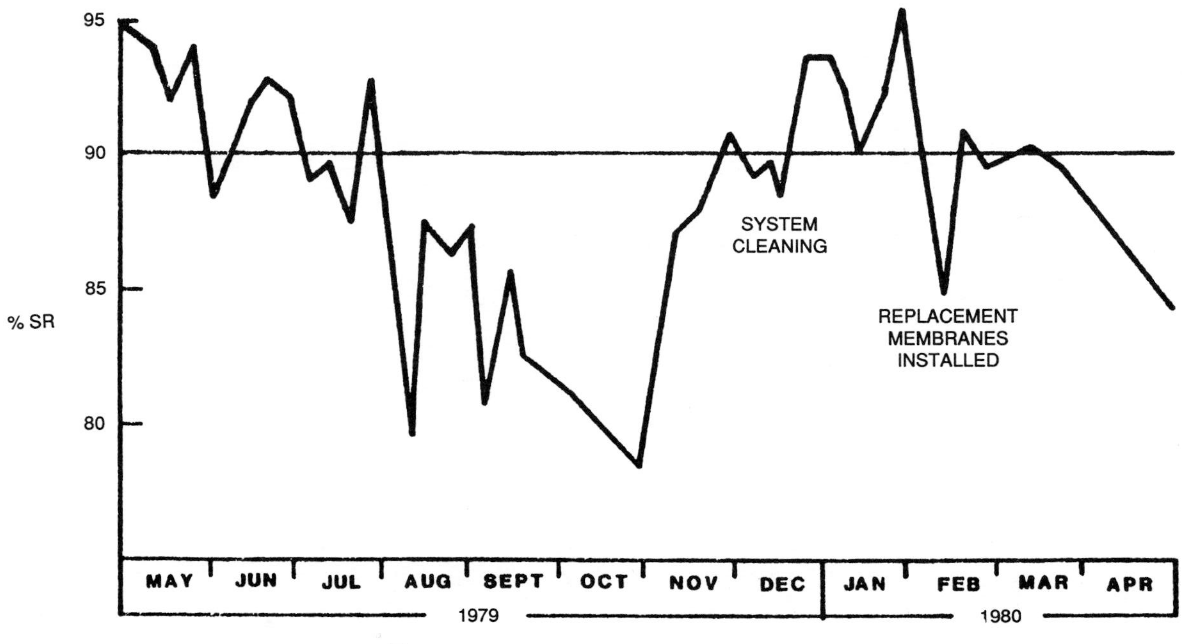

Figure 5.13 Case 2 salt rejection history.

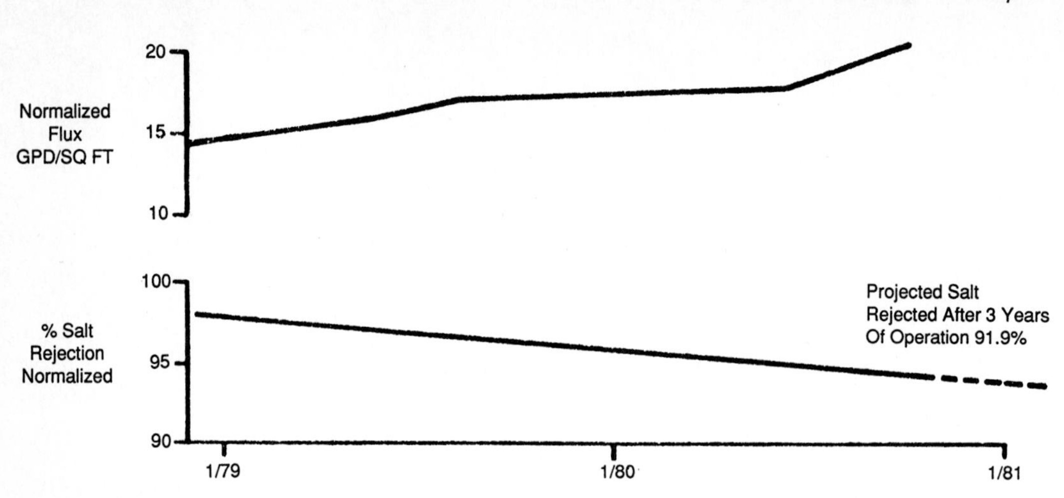

Figure 5.14 Case 3 performance history.

TABLE 5.3 CASE 3 RO PERFORMANCE ANALYSIS

	Average ion rejection (%)				Plant average (%)
	Train A	Train B	Train C	Train D	
Sodium	85.2	88.4	84.6	86.9	86.3
Calcium	96.1	95.8	92.5	95.1	94.9
Chloride	88.4	89.5	84.7	87.9	87.6
Sulfate	98.1	98.9	95.5	97.9	97.6
Silica	79.0	80.1	77.4	83.2	79.0
Salt[a]	88.5	90.4	86.4	88.7	88.5
Average Water Recovery	74.6	74.5	74.4	74.3	74.5
Flux	16.4	21.2	23.0	22.4	20.8

[a] Highest values reported: train A, 93.1; train B, 94.3%; train C, 90.8%; train D, 92.5%; plant average, 92.5%.

Case 4

A Southwest utility uses a man-made reservoir as its source of makeup water. It installed a pretreatment and RO system as viewed in Figure 5.15 with the inlet dissolved solids at 2,000 ppm. After the necessary pretreatment, the water is pumped to two 50 percent booster pumps to two RO trains rated at 216,000 gal/day each at 75 percent recovery. The RO trains use 8-in.-diameter spiral-wound cellulose acetate elements.

The plant started operation in late 1980 and the performance data are shown in Figure 5.16 and Table 5.4. The potable water is presently being supplied by the RO system, and the plant was to be in service in 1984.

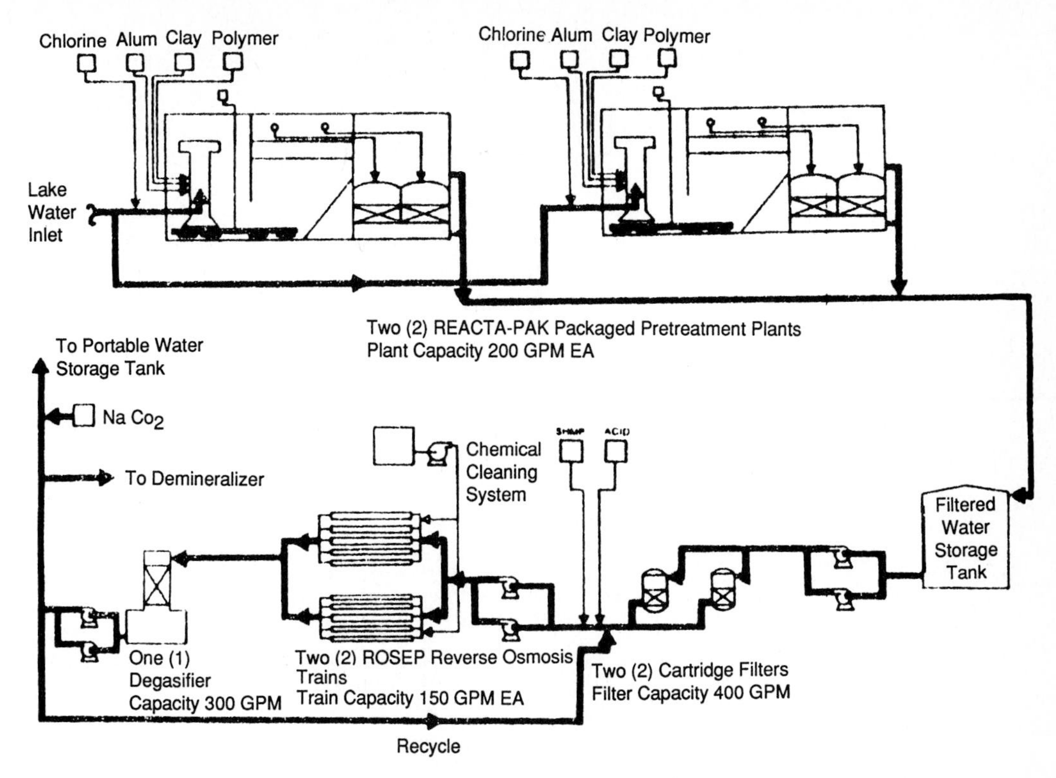

Figure 5.15 Case 4 process diagram.

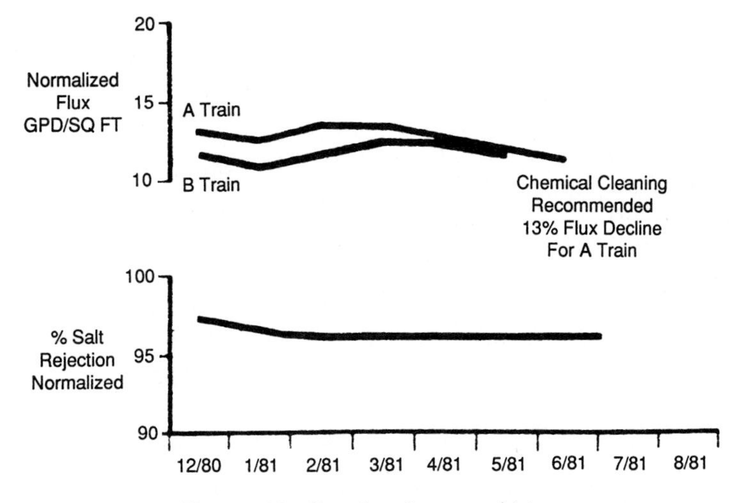

Figure 5.16 Case 4 performance history.

TABLE 5.4 CASE 4 RO PERFORMANCE ANALYSIS

	Average ion rejection (%)		
	Train A	Train B	Plant Average (%)
Sodium	90.7	90.5	90.6
Calcium	97.6	98.4	98.0
Chloride	90.4	90.7	90.6
Sulfate	99.1	99.1	99.1
Silica	80.5	76.0	78.3
Salt	94.0	94.3	94.1
Average Water Recovery	70.0	69.9	69.9
Flux	11.0	10.9	10.9

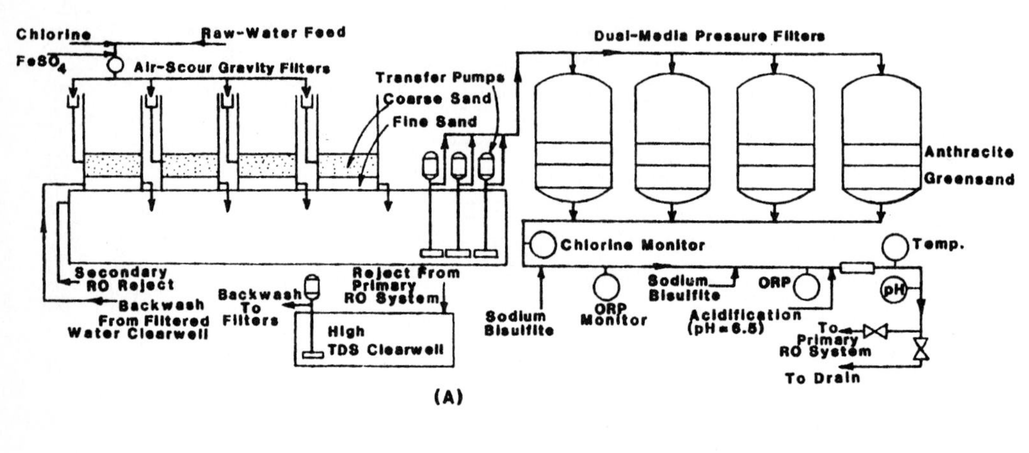

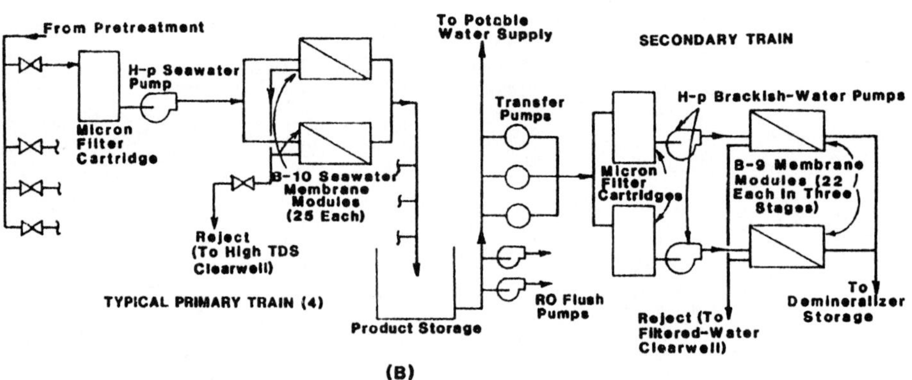

Figure 5.17 (A) Seawater pretreatment removal of suspended solids and organic matter protects membrane modules. (B) Two-stage RO desalting system designed to reduce dissolved solids in 38,000-ppm seawater to 100 ppm.

Case 5

At a utility in Venezuela, RO supplies purified seawater as boiler feed water to a four-unit, 1,600-MW power station. In operation since January 1980, it produces 685,000 gal/day. The total dissolved solids content is reduced from 37,000 to 60 ppm. The length of time between regeneration of the ion-exchange demineralization units has been increased prior to entering the boilers.

The system consists of a pretreatment section, a primary RO seawater section, and a secondary RO unit as shown in Figure 5.17. The pretreated seawater enters the primary seawater RO system, a four-train system, each train producing 196,600 gal/day of treated seawater. Each train consists of a 5-μm-filter cartridge and a high-pressure pump feeding two parallel subunits, each containing 25 Du Pont Permasep B-10 seawater modules for a total of 50 elements per train. The system is designed for 30 percent recovery at 900-psi-rated operating pressure.

The seawater RO product is stored in a 262,000-gal storage tank, which provides the feed to two secondary RO systems, as well as the flushing water required by the primary RO train. Each has a 5-μm-filter cartridge and a high-pressure pump. This output is fed to a single bank of Du Pont Permasep^R B-9 brackish water permeators. These are arranged in three stages, with twelve

TABLE 5.5 CASE 5 PRODUCT WATER ANALYSIS

	RO product	
	Primary	Secondary
Major Cations (mg/l as $CaCO_3$)		
Ca^{2+}	4.5	0.02
Mg^{2+}	22	1.1
Na^+	362	38
K^+	8	0.9
Total	397	40
Major Anions (mg/l as $CaCO_3$)		
Br^-	0.9	0.1
Alkalinity, Total	26	<1
Cl^-	371	41
SO_4^{2-}	11	<1
NO_4^{2-}	<0.2	<0.2
Total	409	41
Specific Conductance (μmho/cm)	864	99
Turbidity (NTU)	<1	<1
Color (apparent color units)	<5	<5
pH	6.5	7.0
Iron (mg/l as Fe)	0.02	0.002
Manganese (mg/l as Mn)	0.003	0.002
TDS at 140°C (mg/l)	490	28

elements in the first stage, seven elements in the second, and three in the third. The secondary train operates at 85 percent recovery. The trains produce 283,900 gal/day or product water each with a TDS of 100 ppm. The only waste stream from the plant is the reject from the seawater system, which is then returned to the ocean. Table 5.5 provides a complete analysis of the product water from the primary and secondary trains.

WATER, WASTEWATER, AND MISCELLANEOUS CASE HISTORIES

With consumer demands for high-quality water increasing, RO has been gaining attention as a means of increasing the supply and quality of potable water. New sources of potable water must be found, since most fresh-water have already been tapped. One alternative would be to treat such potential sources as wastewater, seawater, and brackish water, making them acceptable for human consumption.

The electronics industry has continually demanded higher-quality water due to the increasing complexity of integrated circuits. Water comes in direct contact with the integrated circuit. Components in nearly every manufacturing step, such as rinses after acid etching and cleaning before high-temperature metal deposition. Particulate or metallic ion contamination at any manufacturing step results in a rejected, nonsalable product. Consequently, ultrahigh-purity deionized water must be provided to the electronics manufacturing firm to produce a marketable product.

Case 6

This RO system in Rotunda West, Florida was in operation in early 1972. The feed water is drawn from brackish well water with a TDS content of 5,000 to 7,000 ppm. The design rate of 0.5 mgd is used as the community's potable water supply. The plant consists of six skids containing 8-in. B-9 permeators operating at 50 percent recovery. A fouling index of less than 3 can be maintained with very little pretreatment.

A hollow-fiber aromatic polyamide type system was selected for operation. The permeate is returned to a product storage tank before chlorination and distribution with a 90 percent rejection rate. The chloride and sulfate levels are under 250 ppm as required by the Florida health codes. The composition of the raw feed and the product effluent is shown in Table 5.6.

Case 7

Cape Coral, Florida uses a 3.0-mgd RO plant engineered by Black, Crow, and Eidsness. This town, like many others in Florida, experienced water supply problems because of saltwater intrusion into its groundwater supplies. The brackish feed water is taken from the Hawthorn aquifer and supplies the RO system, which in turn supplies potable water.

TABLE 5.6 ROTUNDA WEST, FLORIDA OPERATING RESULTS

	1	2	2a	3	4
Cations (mg/l as $CaCO_3$)					
Ca^{2+}	272	680	680	10	4
Mg^{2+}	271	1,155	1,155	20	5
Na^+	1,938	4,205	4,205	360	166
H^+ (acidity)					
Total		6,000		390	
Anions (mg/l as $CaCO_3$)					
Alkalinity	183	150	68	30	37
HCO_3^-	183	150	68	30	37
CO_3^{2-}					
OH^-					
PO_4^{3-}					
Cl^-	3,865	5,450	5,450	350	248[a]
SO_4^{2-}	385	400	482	10	10
NO_3^-					
Total		6,000	6,000	390	
Total Hardness (mg/l as $CaCO_3$)		1,795		30	
Alkalinity (mg/l as $CaCO_3$)					
Methyl Orange		150		30	
Phenolphthalein					
Noncarbonate Hardness (mg/l as $CaCO_3$)					
Sodium Alkalinity (mg/l as $CaCO_3$)					
CO_2 (mg/l)	11	11	85	85	10
SiO_2 (mg/l)					
Fe (mg/l)	0.01	0.01	0.01	nil	nil
Turbidity (mg/l)					
Color (mg/l)					
Total Solids (cations + SiO_2)	6,915				470[b]
pH	7.4	7.4	6.2	5.9	6.8

[a] Value estimated from calculation: 250.
[b] Value estimated from calculation: 500.

The system consists of six RO modules, each having a capacity of 0.5 mgd. Each module contains 22 Dowex RO-20K permeators arranged in a two-stage configuration including 16 permeators in the first stage feeding six permeators in the second stage.

At present, the system produces water with a TDS content of 63 ppm. The current 2.5-mgd output can be blended with 0.7 mBd of raw water containing 1,100 to 1,300 ppm of TDS to produce a blended water containing 296 ppm. This is below the 500 ppm of TDS allowed by the present drinking water standards of Florida. The raw and product water qualities are listed in Table 5.7.

TABLE 5.7 CAPE CORAL RO PLANT PERFORMANCE DAB (mg/l)

	TDS	Na	Cl	SO42	Sr
Design Raw Water Quality	2,050	365	743	281	18.1
Actual Acidified Feed	1,234	371	485	234	5.9
Guaranteed Three-Year Quality	145	65	80	30	1.0
Actual Product Water	63	23	24	6	0.34

The plant came on line in March 1977 at a total cost of $4.2 million. Operating costs for the first six months were $0.69/1,000 gal. Total water production costs were approximately $1.00/1,000 gal. These costs were exclusive of membrane replacement, which were estimated to be $0.165/1,000 gal for a three-year life expectancy. Operating equipment data are:

- raw water pumps (Peerless): 2 at 550 gal/min, 1 at 700 gal/min and 3 at 1,000 gal/min
- cartridge filters, 5μm (CUNO): 4 at 1,600 gal/min
- high-pressure pumps (Goulds): 6 at 535 gal/min
- RO modules (Permutit and Dow): 6 at 0.5-mgd output (each with 22 Dowex RO-20K permeators)
- forced-draft degasifiers (Deloach Plastics): 2 at 2,100 gal/min (4 blowers at 5,000 ft^3/min)
- transfer pumps (Peerless): 3 at 4,200 gal/min
- storage reservoir (Crom Corp.): 5.0 mgd
- high-service pumps (Peerless): 2 at 1,750 gal/min, 2 at 3,500 gal/min
- sulfuric acid feed (Permutit): 7,700 lb/day
- hexametaphosphate feed (Permutit): 385 lb/day
- sodium hydroxide feed (Wallace & Tiernan): 625 lb/day
- chlorinators (Fisher-Porter): 210 lb/day
- membrane flushing (Permutit): 150-gal/min. pump
- membrane cleaning (Permutit): 800-gal tank, 120-gal/min pump

Case 8

An electronics rinse-water system uses lake water as its inlet feed supply. The plant went into operation in fall 1979 and supplies 2.0 mgd as product water. The RO plant includes six three-stage trains operating at 85 percent recovery and rated at 413,280 gal/day each. One train is held in standby, making the plant capacity at any time 2,066,400 gal/day. It removes 95 percent of the dissolved solids and virtually all suspended material assisting this electronics manufacturing complex. The RO permeate is pumped to a series of primary and secondary mixed beds where the water is polished to 18 $M\Omega$. The flow diagram is shown in Figure 5.18.

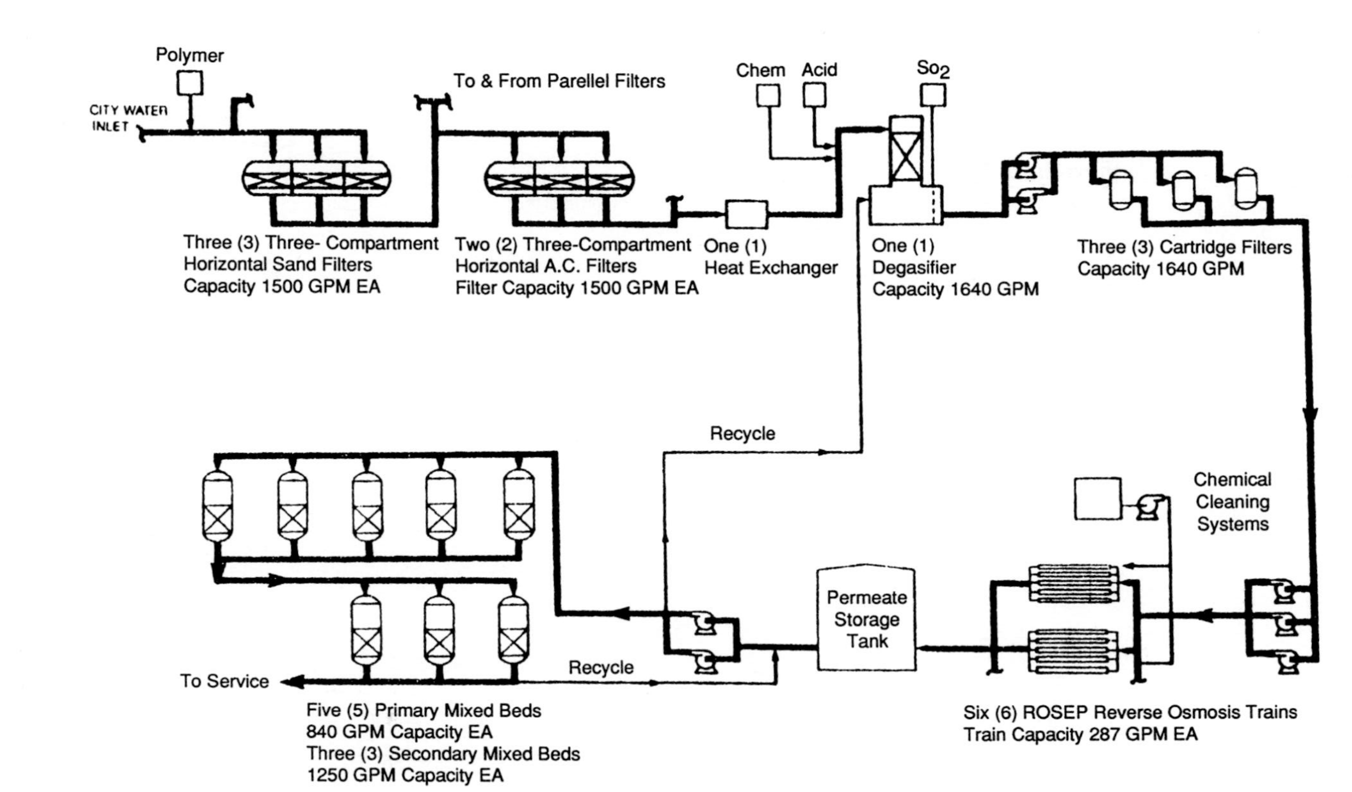

Figure 5.18 Case 8 flow diagram.

Case 9

This installation at the Eastman Kodak plant in Windsor, Colorado was started in July 1975. It is a 1.2-mgd spiral-wound RO plant. There are eight 100-gal/min banks, each containing 216 Model 4101 Roga spiral-wound RO elements assembled in 36 pressure vessels. The 36 pressure vessels are arranged in a three-stage configuration (18 × 11 × 7). The 4101 element is very resistant to fouling; that was one reason for its choice.

Pretreatment consists of chlorination followed by a heat exchanger to maintain a 77°F operating temperature. The water is acidified and passed through sand filters. The fouling index varies from 12 to 18 and the water then passes through 10-μm cartridge filters. The product water is sent to carbon purifiers to remove residual chlorine and then is treated in a two-step demineralizer system. The RO system has a 75 percent overall recovery and a 90 percent salt rejection. The operational results are summarized in Table 5.8.

TABLE 5.8 CASE 9 OPERATING RESULTS

	I[a]	II[b]	III[c]	IV[d]	V[e]
Cations (mg/l as $CaCO_3$)					
Ca^{2+}	16	16	8	8	0.21
Mg^{2+}	6	6	4	4	0.45
Na^+	6	6	6	6	1.5
K^+	1	1	1	1	0.18
Total	29	29	19	19	2.3
Anions (mg/l as $CaCO_3$)					
Alkalinity, Total	17	17	4	6	<1
Alkalinity, Phenolphthalein	0	0	0	0	0
Cl^-	7	4	6	6	<1
SO_4^{2-}	4	7	7	7	<1
NO_3^-	1	1	1	1	<0.5
Total	29	29	18	20	
Iron, total (mg/l as Fe)	0.20	0.20	0.20	0.20	0.05
Manganese, Total (mg/l as Mn)	<0.005	<0.005	<0.005	<0.005	<0.005
Silica, Dissolved (mg/l as SiO_2)	6.9	6.9	6.9	6.9	0.78
Turbidity (NTU)	1.0	1.0	1.0	1.0	0.1
Specific Conductance (μmho/cm)	54	54	34	40	6.4
pH	7.3	7.3	5.5	6.8	6.7
Color, Apparent (color units)	<5	<5	<5	<5	<5
Total Organic Carbon (mg/l as C)	3	2	2	3	1

[a] I = raw water before acidification.
[b] II = influent to sand filters.
[c] III = effluent from sand filters.
[d] IV = effluent from 10-μm filter.
[e] V = RO product.

Case 10

RO can effectively remove viruses from wastewater without additional treatment. Bacterial reductions, however, are not of the same magnitude, and some form of disinfection is necessary before the product water can be put to domestic use. The city of San Diego uses membranes made of cellulose acetate formed into long tubes. The membranes are: one with a pore size of 8 Å, salt rejection of 89 percent to 90 percent and a flux rate of 2 to 25 gal/ft^2 per day; and another with a pore size of 5 Å, salt rejection of 87 percent to 98 percent and a flux rate of 10 to 13 gal/ft^2 per day.

The tubes are arranged in modules containing 8-ft tubes in groups of 18 enclosed in a plastic case. One module is equivalent to 16.7 ft^2 of membrane surface. Figure 5.19 shows a typical module and arrangement and Figure 5.20 shows a flow diagram. This plant produces 15,000 gal/day of product water.

The system was dosed with poliovirus on nine occasions between January and June 1977. The virus recovery is shown in Table 5.9, along with operating parameters presented in Table 5.10. The raw, brine, and product water virus assays are depicted in Table 5.11 and the coliform analysis is shown in Table 5.12. The membranes are less retractive to animal viruses. Disinfection of the product water would ensure its safety from a health point of view.

Case 11

Siemens, a manufacturer of semiconductors in Regensburg, West Germany, uses RO to produce pure water to rinse miniature electronic components. The system was installed in December 1973 and it complements the ion-exchange system. It produces 80,000 gal/day of purified product water.

TABLE 5.9 VIRUS RECOVERY FROM SEEDED RAW SAN DIEGO SEWAGE (JANUARY TO MAY 1977)

	No. of viruses (10^6 pfu/gal)		
Sample date	Theoretical	Observed	Recovery (%)
January 11	1.17	2.47	78.9
January 25	2.06	1.50	92.7
February 16	2.06	6.46	100+
March 1	2.06	3.50	100+
March 17	1.94	4.44	100+
April 6	5.20	6.10	100+
April 15	2.60	5.00	100+
May 3	2.60	7.47	100+
May 19	3.34	3.71	100+

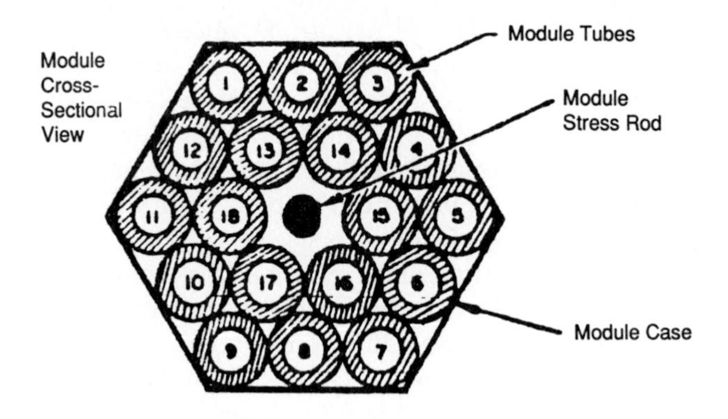

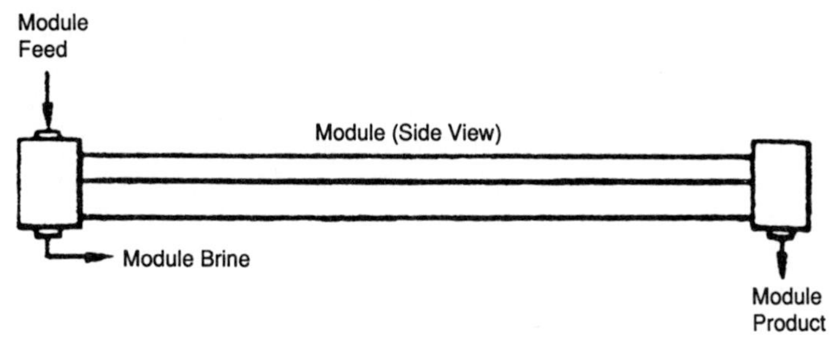

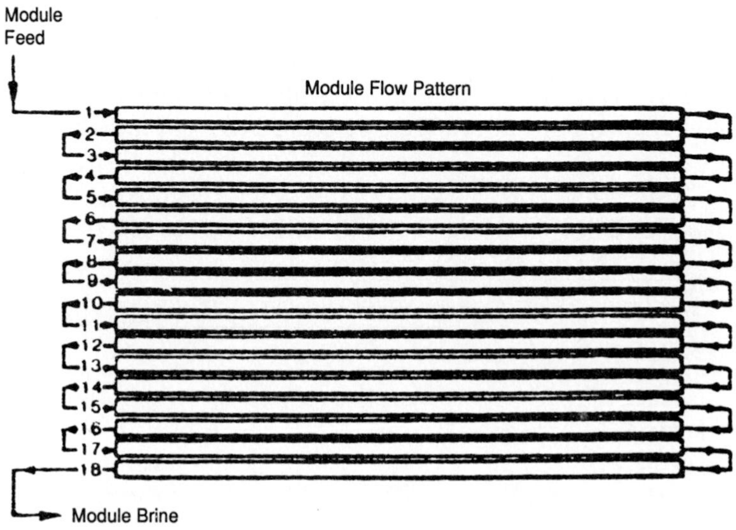

Figure 5.19 Module tube arrangement and tube flow pattern.

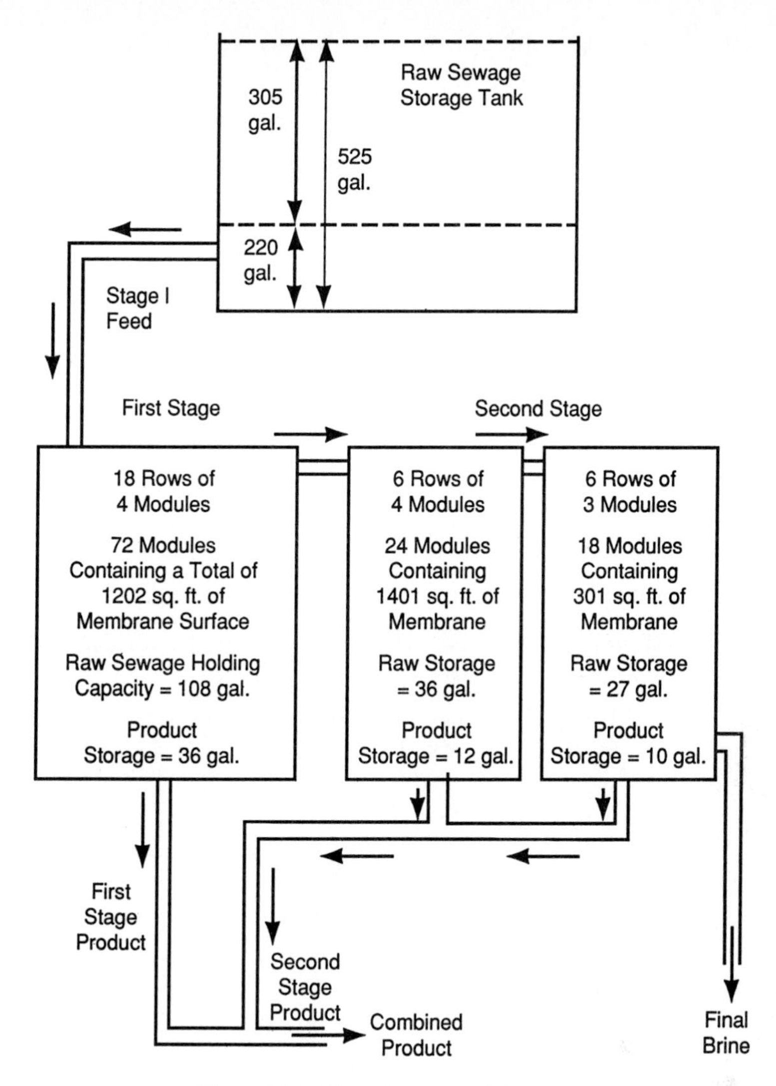

Figure 5.20 Flow diagram of RO system.

City water is the feed supply to the RO with a 400-ppm TDS concentration. The water is low in TDS but high in organic content and alkalinity.

Du Pont aromatic polyamide membranes are used because of their wide pH tolerance. Acid is injected into the feed stream to reduce bicarbonate concentration. A decarbonator removes carbon dioxide formed when the acid is added and raises the pH back to about 7.5. At this pH the "Pourmasep" B-9 permeators reject organics and bicarbonate more readily. Hexametaphosphate is added to prevent scale of calcium salts. Just before the water enters the permeators, it passes through a 5-μm filter.

TABLE 5.10 OPERATING CONDITIONS OF THE RO UNIT DURING VIRUS SEEDLING TESTS (JANUARY TO MAY 1977)

Test date	Sewage		Salt rejection[a] (%)	TDS at sample time[b] (mg/l)			Salt rejection[b] (%)
	Seeded (gal)	Processed (gal)		Raw	Brine	Product	
January 11	525	270	91.4	1,600	3,100	145	90.9
January 25	508	297	91.4	1,750	2,800	117	93.3
February 16	508	300	98.0	2,100	4,300	170	91.9
March 1	514	300	90.4	2,400	4,280	192	92.0
March 17	515	307	90.0	1,900	4,000	182	90.4
April 6	515	307	89.2	2,300	4,600	216	90.6
April 15	515	307	90.8	2,100	4,200	176	91.6
May 3	515	307	91.9	2,200	3,825	227	89.7
May 19	515	307	91.3	3,050	5,500	278	90.9

[a] Rejection rate before tracer salt addition.
[b] After tracer salt addition.

TABLE 5.11 RESULTS OF VIRUS ANALYSIS IN SAN DIEGO, CALIFORNIA RO PLANT (JANUARY TO MAY 1977)

	Raw feed water		Brine		Product	
	Polio-virus[a]	Coli-phage[b]	Polio-virus[a]	Coli-phage[b]	Polio-virus[a]	Coli-phage[b]
January 11	0.247	1.63	1.21	5.70	68	152
January 25	1.50	3.52	2.00	5.70	40	<152
February 16	6.46	35.0	3.80	35.0	29	152
March 1	3.50	35.0	4.50	35.0	12	<152
March 17	4.44	8.74	4.25	16.0	360	<152
April 6	6.10	8.74	6.10	8.74	200	<152
April 15	5.00	1.63	6.60	3.52	10	152
May 3	7.47	8.74	3.12	8.74	<200	152
May 19	3.71	16.3	1.33	35.2	16	152

[a] 10^6 pfu/gal.
[b] 10^6 MPN/gal.
[c] pfu/gal.
[d] MPN/gal.

The RO originally consisted of seven 8-in.-diameter B-9 permeators. In 1975 the capacity was increased to its present level by adding two more. The permeators are series-reject staged in one block. There are three stages, and the reject stream of one stage serves as the feed water for the next. The conversion rate is 80 percent, resulting in a reject concentration factor of five. The working pressure is 400 psi and the product water is further demineralized by a mixed-bed ion-exchange system.

TABLE 5.12 COLIFORN ANALYSIS: SAN DIEGO RO PROCESS WATER (JANUARY TO MAY 1977)

Sample date	Raw feed water (10^6 MPN/100 ml)	Brine (10^6 MPN/100 ml)	Product water (MPN/100 ml)
January 11	2,400	2,400	270
January 25	23	93	26
February 16	93	150	165
March 1	43	43	46
March 17	93	43	33
April 6	43	43	115
April 15	43	240	125
May 3	43	43	1,200
May 19	43	240	240

Case 12

IBM in Mainz, West Germany uses a system incorporating both RO and ion exchange to rinse highly sensitive electronic components. The 245,000 gal/day system began operating in September 1974 and lowered ion-exchange chemical costs and the problem of disposal of regeneration chemicals. The well water has a low TDS (600 ppm) and an SVI below 3.0, but it is very hard and has a high alkalinity. RO reduces the TDS concentration to about 58 ppm. The pretreatment system consists of HCI followed by an activated carbon filter, decarbonator, hexametaphosphate, and a 5-μm filter.

The RO system consists of three blocks of nine 8-in.-diameter B-9 permeators. Each block has three stages of permeators arranged in a five-to-three-to-one series-reject configuration. Conversion rate is 75 percent and the operating pressure is 385 psi. Posttreatment of the product water consists of a continuous mixed-bed ion exchanger, followed by conventional mixed beds.

Case 13

This RO system was installed in Leeds, North Dakota and was designed to produce 100,000 gal/day of potable water. It consists of two modular units, each having six Permasep B-9 permeators. The system was placed in operation in October 1974. The raw water is supplied by two deep wells to the town's water treatment plant.

During pretreatment, potassium permanganate and polyelectrolyte are used to precipitate iron and manganese. Then the water is pumped through sand and coral filters to remove the suspended solids. The water then goes through a weak acid ion-exchange resin softener. Just before entering the RO system, the water passes through cartridge filters to remove any suspended solids.

The RO unit is operated at a 75 percent conversion factor. The product

TABLE 5.13 WATER QUALITY: COMPARISON OF PHS STANDARDS WITH LEEDS DRINKING WATER AND MUNICIPAL WATER BEFORE RO TREATMENT[a]

Characteristic (mg/l)[b]	PHS standards	West well		Well 0.25 mi west of Leeds	
		Mean	Range	Mean	Range
TDS	500	4,200.00	4,180.00–4,220.00	688.00	672.00–704.00
Fixed Solids		3,890.00	3,790.00–3,990.00	516.00	488.00–544.00
Hardness		91.00	89.00–93.00	520.00	500.00–540.00
Sodium		1,555.00	1,550.00–1,560.00	500.00	500.00
pH		8.05	8.00–8.10	7.65	7.30–8.00
Electrical Conductance[c]		6,215.00	6,190.00–6,240.00	1,005.00	950.00–1,060.00
Carbonates		0.00	0.00	0.00	0.00
Bicarbonates		701.00	700.00–702.00	310.00	282.00–338.00
Chloride	250	1,045.00	1,030.00–1,060.00	14.50	8.00–21.00
Sulfate	250	1,250.00	1,240.00–1,260.00	242.00	236.00–248.00
Iron	0.3	2.40	1.90–2.90	4.70	3.20–6.20
Nitrate	45	0.05	0.00–0.10	0.05	0.00–0.10
Fluoride	0.7–4.2	5.15	3.80–6.50	0.10	0.10

[a] Two analyses were done on both wells for all characteristics, except sodium and fluoride, of which there was one analysis for the well 0.25 mi west of Leeds.

[b] Unless otherwise indicated.

[c] μmho/cm.

TABLE 5.14 WATER QUALITY: COMPARISON OF PHS STANDARDS WITH LEEDS MUNICIPAL WATER BEFORE AND AFTER RO TREATMENT[a]

Characteristic (mg/l)[b]	PHS standards	East well		Treated water	
		Mean	Range	Mean	Range
TDS	500	4,148.33	4,030.00–4,260.00	451.00	392.00–506.00
Fixed Solids		3,813.33	3,710.00–3,970.00	404.25	349.00–451.00
Hardness		78.33	62.00–90.00	2.25	0.00–9.00
Sodium		1,533.33	1,490.00–1,570.00	192.25	168.00–207.00
pH		8.08	7.80–8.60	7.23	6.90–7.70
Electrical Conductance[c]		6,350.00	6,060.00–7,160.00	842.50	690.00–992.00
Carbonates		12.00	0.00–72.00	0.00	0.00
Bicarbonates		689.00	608.00–724.00	93.75	86.00–100.00
Chloride	250	1,031.67	1,010.00–1,050.00	139.25	15.00–200.00
Sulfate	250	1,220.00	1,140.00–1,320.00	80.00	48.00–112.00
Iron	0.3	3.98	0.70–11.00	0.28	0.20–0.30
Nitrate	45	0.02	0.00–0.10	0.03	0.00–0.10
Fluoride	0.7–1.2	5.15	3.60–6.50	0.77	0.50–1.00

[a] Six analyses were done on the east well and four on the treated water for all characteristics, except fluoride, of which there were four for the east well and three for the treated water.

[b] Unless otherwise indicated.

[c] μmho/cm.

water is chlorinated and pumped to a water tower for distribution. Tables 5.13 and 5.14 compare the chemical analysis of the Leeds waters with the U.S. Public Health Service (PHS) standards. Installation of an RO system enabled Leeds to meet potable standards. Water users were more satisfied with the general quality of the water and municipal water consumption increased.

Summary

Cases 1 through 4 were systems designed by Ecodyne-Graver Water Division, Union, New Jersey. Case 5 was an RO system designed by Permutit Corporation, Paramus, New Jersey. These five cases reviewed the use of RO supplying boiler feed water. The utility industry has come to rely on RO for this type of application for process and economic reliability. The future industry should move forward to use RO systems in quite a few areas of the power industry.

The zero water discharge concept is becoming a way of life in the electric industry. Wastewater recovery using RO is becoming very evident. Wet sulfur dioxide scrubber waste is being treated by RO and recycled back for use in other areas of the plant. Further expansion of the technology can only increase the use of the RO throughout the utility industry.

Cases 6, 7, and 13 reviewed installations of RO to reduce TDS levels to obtain a potable water supply for three communities. Cases 8, 11, and 12 show the use of RO to produce ultrapure rinse water for the electronics industry. The photography industry uses RO for water use as reviewed in Case 9. Wastewater treatment for virus reduction is a possible consideration, as seen in Case 10.

With contamination of groundwater supplies and limited amount of surface water in many parts of the world, use of RO in the production of potable water supplies has been proved and will continue to be on the upswing in years to come. Many photographic, electronic, and utility firms use this process to reduce overall cost of demineralization.

The technology of RO has increased dramatically over the past ten years and mechanical problems, such as O-ring failure, stub-tube modifications, and membrane material, have virtually been eliminated. RO membranes are usually purchased from one manufacturer and a vendor incorporates the hardware into a total treatment system. The pretreatment system, which can cause problems with the inlet feed conditions to an RO package, will be the largest problem for engineers in future designs. Further information is given in the literature.

REFERENCES

1. Technical bulletins, product literature and photographs, Permutit Company.
2. Technical bulletins and product literature, Ecodyne-Graver Water Division.

3. Vera, I., in *Official Proceedings—The 42nd International Water Conference*, Pittsburgh, PA, October 25–27, 1981, pp. 35–45.
4. Askim, M. C., and C. M. Janecek, "Reverse Osmosis Can Increase Supply and Quality of Potable Water," *Water Sew. Works* (October 1976), pp. 76–79.
5. Carr, I. T., Jr., and J. D. Beffert, in *Proceedings of the American Water Works Association Annual Conference, Part 2*, pp. 1003–1012. Denver, CO: American Water Works Association, 1979.
6. Cooper, R. C., and D. Straube, *Water Sew. Works*, April 30, 1979, pp. 162–164, 166–167.
7. Shields, C. P., *Desalination*, 28(3) (1979), 157–179.
8. Kaup, E. C., "Design Factors in Reverse Osmosis," *Chem. Eng.*, April 2, 1973, pp. 46–55.
9. Kihustler, W., *Basic Reverse Osmosis Theory and Operation—A Master's Project*. Newark, NJ: New Jersey Institute of Technology, 1978.
10. Kosarek, L. J., "Purifying Water by Reverse Osmosis—Part 1," *Plant Eng.*, August 9, 1979, pp. 103–106.
11. Kremen, S. S., "Reverse Osmosis Makes High Quality Water Now," *Environ. Sci. Technol.*, April 1, 1975, pp. 314–318.
12. Sleigh, J. H., and R. L. Truby, "Cleaning Water by Reverse Osmosis," *Plant Engr.*, December 12, 1974, pp. 83–86.
13. Westbrook, G., and L. Wirth, "Water Management at Power Plants," *Environ. Sci. Technol.*, 11(2) 140–143, 1973.
14. Baxter, A. G., M. E. Bednaw, T. Matsura, and S. Seurirajan, *Chem. Eng. Commun.*, 10(4–5) 471–483, 1976.
15. Heizer, R. T., and C. E. Plock, in *Proceedings of the Sixth Annual Industrial Pollution Conference*, pp 103–111. McLean, VA: Water and Wastewater Equipment Manufacturers Association, 1978.

6

Selection and Sizing of Prefilter/ Final Filter Systems

Filtration separates fluids (whether liquid or gaseous) from their contaminating particulate matter. In the process, the solid particles being removed are collected by the filter, which becomes progressively blocked. The successive diminution in effective filtration area results in ever diminishing flow rates for constant applied differential pressure ΔP, and in escalating ΔP for the operations where flow rates are constant. Ultimate blockage of the filter, however practically defined, determines the throughput (the total liquid volume that is filtered).

Fluid flow through a filter depends on many factors. The cross-sectional dimensions of the filter passageways are of obvious influence, since for near-micrometer and submicrometer orifices, flow varies as the fourth power of the radius. Flow through such pores varies inversely with the viscosity of the fluid, which in turn varies inversely with temperature. For most fluids, flow through the filter pores also varies directly with ΔP. The longer the pore passageways and the more complex their convolutions, the greater their impediment to fluid flow. Hence, the thinner the filter, the greater its rate of flow. Obviously, the more numerous the pore paths, the greater the rate of flow. Therefore, the total porosity of the filter and the extent of filter surface directly govern the flow properties of a filter. Time is a factor in any practical consideration of filtration; hence, flux is the expression that defines the volume flow through a unit area of a given filter per unit of time, at a given ΔP.

This chapter presents a simple mathematical treatment of system sizing based on assessing flow decay, the diminution in the rate of flow of a filter that is caused by the buildup of retained particulate matter. This treatment provides a uniform method of system sizing that can be used by anyone familiar with the basics of filtration and will result in an orderly approach to the problem, yielding consistent and accurate solutions. Background concepts are reviewed. The discussion is geared toward cartridge filters, since larger cartridge systems are the most difficult to size. However, the methods are also applicable to the simpler disk systems.

INFORMATIONAL REQUIREMENTS

The first step in designing and sizing a filtration arrangement is to obtain detailed information on the important attributes of the application, fluid, and system. Application attributes include:

1. Purpose of filtration, e.g., sterilization, clarification, bacterial reduction, or particulate removal.
2. Final filter pore size, if not inherent in the purpose of the filtration, e.g., sterilizing filtration would require an absolute 0.2-μm-rated membrane.
3. Sterilization/sanitization, including methods, frequency, and conditions.
4. Integrity testing methods.
5. Pretreatments/prefiltration of the fluid, e.g., deionization (DI), reverse osmosis (RO), or roughing filters.
6. Prior experience with the solution, e.g., filters used, flow rates, and throughputs.

Fluid attributes include:

1. Chemical composition, providing information on the properties of the fluid, as well as any compatibility issues with the materials to be used.
2. Fluid viscosity, which directly affects flow rates.
3. Surface tension, which can affect the results of integrity testing.
4. Fluid temperature, affecting viscosity and applicable material properties.

System conditions include:

1. Flow rate required, whether average, constant, or peak.
2. Pressure available, including inlet pressure, initial (clean) differential pressure, and maximum (dirty) differential pressure.
3. Throughput/change frequency, whether a function of batch size, operating life, or differential pressure increase.

4. Equipment requirements, including provisions for peak versus average demand, future increases in batch size or flow rate, and spatial limitations.

DIFFERENTIAL PRESSURE

Insertion of a filter into a fluid stream creates an impediment to flow. To overcome this resistance to flow, a higher pressure must be applied upstream of the filter. The difference in the pressure levels between the upstream and downstream sides of the filter is the applied differential pressure ΔP. It is the driving force of the filtrative operation. ΔP bears a straight-line relationship to the rate of fluid flow through the filter. There are several ways in which pressure may be defined:

- Inlet pressure is the pressure entering the inlet (upstream) side of the filters, also called upstream pressure or line pressure.
- Outlet pressure is the pressure exiting the downstream side of the filter, also called downstream pressure.
- Gauge pressure is the pressure measured by a pressure gauge, pressure above atmospheric (ambient) pressure, symbolized psig.
- Absolute pressure is the pressure above an absolute vacuum, which is 14.7 psi (one atmosphere) above gauge pressure, symbolized pisa.
- Differential pressure is the difference between inlet pressure and outlet pressure, also called ΔP, or psid, or pressure drop.

Differential pressure, not inlet or line pressure, controls the flow rate through the filter.

Inlet pressure is usually expressed as gauge pressure. In a closed system (Figure 6.1), gauges will be used before and after the filter. The applied differential pressure is, therefore, the difference between the two gauge pressures.

An open or vented system is a special case of a closed system. The same calculation holds true for each system. However, in an open or vented system (Figure 6.2), the downstream pressure is equal to ambient pressure. Thus,

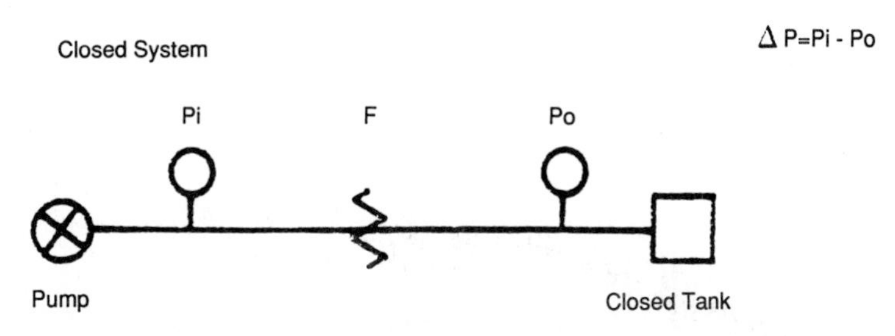

Figure 6.1 ΔP for a closed system. ΔP = Pi − Po.

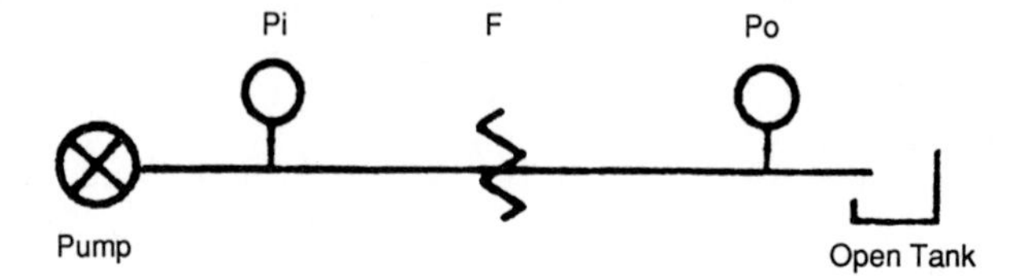

Figure 6.2 ΔP for an open system. $\Delta P = \text{Pi} - Po$. However, Po is at ambient pressure; therefore, $\Delta P = \text{Pi}$.

when gauge pressure is used, the downstream pressure is zero, so the differential pressure is equal to line pressure.

Parallel Filtration

The selected filter arrangements will be composed of standard filter units. Thus, as many cartridges or disks of whatever sizes will be joined to give the required effective filtration area. Where many such cartridges are required, they will be contained in a multicartridge holder.

The flow pattern within such a holder will involve the simultaneous passage of fluid through the filters. Such an arrangement is parallel flow (Figure 6.3). Regardless of the number of parallel or replicating flow paths, ΔP remains the same. However, the total flow rate and ΔP will be equally distributed across each filter. For a given flow, the total ΔP will be reduced proportionately by the number of filters involved.

Series Filtration

In parallel flow, the total ΔP does not depend on the number of filters operating in parallel. That is not the case in a series arrangement, wherein the fluid flows first through one filter, then through another, as is classically the

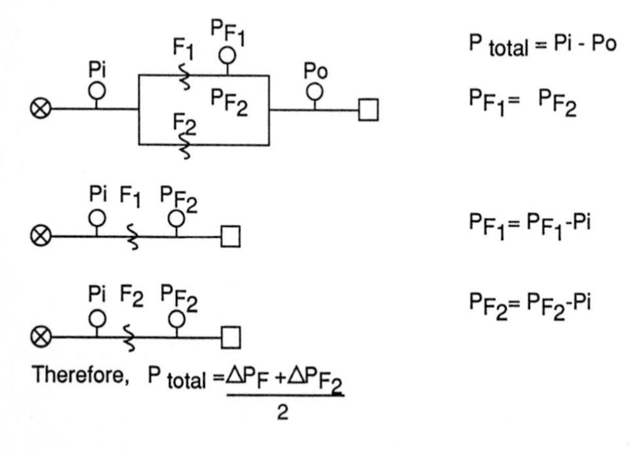

Figure 6.3 ΔP for parallel filtration.

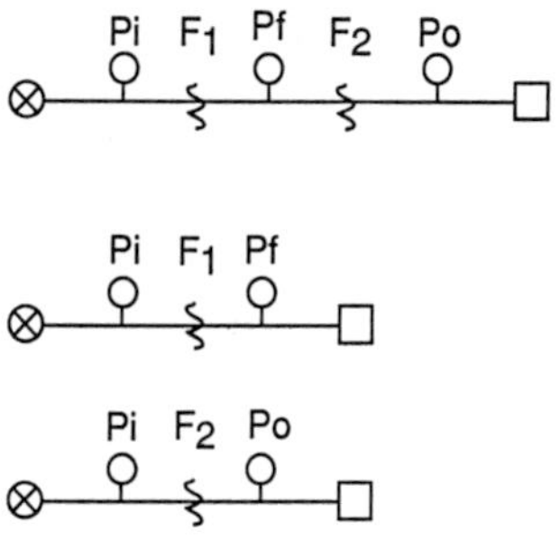

Pi F1 Pf F2 Po

$$P_{total} = Pi - Po$$

Pi F1 Pf

$$P_{F_1} = Pi - Pf$$

Pi F2 Po

$$P_{F_2} = Pf - Po$$

Therefore, $P_{total} = P_{F_1} + P_{F_2} = (Pi-Pf)+(Pf-Po) = Pi-Po$

Figure 6.4 ΔP for series filtration.

case for prefilter/final filter patterns. Here the total ΔP is the sum of the ΔP across each individual filter (Figure 6.4).

Because of the additive nature of each ΔP in series filtration, addition of prefilters should not be made without good reason. The unnecessary presence of filters in such a system is wasteful both of material and of pumping energy. More importantly, the unnecessary ΔP increment might, given the finite rating of any pumping system, limit complete use of the filters.

DIFFERENTIAL PRESSURE LEVELS

It is useful in filtration operations that the rate of fluid flow is directly proportional to ΔP. Accordingly, higher levels of ΔP will result in more rapid rates of filtration. However, the shape of the flow-decay curve may also depend on the differential pressure because elevations in the level of that driving force may increasingly promote the rate of filter blockage and, hence, the throughput.

It is almost universally more advantageous to use as low a ΔP as possible in process filtration. Filter economics will be enhanced as an expression of filter longevity. Throughput volume will usually be greater, and the rates of flow will compensate in the extent of their duration for their lower volumes per unit time. The preceding will especially be true, the greater the degree of particle loading in the fluid.

The longevity of membrane filters is promoted by the use of lower ΔP. At low levels, the particulate matter being removed forms a diffuse polarized layer adjacent to the filter face. The fluid being filtered finds little resistance in permeating this diffuse layer. At higher ΔP levels, the polarized layer becomes impacted, offering an increased resistance to fluid flow (Figure 6.5). Impaction of the removed particulate matter on the filter surface will unduly block fluid flow. This is particularly possible where gelatinous or otherwise deformable depositions are involved. Such an occurrence may more than

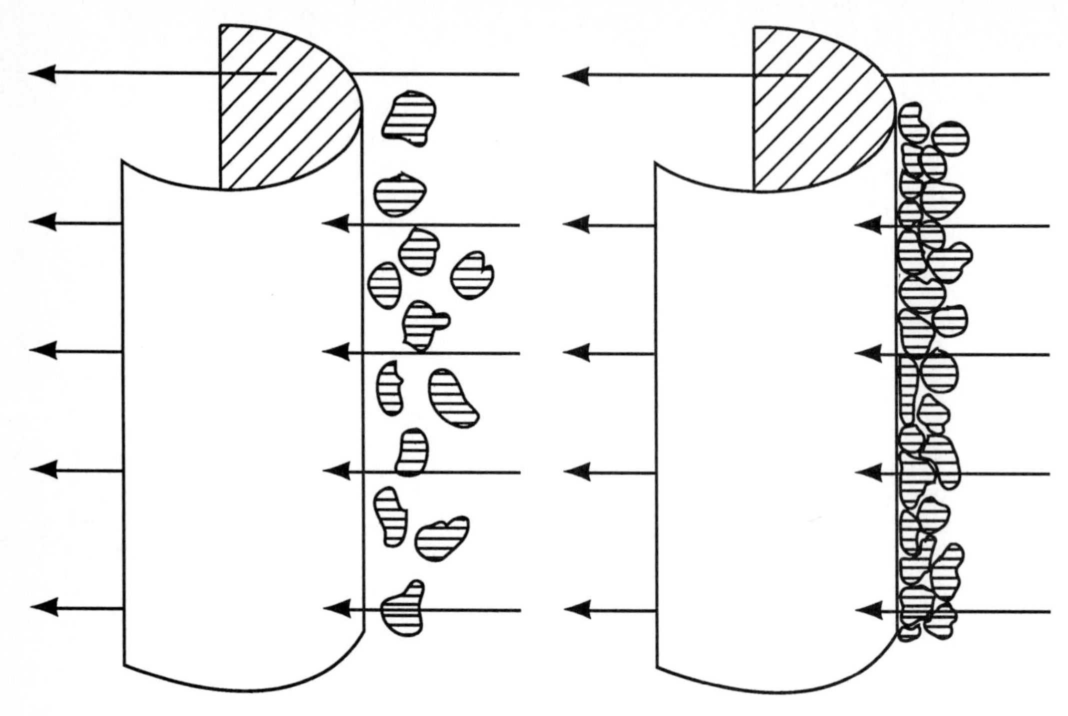

Figure 6.5 Effect of differential pressure on throughput volume: economics of filter longevity.

offset the advantages to rates of flow that derive from the use of higher ΔP. A concomitant decline in throughput may also result.

For membrane filters, surety of particle retention is mostly a result of sieve-type capture; that is, particles larger than the pore size are retained on the surface of the membrane filter. This retention characteristic is virtually unaffected by ΔP. However, when the particles are small enough to enter the filter's pores, and when their arrest then depends on their making contact with the pore walls, the probability of their capture is inversely proportional to ΔP. The probability of particle/pore wall encounter depends on the residence time of the particle in the pore. This primarily depends on the fluid velocity in the pore, which is a direct function of ΔP. Therefore, except where the particle-size distribution completely exceeds the largest of the filter pores, the efficiency of particle capture will vary inversely with ΔP. Cognizance of this effect is important to assure the efficiency of depth filters regardless of particle size.

As a practical measure, higher rates of flow and greater throughput volumes should be sought in process filtration contexts through using larger effective filtration areas rather than by employing elevated ΔP.

FILTER BLOCKING

As particulates are removed, they become deposited on or within the filter medium, occluding the pores. As filtration proceeds, the filter becomes increasingly loaded with particulates and becomes increasingly blocked. Thus, the effective filtration area progressively decreases.

Flow rate (F) is related directly to ΔP and effective filtration area (EFA). For a constant flowrate, ΔP and EFA bear an inverse relationship. It follows that in actual filtration, changing one of these parameters alters at least one of the others. For example:

F	ΔP	EFA
Constant	Increase	Decrease
Constant	Decrease	Increase
Increase	Constant	Increase
Decrease	Constant	Decrease
Increase	Increase	Constant
Decrease	Decrease	Constant

These relations are proportionate. Changing one parameter by some percentage or multiple will necessitate a change in a second parameter by the same percentage or multiple. This follows from the knowledge that F per ΔP per EFA is a given value.

If F is to remain constant and ΔP doubles, then EFA must be half its initial value because of filter blocking by the removed contaminants. Alternatively, if the ΔP remains constant, and EFA doubles, F will double. For example:

Condition	EFA	F	ΔP
Initial Area	5 ft^2	10 gpm	10 psi
Double Area	10 ft^2	10 gpm or 20 gpm	5 psi or 10 psi
Half Area	2.5 ft^2	10 gpm or 5 gpm	20 psi or 10 psi
Quarter Area	1.25 ft^2	10 gpm or 2.5 gpm	40 psi or 10 psi

Using these proportions, increases in the differential pressure of an operating filtration system indicate not only that the filter is becoming blocked, but also show how much of the filter is unblocked and still available. Thus:

Multiple of ΔP	% Blocked
ΔP (initial clean condition)	0
2 ΔP	50
4 ΔP	75
8 ΔP	87

This demonstrates that for each time the differential pressure doubles, the available filtration area is halved. Expressed mathematically:

$$\%\ EFA \text{ Available} = \frac{100\%}{\text{multiple of clean } \Delta P}$$

$$\%\ EFA \text{ Blocked} = 100\% - \frac{100\%}{\text{multiple of clean } \Delta P}$$

When plotted graphically, the second equation provides a curve whose slope is increasing at a steadily diminishing rate (Figure 6.6). This curve il-

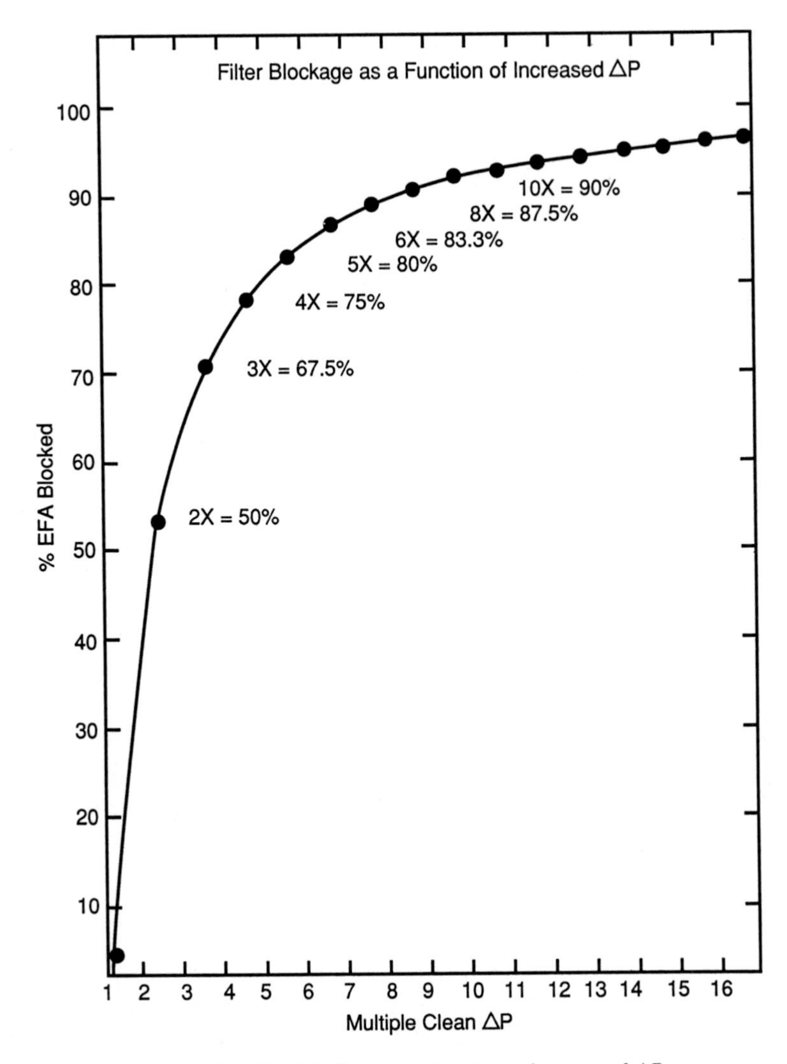

Figure 6.6 Filter blockage as a function of increased ΔP.

lustrates the law of diminishing returns with the maximum efficacy at 80 percent filter blockage. It is for this reason that filters are usually replaced at this point: either where, at constant pressure, the flow has diminished to 80 percent of its original rate, or where, at constant rate of flow, the pressure has increased fivefold.

In the fivefold multiplication of the original ΔP to the point of diminishing returns on the flow-decay curve lies the economic optimization of filter use previously discussed. To permit its achievement, the original ΔP must be as low as possible.

Whether necessitated by the use of a long series of prefilters/final filters, or by a desire for a high initial flow rate, an elevated initial ΔP is sometimes used. In such cases, the possibilities of manyfold pressure increases may be forfeited. Restrictions on the extent of pressure increase may also be imposed by limitations in the capacity of the system's pump. In that event, the filter system will not be usable to 80 percent of its capacity.

Different characteristics may be desired in a filter system, depending on its operational purposes. In continuous filtration, or where the batches are very large, extreme durability of the filter system is usually sought to minimize interim filter changes. These changes are always disruptive and time consuming. Their avoidance is always desired.

Savings in filtration time are best achieved by increase of the system's EFA. Parallel filter arrangements may be used advantageously to this end. Where batch filtration is used, it is helpful if the filter's exhaustion coincides with completion of the batch.

Where the batch size is so small as not to deplete the filter capacity, and where the filter is not reused as a precaution against batch cross-contamination, the operation offers the possibilities of using higher initial ΔP to attain peak flow rates or to reduce processing times. In such cases, compromise of filter longevity becomes less important. The filter pore-size rating must, however, demonstrably ensure against diminution in particle capture efficiency at higher ΔP.

USE OF FLOW-DECAY MEASUREMENTS

Filtration comprises one of the experimental branches of the physical sciences. Thus, how rapidly a filter will block during the filtration of a given fluid can be ascertained only by actual trial. Obviously, where system design involving large volumes of fluids is involved, some prior indication of the fluid filterability is desirable to avoid costly mismatching of process expectations and performance.

Flow-decay measurements provide a means to this end. Filtration is performed using filters of such small areas and/or such volumes of the subject fluid as will reveal in microcosm the blocking experience of that filter/fluid combination. The rate of flow and throughput will be elucidated. From the known effective filtration area of the small filter and the volume flow it yielded, an arithmetical proportion will reveal the EFA necessary to complete processing of specified large volumes of that particular fluid. If more rapid rates of filtration are required, a suitable multiplication of the EFA can be deduced.

Final Filters

In this exercise, the first filters to be considered are the final filters through which the fluid is to receive its eventual filtrative cleansing; the filters are selected on the basis of their pore-size ratings, composition, and so on to give the treated fluid the ultimate degree of purification sought for it in the filtrative step.

The pore-size rating selected for the final filter will be in accordance with the purposes of the filtration. Thus, clarification/visible particle removal may employ a 5-μm-rated membrane; bacterial reduction may use a 0.45-μm-rated filter; sterilization modeled by Pseudomonas diminuta will involve a 0.2-μm-rated filter; while given particle removals could conceivably require membranes of even finer pore size.

Membrane filters are usually used as final filters because their narrow pore-size distributions allow complete retention of particulate matter. When correctly selected for an application, the largest pores of a membrane's pore-size distribution will be smaller than the smallest particles of the particle-size distribution they are required to retain. The result is absolute particle capture, as evidenced by surface or screen retention. Surface retention occurs because the particles are too small to enter the membrane pores.

Regrettably, surface retention, the consequence of the desired absolute particle capture, may prematurely block the membrane surface. Especially where fluids involving high particulate contents are involved, prefilters must be employed to remove the bulk of particles, thus sparing the final filter and enhancing its longevity to the point of economic feasibility for the entire filter system.

Prefilters

Prefilters are usually of the depth-type construction, although membranes of appropriate pore size may also serve. In any case, the larger pores present in the depth-type constructions as a result of broader pore-size distributions invite penetration by particles. Particle arrest occurs largely within the pores of the prefilter, using its vast inner surfaces. Hence, the prefilter is able to accommodate large loadings of particles. Less than complete particle retention is a concomitant of prefilter pore penetration. Some particles pass completely through the prefilter because of its broader pore-size distribution. These particles emerge to confront the final filter. The load on the final filter is, however, reduced by the prefilter, and the useful life of the final filter is thereby increased.

A balance of retentivity and permeability for the prefilter is needed so that the burden of particulates it assumes significantly prolongs the life of the final filter, while the material the prefilter avoids trapping keeps its own life from becoming unduly abbreviated.

When flow-decay studies on the final filter alone show too large a drop in flow rate resulting from too rapid blockage of the membrane surface by high particulate loading, the need for a prefilter is indicated. As illustrated in Figure 6.7, a final filter by itself has a given level of flow sufficiency. In cases where this is not enough, coupling the final filter with a prefilter prolongs the period of meaningful flow rates. A second prefilter, protective of the first, can give an additional period of useful life to the filtration system. In these situations, the area under each appropriate curve represents the total throughput volume.

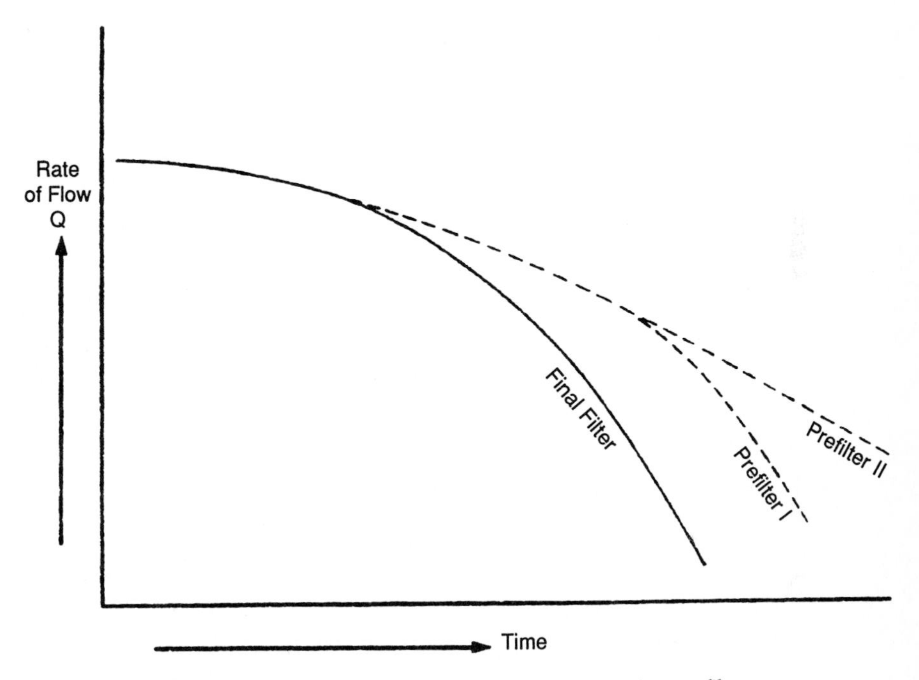

Figure 6.7 Prolongation of final filter longevity using prefilters.

OPERATIONAL CONSIDERATIONS

A flow-decay study is the most cost-effective method for selecting filter types, pore sizes, and filter areas, particularly in a new system. Flow-decay tests can be performed quickly and easily in a matter of hours using small disk filters, whereas the option usually will require experimentation with cartridge filters for days or weeks. When completed, a flow-decay test will provide fairly accurate data leading to the choice of proper prefilter/final filter combinations and the proper ratio of filter areas.

The purpose of a flow-decay study is to provide information on the particle distribution of the fluid to be filtered, and to provide information on the interaction between the fluid and the filter media. There are various ways of conducting a flow-decay test, and without dwelling on specific procedures, a general approach is suggested.

Begin with the Final Filter

The final filter will almost always be specified, at least indirectly, by the application, fluid, and system conditions so that the starting point becomes clear. In the rare case where the choice of final filter is not clear, preliminary testing should be performed to define this. The flow-decay measurement is performed first on this, the ultimate filter.

Add Prefilters in Ascending Pore-Size Order

After obtaining initial data on the unprotected final filter, a prefilter of the next logical pore size should be added and the test rerun on that prefilter. Generally, the prefilter first tried should be of a pore rating somewhat greater than that of the final filter to remove particles that may just fit, and therefore completely block, its pores. One progressively tries prefilters of increasingly larger pore ratings.

Plotting the flow-decay data is not essential. It will, however, provide a useful visual aid. Within the interval on the graph that represents the increased margin of flow of the prefilter over the final filter lies the possibility of improving the flow of their combinative arrangement (Figure 6.8). If the prefilter flow curve showed no decay, or very little, it would indicate an absence of significant particle retention by the prefilter. Although the margin in flows between the prefilter and final filter would then be large, it would offer no promise of improvement. That particular prefilter would not protect the final filter. The combination of both filters would show essentially the flow-decay characteristics of the final filter alone.

If the prefilter flow-decay pattern is rather close to that of the final filter, then the brevity of the interval between the two flows offers only small possibilities for improvement. In effect, the prefilter is itself too similar to the final filter in its particle capture and flow propensities to offer improvement.

It is impossible to predict, short of actual experimental trial, how much of the potential improvement interval between the flow-decay curves for the prefilter and final filter will translate into benefit for their combination.

If the chosen prefilter has throughput similar to the tested final filter, then its pore-size rating is too small. A larger pore-size prefilter is indicated.

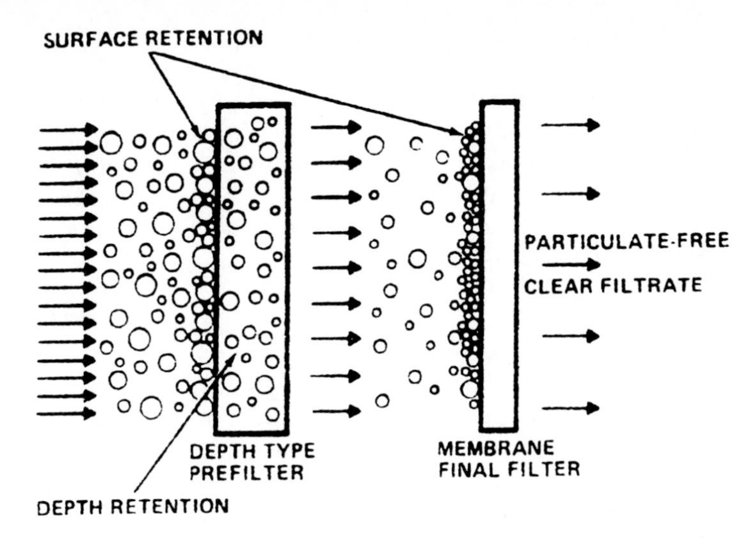

Figure 6.8 Prefilter/final combination.

If the chosen prefilter does not adequately protect the final filter, as evidenced by a very high flow rate, then the prefilter has a pore-size rating that is too large.

Use Separate Holders for Prefilters and Final Filters

Flow-decay tests are sometimes conducted by placing simultaneously as many different filter/prefilter disks in a holder as will fit while still providing adequate sealing and freedom from interference. This method is simpler than using multiple holders but will yield fewer useful data. By using a single disk holder, the flow-decay test will by experimental design yield equivalent areas for the prefilters and final filter. This limits the utility of the flow-decay test to providing information on the proper pore size and material selection. It will not provide any information on the ratio of filter areas. The suggested method is to use a separate filter holder for each simulated filter, to obtain data on the individual units, collect the effluent, and filter solution through the next filter unit in series. Using this approach will generate data both on proper pore-size rating selection and relative prefilter/final filter areas.

Simulate Cartridge Configurations

When a serial or double-layer cartridge filter is being considered, this construction must be reflected in the flow-decay test. In this case, the multiple layers should be placed in the same disk holder, since this will automatically yield the equivalent areas inherent in the serial cartridge design.

Use Actual Operating Parameters

As stated, the flow-decay test should be performed as the ΔP and fluid temperatures to be used in the final system. This will allow the filter/fluid interaction best to stimulate operating conditions. Changing these parameters, such as the common method of accelerating the flow decay by raising the pressure, may adversely affect the accuracy of the data due to factors such as particle impaction.

Monitor the Test

The flow decay will be monitored by recording data at preselected time intervals. The intervals may be modified as the decay progresses. It is suggested that either flow rate versus time (in a constant ΔP system) or ΔP versus time (in a constant flow rate system) be recorded. The test will be completed when the flow is 20 percent to 33 percent of the initial rate, or when ΔP reaches three to five times its initial value. This corresponds to a flow decay of 67 percent to 80 percent. Either of these data sets can be translated into throughput.

Larger Filter Areas Are Preferred

Flow decays are most commonly performed on 47-mm disks. However, smaller disks are used on clean solutions, such as deionized water, to shorten the time interval to filter blockage. Use of a smaller disk, decreasing filtration area, and reduced testing times will sacrifice accuracy. The largest filtration area that is reasonable with respect to accuracy, time, and expense should be considered. In some cases, a flow decay conducted on small cartridge filters may be appropriate.

Other Variables

In addition to EFA and operating conditions (mentioned previously), two other factors can seriously affect the accuracy of the tests. These are changes in particle distribution/burden/cleanliness of the fluid and fluid/system dynamics. To account for these variables, filtration engineers often will "oversize" the calculation by a factor of 1.5 to 2. It is suggested that any appropriate correction of this type be factored into the equation after completing system sizing, rather than after the flow-decay test.

Design Options

When the system is sized, its effective filtration area will usually reflect the most arduous demands, such as the peak flow rates. The accommodation of these conditions obviously requires a larger filter area than would an average or more modest constant rate of flow. There are engineering alternatives to such problems. Thus, the use of a storage vessel, where such a device can be tolerated and properly maintained, may permit system sizing on the basis of average rates of flow. The filter requirements would then be diminished.

Where the particle content is not excessive, the system may be based on the final filters alone. However, prefilters may serve, even here, to decrease the final filter area. Prefilter action serves to delay blockage of the final filter, thus influencing that filter's rates of flow and throughput. Prefilters are less expensive than final membrane filters. An economic advantage may be had from their employment.

Cartridge Scaleup

The final step in the flow-decay process is to scale up the results of cartridges. First, determine throughput per cartridge:

$$\frac{\text{throughput per disk}}{\text{cm}^2 \text{ EFA per disk}} \times \frac{6.45 \text{ cm}^2}{\text{in.}^2} \times \frac{144 \text{ in.}^2}{\text{ft}^2} \times \frac{\text{ft}^2 \text{ EFA}}{\text{cartridge}} = \text{throughput per cartridge}$$

Second, determine clean flow rate per cartridge at ΔP:

$$\frac{\text{clean flow per disk}}{\text{cm}^2 \text{ EFA per disk}} \times \frac{6.45 \text{ cm}^2}{\text{in.}^2} \times \frac{144 \text{ in}^2}{\text{ft}^2} \times \frac{\text{ft}^2 \text{ EFA}}{\text{cartridge}} = \text{clean flow per cartridge}$$

Third, determine clean flow rate per unit ΔP:

$$\frac{\text{clean flow per cartridge}}{\text{flow-decay pressure}} = \text{clean flow per unit } \Delta P$$

Fourth, determine average flow rate per cartridge:

$$\frac{\text{throughput per cartridge}}{\text{flow-decay time}} = \text{average flow rate per cartridge}$$

At any convenient time, units may be converted to those specified in the system parameters, e.g., 1 liter = 10^3 ml; 1 gal = 3.79 liters, 1 ft^3 = 28.32 liters.

Flow-decay tests are conducted almost exclusively at constant pressure with variable flow. However, many applications require constant flow with variable pressure. In a constant pressure-variable flow application, average flow rate is used for sizing. In a constant flow-variable pressure application, clean flow rate is used for sizing and clean ΔP is adjusted accordingly.

OPERATIONAL STEPS OF SYSTEM SIZING

Liquid Systems

Determine the maximum available ΔP:

$$\text{available } \Delta P = \text{inlet pressure} - \text{outlet pressure}$$

When the system is open or vented and flowing to ambient pressure, the available ΔP is equal to the inlet pressure.

Select the initial clean ΔP. Starting the filtration at the maximum available ΔP in a constant flow-variable pressure application leaves no room for additional pressure increases. As soon as the flow decreases to 80 percent of its initial value, the filter must be discarded. If the initial clean ΔP is not specified by the user, and only the maximum available ΔP is known, the clean ΔP should be chosen as 20 percent to 33 percent of the maximum available ΔP. Only for a constant pressure-variable flow application, such as batch processing, should clean ΔP be selected to equal the maximum available ΔP.

Determine flow rate per unit EFA at the clean ΔP:

$$\text{flow rate per } EFA = \text{flow rate per EFA at unit } \Delta P \times \text{clean } \Delta P$$

Flow rate at a specific clean ΔP may also be determined directly from the manufacturer's flow rate graphs.

Correct flow rate for fluid viscosity:

$$\text{corrected flow rate} = \frac{\text{flow rate per EFA at clean } \Delta P}{\text{viscosity in centipoise}}$$

Determine number of filtration units required in terms of cartridges or other increments of EFA:

$$\text{number of filtration units} = \frac{\text{system flow-rate requirement}}{\text{corrected flow rate per } EFA \text{ at clean } \Delta P}$$

Recalculate actual clean ΔP of final system:

$$\text{actual clean } \Delta P = \frac{\text{system flow-rate requirement}}{\text{flow rate per unit EFA at unit } \Delta P \times \text{number of units}}$$

Determine the percentage of the filter blocked at the change point:

$$\text{multiple of } \Delta P = \frac{\text{maximum available } \Delta P}{\text{actual clean } \Delta P}$$

$$\%\text{EFA blocked} = 100\% - \frac{100\%}{\text{multiple of } \Delta P}$$

Review the system for possible competing factors of flow rate, throughput, and economics.

Oversizing the System

There is always uncertainty regarding the accuracy of extrapolations made from small trials to larger operations. Flow-decay measurements are no exception. It is common practice, therefore, particularly for engineers representing filter manufacturers, to oversize such systems by factors of 1.5 to 2. Their intention is the laudable one of avoiding costly underdesign, which entails interim filter changes and insufficient housing capacity. Far more economical, it is commonly reasoned, to spend unnecessarily (but not extravagantly so) for added filters than to be saddled with an installed system insufficient for its purposes.

However justified the practice, it should not serve as a substitute for responsible conclusions being drawn from careful flow-decay determinations. Decisions on oversizing should in every case become fixed in consultative discussions with the end user.

It is possible that the degree of overdesign may be stipulated by the filter holder hardware. Thus, if a water purification system required five 10-in. cartridges in parallel to deliver the necessary flow, a seven-cartridge housing would be indicated (assuming five- or six-cartridge housings were not available in the market). Automatically, the five-cartridge system would become oversized by two cartridges or 1.4 times. Here the degree of overdesign would be permitted to rest, particularly if the next larger available housing was for 14 cartridges.

However uncertain its numerical prescription, the practice of oversizing is prudent. It provides against the vagaries of solution variability and seeks to avoid the interruption of critical operations.

High Flow Rates

Using filters with a large pore size and high flow rate or designing systems with high ΔP can yield very high theoretical flow rates. Based on extrapolating graphs, it may appear that a high flow-rate system (e.g., 100 gal/min) could be designed with one or two 10-in. cartridges. This design, however, would not take into account the resistance to flow of the filter housings.

Filter manufacturers publish graphs of flow rates that are fairly linear within a range. Within this range, most of the ΔP drives the fluid through the filter, and only a small portion is used to overcome the flow resistance of the housing. When the recommended flow rate is exceeded, however, deviation can occur.

First, filter efficiency and throughput may be decreased due to high-velocity impaction of particles on the filter medium. Additionally, ΔP will increase in a curvilinear fashion at a level higher than predicted by straight-line extrapolation. While the filter may still behave in a linear manner, the housing inlet, cartridge core, and outlet become increasingly restrictive to fluid flow because of the relatively small orifices involved. Also, the increased frictional losses and turbulence will generate heat. This can have a destabilizing influence on certain solutions and/or components.

Based on these restrictions, certain filter flow-rate limitations can be recommended:

1. When possible, maintain a flow rate of 1 to 3 gal/min/ft^2 EFA
2. The maximum flow rate should be 5 gal/min/ft^2 EFA
3. The filter cartridge-to-housing interface will normally be the most restrictive area to flow. The flow per interface should be maintained at 5 to 15 gal/min when possible.
4. The maximum flow rate per interface should be 25 gal/min. This implies that a single cartridge housing should have a flow rate not exceeding 25 gal/min whether the cartridge length is 10, 20, or 30 in. If a flow rate of 25 gal/min can be achieved with a 10-in. cartridge, increased lengths should be used to increase throughput rather than increase flow rate. Equivalent maxima for three-round and seven-round housings should be 75 and 175 gal/min, respectively.

These guidelines should prove acceptable for the majority of applications.

Prefilter/Final Filter Combination

Ideally, system sizing should always be completed using test data from a flow-decay study. This becomes increasingly important in large systems where one or more prefilters must also be sized. Unfortunately, performing flow-decay studies is not always possible and reliance on other sources of information such as a prior experience with the product, its probable cleanliness, and considerations of its particle-size distribution assumes a greater role.

In case of a complete absence of data, other approaches can be used:

1. Size the prefilter and final filter systems for the same clean ΔP. This approach assumes that there is a uniform particle distribution that will result in concurrent prefilter and final filter blocking. Since prefilters generally have a lower ΔP per unit flow rate than final filters, this assumption will result in the use of fewer prefilters than final filters.
2. Size for the same number of prefilters and final filters. This approach simplifies inventory and hardware requirements. Generally this will result in a lower ΔP for the prefilter group than for the final filter section.
3. Size two prefilters for every final filter. This approach follows from the fact that the function of a prefilter is to protect a final filter and that it will therefore accumulate greater particle loadings. Prefilters are also

considerably less expensive than final filters and thus can be discarded with less economic loss.

4. Prefilter/final filter ratios may be sized respecting a desire for system versatility, which may dictate that the same housing be used for each. System design should include provisions for peak load versus average demand. Future plans for batch size or flow rate scaleup should also be considered.

There is merit to each of these approaches, although the sizing solutions are less than optimal. It should be remembered that, ideally, prefilters and final filters should block concurrently to maximize area use and minimize change frequencies.

These pertinent concepts should be kept in mind regardless of whether flow decay or some other method is used to size the system:

1. Determine ratio of prefilters (PF) to final filters (FF):

$$\text{number of PF per FF} = \frac{\text{FF throughput per unit EFA}}{\text{PF throughput per unit EFA}}$$

   This calculation assumes data from a flow-decay assay. If an arbitrary ratio is chosen, that ratio should be substituted.

2. Determine maximum available ΔP:

$$\text{available } \Delta P = \text{inlet pressure outlet pressure}$$

3. Determine initial clean ΔP:

$$\text{available } \Delta P = \text{inlet pressure} - \text{outlet pressure}$$

4. Determine ΔP of corrected unit EFA of combined assembly to meet system requirements:

   clean ΔP per flow rate per unit EFA combination

$$= \frac{\text{system flow requirement}}{\text{FF flow/EFA/unit } \Delta P} + \frac{\text{system flow requirement}}{\text{PF flow/EFA/unit } \Delta P \times \text{PF/FF ratio}}$$

$$= \text{clean } \Delta P \text{ of unit combined assembly}$$

5. Correct clean ΔP for viscosity:

$$\text{clean } \Delta P \text{ of unit combined assembly} \times \text{viscosity}$$
$$= \text{corrected clean } \Delta P \text{ of combined assembly}$$

   A viscosity correction will not be needed if the flow decay was conducted automatically accounting for viscosity.

6. Determine number of filtration unit assemblies required:

$$\text{number of unit assemblies} = \frac{\text{unit assembly combined } \Delta P}{\text{system clear } \Delta P}$$

$$\text{number of unit assemblies} \times \text{number of FF per assembly} = \text{number of FF cartridges}$$

$$\text{number of unit assemblies} \times \text{number of PF per assembly} = \text{number of PF cartridges}$$

7. Recalculate actual clean ΔP of system:

$$\frac{\text{system flow measurements}}{\text{FF flow/unit EFA/unit } \Delta P \times \text{number FF cartridges}} = \text{FF clean } \Delta P$$

$$\frac{\text{system flow measurements}}{\text{PF flow/unit EFA/unit } \Delta P \times \text{number PF cartridges}} = \text{PF clean } \Delta P$$

$$\text{FF clean } \Delta P + \text{PF clean } \Delta P = \text{actual clean } \Delta P \text{ of system}$$

8. Determine the percentage of the filter blocked at the change point:

$$\text{multiple of } \Delta P = \frac{\text{maximum available } \Delta P}{\text{actual clean } \Delta P}$$

$$\%\ \text{EFA blocked} = 100\% - \frac{100\%}{\text{multiple of } \Delta P}$$

9. Review the system for possible competing factors of flow rate, throughput, and economics.

COMPRESSED GAS

First, determine the maximum available ΔP:

$$\text{available } \Delta P = \text{inlet pressure} - \text{outlet pressure}$$

Second, determine the initial clean ΔP:

$$\text{clean } \Delta P = \frac{\text{available } \Delta P}{3 - 5}$$

Third, convert the flow rate to actual cubic feet per minute (acfm) from standard cubic feet per minute (scfm).

$$\text{acfm} = \frac{\text{system flow requirement} \times \text{one atmosphere}}{\text{inlet pressure} + \text{one atmosphere} - \text{clean } \Delta P}$$

Standard cubic feet per minute (scfm) is the flow rate or volume at ambient pressure (one atmosphere). Unlike a liquid, gases are highly compressible. The ideal gas law shows that at constant temperature a given volume of gas will be compressed as an inverse function of applied pressure, that is, increase pressure/decrease volume. Therefore, a given flow rate of gas measured at ambient pressure will occupy less volume when compressed but will remain the same quantity and the compressed flow in actual cubic feet per minute (acfm) should be used for filter sizing. Some applications may specify different units, and the following conversion factors may be used: 1 ft^3 = 28.3 liters, 1 atm = 14.7 psig = 0 pisa = 760 mm Hg = 30 in. Hg = 1.0 kg/cm^2 = 1.0 bars.

Fourth, determine the flow rate per unit EFA at the clean ΔP:

$$\text{flow rate per EFA} = \text{flow rate per EFA at unit } \Delta P \times \text{clean } \Delta P$$

Fifth, determine the number of filtration units required:

$$\text{number of filtration units} = \frac{\text{system flow requirement (acfm)}}{\text{flow rate per unit EFA at clean } \Delta P}$$

Sixth, recalculate actual clean ΔP of final system:

$$\text{actual clean } \Delta P = \frac{\text{system flow-rate requirement}}{\text{flow rate per unit EFA per unit } \Delta P \times \text{number of units}}$$

Seventh, determine the percentage of the filter blocked at the change point:

$$\text{multiple of } \Delta P = \frac{\text{maximum available } \Delta P}{\text{actual clean } \Delta P}$$

$$\%\text{ EFA blocked} = 100\% - \frac{100\%}{\text{multiple of } \Delta P}$$

Eighth, review the system for possible competing factors of flow rate, throughput, and economics.

Compressed gas systems rarely require prefiltration. However, particularly in compressed air systems, care should be taken to assure that the gas is dry and oil free through pretreatment with a dryer and oil coalescer.

EXAMPLE PROBLEMS

Example 6.1: Flow Decay

Applications:

1. purpose of the filtration: removal of submicrometer particles contained in water
2. final filter pore size: 0.2-μm-rated

3. sterilization/sanitization: weekly sanitization with 85°C distilled water
4. integrity testing methods: none required
5. pretreatments/prefiltration tap water treated with chlorine and passed through activated carbon and deionizing beds to 1 MΩ resistivities
6. prior experience with the solution: filtration through 1-μm string-wound cartridge, which does not block.

Fluid conditions:

1. chemical composition: water with fairly low bacteria counts and a high level of submicrometer particles
2. viscosity: 1cP
3. surface tension: not required
4. fluid temperature: ambient

System conditions:

1. flow rate: 10 gal/min
2. pressure: 30 psig inlet filtering to ambient
3. throughput/change frequency: as needed
4. equipment requirements: none

The first step in the design process is to conduct a flow-decay determination. Since 0.2-μm-rated filtration and flow rate are major requirements, with integrity testing and sterilization not required, the Gelman Acroflow II and Acroflow Super E cartridges are considered for the application.

Gelman Acropor AN-200 (0.2-μm-rated) is placed into a 47-mm holder as this is the final layer in both cartridges. The test is conducted at the user's operating conditions. Flow-rate readings at 5-min intervals showed:

Time (min)	Flow rate (ml/min)	Throughput (ml)
Start	200	
5	200	1,000
10	100	1,800
15	90	2,200
20	70	2,600
25	50	2,900
30	40	3,200

Flow decay of 80 percent was reached and the test was terminated.

Next, Acropor AN-450 (0.45-μm-rated) was added to a new AN-200 disc in the same holder to simulate the Acroflow II cartridge.

Time (min)	Flow rate (ml/min)	Throughput (ml)
Start	140	
5	130	700
10	110	1,300
15	100	1,800
20	80	2,300
25	70	2,600
30	60	3,000
35	50	3,200
40	40	3,500

Flow decay of 80 percent was reached and the test was terminated.

This second test shows a decrease in initial flow rate due to the increased pressure drop incurred by adding the 0.45-μm-rated filter. Throughput was increased approximately 10 percent.

For a third test, AN-800 (0.8-μm-rated) is tried with a new AN-200 in the same holder. This simulates the configuration for Gelman's Acroflow Super E.

Time (min)	Flow rate (ml/min)	Throughput (ml)
Start	170	
5	150	800
10	140	1,500
15	130	2,200
20	110	2,800
25	100	3,300
30	90	3,800
35	70	4,200
40	50	4,500
45	40	4,700

Flow decay of 80 percent was reached and the test was terminated.

The AN-800/AN-200 combination has provided a 40 percent increase in throughput as compared to the AN-200 alone. Initial flow rate was not decreased as much by the AN-800 prefilter as the AN-200.

Analysis of the three data sets indicates that most of the particles that blocked the 0.2-μm-rated filter also blocked the 0.45-μm-rated membrane. However, using the 0.8-μm-rated layer appears to provide a good separation of the particle distribution.

Thus, the final filter will be the Acroflow Super E, a cartridge comprising a 0.2-μm-rated final membrane with an integral 0.8-μm-rated membrane as a prefilter. However, the situation may not yet be optimized. The next step, then, is to choose a prefilter. The pore-size rating of the prefilter should be somewhat greater than 0.8 μm. The next choice is to try a prefilter with a pore-size rating of 1 μm. Two possibilities exist in the Gelman line: the Acroflow II, 1.2-μm-rated membrane, and the Preflow 200, 1-μm-rated depth-type. The Acroflow II contains a membrane filter and has the inherent advantages of absolute retention and no fibrous materials. The Preflow 200 contains a nominal fiberglass medium with the depth filter characteristic of high dirt-holding capacity. Since the application does not proscribe the use of a fibrous filter, the high dirt-holding capacity of glass fiber is seen to outweigh its disadvantages. The Preflow 200 is chosen.

A 47-mm disk of the Type AIE Glass used in the Preflow 200 is placed in a holder. A second holder is prepared with AN-800/AN-200 to simulate the Super E.

Now a flow-decay study is conducted on the A/E Glass alone. The effluent solution is collected in a clear container.

Time (min)	Flow rate (ml/min)	Throughput (ml)
Start	800	
5	800	4
10	700	8
15	600	11
20	600	14
25	500	17
30	400	19
35	300	21
40	250	22
45	200	23
50	160	24

Flow decay of 80 percent was reached and the test was terminated.

The collected 1-μm-filtered solution is now filtered through the AN-800/AN-200 combination.

Time (min)	Flow rate (ml/min)	Throughput (ml)
Start	170	
5	160	830
10	150	1,600
15	150	2,400
20	140	3,100
25	130	3,800
30	110	4,400

Time (min)	Flow rate (ml/min)	Throughput (ml)
35	100	4,900
40	80	5,300
45	70	5,700
50	60	6,000
55	50	6,300
60	40	6,500

Flow decay of 80 percent was reached and the test was terminated.

The A/E Glass prefilter has increased the throughput of the AN -800/AN-200 combination by 40 percent. Together the A/E Glass and AN-800 have provided an aggregate increase in 100 percent or doubled the throughput of the 0.2-μm-rated AN-200 membrane.

The fact that the 1-μm string-wound cartridge has not clogged in the application indicates that continuing testing to larger pore-size filters is unnecessary. However, as evidenced by the 1-μm A/E glass having a definite clogging factor, pore-size ratings by industrial depth-type medium manufacturers may not be the same as the ratings by membrane manufacturers. In this case, trial of a 3-μm membrane demonstrated no appreciable flow decay after two hours and was considered ineffective in further prolonging the life of the 0.2-μm-rated filter.

Using the final flow-decay data and equations, the theoretical flow rate and throughput for each cartridge may be calculated.

Acroflow Super E. 10-in. (AN-800/AN-200):

1. Determine throughput per cartridge:

$$\frac{\text{throughput per disk}}{\text{cm}^2\text{ EFA per disk}} \times \frac{6.45\text{ cm}^2}{\text{in}^2} \times \frac{144\text{ in}^2}{\text{ft}^2} \times \frac{\text{ft}^2\text{ EFA}}{\text{cartridge}} = \text{throughput per cartridge}$$

$$\frac{6500\text{ ml throughput}}{10\text{ cm}^2\text{ EFA/47 mm disk}} \times \frac{6.45\text{ cm}^2}{\text{in.}^2} \times \frac{144\text{ in.}^2}{\text{ft}^2} \times \frac{6.5\text{ ft}^2\text{ EFA}}{\text{cartridge}} \times \frac{\text{liter}}{10^3\text{ ml}} \times \frac{\text{gal}}{3.79\text{ liter}} = 1{,}037\text{ gal throughput per 10-in. Super E}$$

2. Determine clean flow rate per cartridge at ΔP:

$$\frac{\text{clean flow per disk}}{\text{cm}^2\text{ EFA per disk}} \times \frac{6.45\text{ cm}^2}{\text{in.}^2} \times \frac{144\text{ in.}^2}{\text{ft}^2} \times \frac{\text{ft}^2\text{ EFA}}{\text{cartridge}} = \text{clean flow per cartridge}$$

$$\frac{170 \text{ ml/min. flow}}{10 \text{ cm}^2 \text{ EFA/47 mm disk}} \times \frac{6.45 \text{ cm}^2}{\text{in.}^2} \times \frac{144 \text{ in.}^2}{\text{ft}^2} \times \frac{6.5 \text{ ft}^2}{\text{Super E}}$$

$$\times \frac{\text{liter}}{10^3 \text{ ml}} \times \frac{\text{gal}}{3.79 \text{ liter}} = 27 \text{ gal/min clean flow per 10-in. Super E}$$

3. Determine clean flow rate per unit ΔP:

$$\frac{\text{clean flow per cartridge}}{\text{flow-decay pressure}} = \text{clean flow per unit } \Delta P$$

$$\frac{27 \text{ gal/min clean flow}}{30 \text{ psi}} = 0.9 \text{ gal/min-psi per Super E}$$

4. Determine average flow rate per cartridge:

$$\frac{\text{throughput per cartridge}}{\text{flow-decay time}} = \text{average flow rate per cartridge}$$

$$\frac{1{,}037 \text{ gal throughput}}{60 \text{ min}} = 17 \text{ gal/min average flow per 10-in. Super E}$$

Preflow 200, 10-in. (A/E Glass)

1. Determine throughput per cartridge:

$$\frac{24 \text{ liters throughput}}{10 \text{ cm}^2 \text{ EFA/47-mm disk}} \times \frac{6.45 \text{ cm}^2}{\text{in.}^2} \times \frac{144 \text{ in.}^2}{\text{ft}^2} \times \frac{4.5 \text{ ft}^2 \text{ EFA}}{\text{preflow}}$$

$$\times \frac{\text{gal}}{3.79 \text{ liter}} = 2{,}650 \text{ gal throughput per 10-in. Preflow-200}$$

2. Determine clean flow rate per cartridge at ΔP:

$$\frac{800 \text{ ml/min flow}}{10 \text{ cm}^2 \text{ EFA/47-mm disk}} \times \frac{6.45 \text{ cm}^2}{\text{in.}^2} \times \frac{144 \text{ in.}^2}{\text{ft}^2} \times \frac{4.5 \text{ ft}^2 \text{ EFA}}{\text{preflow}}$$

$$\times \frac{\text{liter}}{10^3 \text{ ml}} \times \frac{\text{gal}}{3.79 \text{ liter}} = 88 \text{ gal/min clean flow per 10-in. Preflow 200}$$

3. Determine clean flow rate per unit ΔP:

$$\frac{88 \text{ gal/min clean flow}}{30 \text{ psi}} = 2.9 \text{ gal/min-psi per 10-in. Super E}$$

4. Determine average flow rate per cartridge:

$$\frac{2{,}650 \text{ gal throughput}}{50 \text{ min}} = 53 \text{ gal/min average flow per 10-in. Preflow}$$

Given the eventuality that systems will be designed without data being available from a now-decay exercise, Examples 6.2 and 6.3 will use catalog values. Only a final filter will be sized for simplicity. Example 6.4, the flow-decay data, will be used for both prefilter and final filter.

Example 6.2: Liquid System

This example is for a deionized water system flowing into a vented tank at 20 psig line pressure.

1. Determine the maximum available ΔP:

$$\text{available } \Delta P = \text{inlet pressure} - \text{outlet pressure}$$
$$= 20 \text{ psig} - 0 \text{ psig} = 20 \text{ psig}$$

2. Determine initial clean ΔP: A constant pressure variable flow-rate condition will be assumed, so that clean ΔP is equal to the maximum available differential pressure of 20 psig. If the application were for constant flow-variable pressure, 20 percent to 33 percent of the maximum available ΔP would have been chosen.
3. Determine flow rate per unit EFA at the clean ΔP:

$$\text{flow per EFA} = \text{flow per EFA at unit } \Delta P \times \text{clean } \Delta P$$
$$= 1 \text{ gal/min/psi} \times 20 \text{ psi}$$
$$= 20 \text{ gal/min per Super E at 20 psi}$$

This calculation is conducted using the slope of the Super E flow-rate curve from catalog data. The result may also be determined directly by reading the flow rate at 20 psi on the graph.

4. Correct the flow rate per fluid viscosity:

$$\text{corrected flow rate} = \frac{\text{flow rate per EFA at clean } \Delta P}{\text{viscosity}}$$
$$= \frac{\text{20 gal/min per cartridge at 20 psig}}{\text{ICP viscosity}}$$
$$= 20 \text{ gal/min per cartridge}$$

5. Determine number of filtration units required:

$$\text{number of filtration units} = \frac{\text{system flow-rate requirement}}{\text{corrected flow per EFA at clean } \Delta P}$$
$$= \frac{\text{10 gal/min system requirement}}{\text{20 gal/min per 10-in. Super E}}$$
$$= 0.5 \text{ 10-in. Super E needed}$$

The system chosen will consist of one 10-in. Acroflow Super E cartridge.

6. Recalculate actual clean ΔP:

actual clean ΔP

$$= \frac{\text{system flow-rate requirement}}{\text{flow rate per unit EFA at unit } \Delta P \times \text{number of units}}$$

$$= \frac{\text{10 gal/min system}}{\text{1 gal/min per cartridge per psig} \times \text{1 cartridge}} = \text{10 psig}$$

7. Determine percentage of filter blocked at change point:

$$\text{multiple of } \Delta P = \frac{\text{maximum available } \Delta P}{\text{actual clean } \Delta P} = \frac{\text{20 psig}}{\text{10 psig}} = \text{2X multiple } \Delta P$$

$$\%\text{ EFA blocked} = 100\% - \frac{100\%}{\text{actual clean } \Delta P}$$

$$= 100\% - \frac{100\%}{\text{2X multiple}} = 50\%\text{ EFA blocked}$$

8. Review the system for possible competing factors of flow rates, throughput, and economics.

An initial premise of this example was that it involves a constant pressure-variable flow application. Under this condition, steps 6 and 7 are not needed in that the pressure will be 20 psig and the initial flow rate will be 20 gpm. The filter will be used until it stops flowing.

One alternative recommendation might be to begin at 10 psi and increase the pressure as the filter blocks and flow rate declines. Under these conditions, 50 percent of the filter will be blocked at the change point as shown previously.

A second alternative would be to recommend a 20-in. cartridge. The clean ΔP would then be 5 psi. This allows for a multiple of 4 in pressure increase resulting in 75 percent filter blockage at the change point.

Example 6.3. High Flow-Rate System

This example designs a system for filtration of an aqueous solution into a high-flow filling machine. The pump upstream of the filter is producing 30 psig. There is a pressure requirement of 10 psig downstream into the filling machine. A constant flow rate of 50 gal/min is needed. The filter is the 5-μm Acroflow 11.

1. Determine the maximum available ΔP:

$$\text{30 psig inlet} - \text{10 psig outlet} = \text{20 psig max. available } \Delta P$$

2. Determine the initial clean ΔP: to allow for filter blockage, a clean ΔP of 25 percent of the maximum available, or 5 psig, will be chosen.

3. Determine the flow rate per unit EPA at the clean ΔP:

$$10 \text{ gal/min/psi} \times 5 \text{ psi} = 50 \text{ gal/min per cartridge at } 5 \text{ psi}$$

4. Correct flow rate for fluid viscosity: since the fluid is aqueous, the viscosity is equal to one and no correction is needed.
5. Determine the number of filtration units required:

$$\frac{50 \text{ gal/min system requirement}}{50 \text{ gal/min per 10-in. cartridge}} = 1 \text{ Acroflow II needed}$$

Theoretically, the system will work with one 10-in. cartridge. However, at this flow rate, the velocity through the filter is too great and will adversely affect filter performance and life. Since the cartridge has a 4.5-ft^2 filtration area, the flow rate should be a maximum of 23 gal/min. and preferably, 5–14 gal/min. Therefore, we must have at least three 10-in. cartridges and preferably four 10-in. cartridges.

Following this line of thought, one 30-in. or 40-in. cartridge should be sufficient. However, there is a second limitation on flow rate through the cartridge/housing interface and housing outlet. The flow rate per interface should be 25 gal/min maximum and preferably, 5 to 15 gal/min. A 30-in. or 40-in. cartridge will not suffice and a three-round housing, or four single housings in parallel, must be used.

While four assemblies are preferred, there is no four-round housing available. Consultation with the user indicates that four single housings in parallel is not feasible for the installation.

Thus, limitations in equipment, space, and connections dictate that a three-round housing with three cartridges in parallel will be selected. The flow rate per cartridge will now be 17 gal/min.

6. Recalculate actual clean ΔP:

$$\frac{50 \text{ gal/min system requirement}}{10 \text{ gal/min per cartridge per psi} \times 3 \text{ cartridges}} = 1.7 \text{ psi}$$

7. Determine percentage of filter blocked at change point:

$$\frac{20 \text{ psig available } \Delta P}{1.7 \text{ psi clean } \Delta P} = 12\text{X multiple } \Delta P$$

$$100\% - \frac{100\%}{12\text{X Multiple}} = 92\% \text{ EFA blocked}$$

8. Review the system for competing factors of flow rate, throughput, and economics.

The clean ΔP is much lower than strictly needed. However, equipment and engineering considerations have dictated the final configuration. While the cost is high in terms of filter use, this system virtually assures that the filters will perform well without blocking throughout the service life.

Example 6.4: Prefilter/Final Filter Combination

In this example, the flow-decay data obtained in Example 6.1 are used.

- Flow-rate requirement = 10 gal/min
- maximum available ΔP = 30 psig
- Acroflow Super L., 10 in.: throughput = 1,037 gal, clean flow rate = 27 gal/min at 30 psi (theoretical) and clean flow rate = 0.9 gal/min-psi
- Preflow 200, 10 in.: throughput = 2,650 gal, clean flow rate – 88 gal/min at 30 psi (theoretical) and clean flow rate = 2.9 gal/min/psi

1. Determine ratio of prefilters to final filters:

$$\text{number of PF per FF} = \frac{\text{FF throughput per unit EFA}}{\text{PF throughput per unit EFA}}$$

$$= \frac{\text{1,037 gal per 10-in. Super E}}{\text{2,650 gal per 10-in. Preflow}}$$

$$= \text{0.4 Preflows per Super E}$$

The unit EFA has been chosen as one 10-in. cartridge in each preceding case. However, ft^2 or other units may also be used.

2. Maximum available ΔP = 30 psi.

3. Initial clean ΔP = 30 psi/4X multiple = 7.5 psi.

4. Determine ΔP of corrected unit EFA of combined assembly to meet system requirement:

$$\frac{\text{system requirement}}{\text{FF flow/EFA/unit } \Delta P} + \frac{\text{system requirement}}{\text{FF flow/EFA/unit } \Delta P \times \text{PF/FF ratio}}$$

$$= \text{clean } \Delta P \text{ of unit combined assembly}$$

$$\frac{\text{10 gal/min system}}{\text{0.9 gal/min/cartridge/psi}} + \frac{\text{10 gal/min}}{\text{2.9 gal/min/cartridge/psi} \times \text{0.4 PF/FF}}$$

$$= \text{11.1 psi FF} + \text{8.6 psi PF} = \text{19.7 psi unit combined assembly}$$

5. Correct clean ΔP viscosity:

$$\text{clean } \Delta P \text{ of unit combined assembly} \times \text{viscosity}$$

$$= \text{corrected clean } \Delta P \text{ of combined assembly}$$

No correction is needed since conducting the flow decay automatically includes the viscosity correction.

6. Determine the number of filtration unit assemblies required:

$$\text{number of unit assemblies} = \frac{\text{unit assembly combined } \Delta P}{\text{system clean } \Delta P}$$

$$\frac{\text{19.7 psi unit combined assembly}}{\text{7.5 psi clean system } \Delta P} = \text{2.6 combined assemblies}$$

$$\text{2.6 assemblies} \times \text{1 Super E per assembly} = \text{2.6 10-in. Super E cartridges}$$

$$\text{2.6 assemblies} \times \text{0.4 preflow per assembly} = \text{1.1 10-in. Preflow cartridges}$$

Since the cartridges must be complete units, a 30-in. Super E would be selected. A 20-in. Preflow 200 would also be recommended to compensate for the added final filter area so that the prefilter will not block before the final filter. The unused area in the prefilter is acceptable given the low cost of the prefilter when compared to the final filter.

7. Recalculate actual clean ΔP of system:

$$\frac{\text{10 gal/min system}}{\text{0.9 gal/min/cartridge/psi} \times \text{3 cartridges}} = \text{3.7 psi 30-in. Super E}$$

$$\frac{\text{10 gal/min system}}{\text{2.9 gal/min/cartridge/psi} \times \text{2 cartridges}} = \text{1.7 psi 20-in. Preflow}$$

$$\text{3.7 psi} + \text{1.7 psi} = \text{5.4 psi actual clean } \Delta P \text{ of system}$$

8. Determine percentage of filter blocked at change point:

$$\frac{\text{30 psi max, available } \Delta P}{\text{5.4 psi clean } \Delta P} = 5.6 \times \text{ multiple } \Delta P$$

$$100\% - \frac{100\%}{5.6 \times \text{ multiple}} = 82\% \text{ blocked}$$

9. Review the system for possible competing factors of flow rate, throughput, and economics.

The designed system results in greatly "oversized" prefilters to compensate for a mildly oversized final filter. Given the limitations of 10-in., 20-in., and 30-in. units, a 30-in. final filter must be used. As indicated, the 20-in. prefilter assures that premature blocking will not occur. Economically, this is reasonable.

However, other alternatives are available to the user. A 10-in. Preflow could be selected along with the 30-in. Super E, increasing the clean ΔP to 7.2 psi. This theoretically will allow 76 percent filter use before the ΔP reaches 30 psi and necessitates filter change. The disadvantage to this approach is that the prefilter may block prematurely, disallowing full use of the final filter.

Another alternative design would be to use a 20-in. unit for both the prefilter and final filter. This has the advantage of simplifying hardware and providing versatility. For this case, the clean ΔP will be 7.3 psi, allowing 75 percent filter blocking. However, the final filter may block prematurely, disallowing full use of the prefilter.

Example 6.5: Compressed Gas

A compressed air system is required to deliver 500 scfm of sterile air into a fermenter. Line pressure at the fermenter is 50 psig. ΔP is specified to be 1 psi clean and 5 psi the maximum allowable. The cartridge used is the Gelman Acrovent.

1. Determine the maximum available ΔP: minimum ΔP is specified as 5 psi.
2. Determine the initial clean ΔP: clean ΔP is specified as 1 psi.
3. Convert the flow rate to acfm from scfm:

$$\text{acfm} = \frac{\text{system flow requirement} \times \text{one atmosphere}}{\text{inlet pressure} + \text{one atmosphere} - \text{clean } \Delta P}$$

$$\frac{500 \text{ scfm} \times 14.7 \text{ psia}}{50 \text{ psig} + 14.7 \text{ psig} - 1 \text{ psid}} = 115 \text{ acfm}$$

4. Determine flow rate per unit EFA at the clean ΔP:

$$\text{flow rate per EFA} = \text{flow rate per EFA at unit } \Delta P$$

$$\times \text{ clean } \Delta P \frac{20 \text{ cfm/psi}}{10\text{-in. Acrovent}} \times 1 \text{ psi } \Delta P = 20 \text{ cfm per 10-in. Acrovent}$$

5. Determine the number of filtration units required:

$$\text{number of filtration units} = \frac{\text{system flow requirement (acfm)}}{\text{flow rate/unit EFA at clean } \Delta P}$$

$$\frac{115 \text{ acfm}}{20 \text{ cfm/Acrovent at 1 psi}} = 5.8 \text{ 10-in. Acrovents}$$

Reviewing available equipment, a seven-place multicartridge housing is found to be the proper holder. Therefore, seven 20-in. Acrovents will be used.

6. Recalculate actual clean ΔP of the final system:

$$\text{actual clean } \Delta P = \frac{\text{system flow-rate requirement}}{\text{flow rate/unit EFA/unit } \Delta P} \times \text{number of units}$$

$$\frac{115 \text{ acfm}}{20 \text{ cfm/Acrovent/psi} \times 7 \text{ Acrovents}} = 0.8 \text{ psi clean } \Delta P$$

7. Determine the percentage of the filter blocked at the change point:

$$\frac{5 \text{ psi maximum available } \Delta P}{0.8 \text{ psi actual clean } \Delta P} = 6.3 \times \text{ multiple}$$

$$100\% - \frac{100\%}{6.3 \times \text{ multiple}} = 84\% \text{ EFA blocked}$$

8. Review the system for competing factors of flow rate, throughput, and economy.

Compressed gas filters are the one case in pharmaceutical applications where the filter will be sterilized repeatedly and placed in the system. When the system is well maintained to produce dry, oil-free air, the burden of viable and nonviable particulates removed by the filter will be small. Also, unlike liquid systems, there is no risk of pyrogens. Thus, compressed air filters are rarely replaced because of filter blockage. Typically, they are replaced as a part of a routine maintenance schedule, or for integrity failure as a consequence of too many sterilization cycles.

A a result of these factors, point-of-use filters in a high-quality compressed air system usually do not require significant "oversizing."

CONCLUSION

As is common to situations where experimental procedures are involved, different practitioners advocate somewhat different approaches to the subject. Some years ago, an effort was made to computerize the complexities involved in prefilter/final filter selection and sizing, no doubt to free the exercise from its experimental trial and error requirements. The effort, as judged from its apparent application absence, was not a success.

Whether by the method advanced here, or by any of the several possible variations, prefilter/final filter selection and sizing can be performed successfully. The results will yield filter arrangements that are suitable and practical in terms of meeting both the technical and economic requirements of such systems. The literature [5] provides further discussion.

REFERENCES

1. Cole, J. C., and R. H. Shumsky, "Filter Sizing for Economics: Liquid Systems," *Pharm. Technol.*, 1(11) (1977), 39–41.
2. Trasen, B., "Designing Microfiltration Systems," *Pharm. Technol.*, 15(11) (1981), 62–69.
3. Johnston, P. R., and D. L. Beals, "Constant-Pressure Liquid Filtration: Mathe-

matical Models for Fast Plugging of the Filter Medium," *J. Testing Eval.*, 8(2) (1980), 57–62.

4. MELTZER, T. H., and R. C. LUKASZEWICZ, "Filtrative Sterilization with Porous Membranes," In *Quality Control in the Pharmaceutical Industry*, vol. 3. New York: Academic Press, Inc., 1979.
5. LUKASZEWICZ, R. C., P. R. JOHNSTON, and T. H. MELTZER, "Prefilters/Final Filters: A Matter of Particulate/Pore Size Distribution," *J. Parenteral Sci. Technol.*, 35 (1981), pp. 40–47.

7

Filter Aids and Filter Media

Filter aids and/or flocculants are employed to improve the filtration characteristics of hard-to-filter suspensions. A filter aid is a finely divided solid material, consisting of hard, strong particles that are, en masse, incompressible. The most common filter aids applied as an admix to the suspension are diatomaceous earth, expanded perlite, Solkafloc, fly ash, or carbon. Filter aids build up a porous, permeable, and rigid lattice structure that retains solid particles and allows the liquid to pass through. These materials are applied in small quantities in clarification or in cases where compressible solids have the potential to foul the filter medium. Typical applications for filter aids are described in the literature.

Filter aids may be applied in one of two ways. The first method involves the use of a precoat filter aid, which can be applied as a thin layer over the filter before the suspension is pumped to the apparatus. A precoat prevents fine suspension particles from becoming so entangled in the filter medium that its resistance becomes excessive. Further, it facilitates the removal of cake at the end of the filtration cycle.

The second application method involves incorporation of a certain amount of the material with the suspension before introducing it to the filter. The addition of filter aids increases the porosity of the sludge, decreases its compressibility, and reduces the resistance of the cake. In some cases the filter aid displays an adsorption action, which results in particle separation of sizes down to 0.1 μm. The adsorption ability of certain filter aids, such as bleached earth and activated charcoals, is manifest by a decoloring of the suspension's liquid phase. This practice is widely used for treating fats and oils.

Filter aids are also used in processing sugar, beer, wine, gelatin, antibiotics, glycerol, solvents, synthetic tars, caustic, sulfur and uranium salts, for water cleaning, and in preparing galvanic solutions. The properties of these additives are determined by the characteristics of their individual components. For any filter aid, size distribution and the optimal dosage are of great importance. Too low a dosage results in poor clarity; too great a dosage will result in the formation of very thick cakes. In general, a good filter aid should form a cake having high porosity (typically 0.85 to 0.9), low surface area, and good particle-size distribution. An acceptable filter aid should have a much lower filtration resistance than the material with which it is being mixed. It should reduce the filtration resistance by 67 percent to 75 percent with the addition of no more than 25 percent by weight of filter aid as a fraction of the total solids.

The addition of only a small amount of filter aid (e.g., 5 percent of the sludge solids) can actually cause an increase in the filtration resistance. When the amount of filter aid is so small that the particles do not interact, they form a coherent structure, and resistance may be affected adversely.

FILTER AID REQUIREMENTS

Filter aids are evaluated in terms of the rate of filtration and clarity of filtrate. Finely dispersed filter aids are capable of producing clear filtrate; however, they contribute significantly to the specific resistance of the medium. As such, applications must be made in small doses. Filter aids comprised of coarse particles contribute considerably less specific resistance; consequently, a high filtration rate can be achieved with their use. Their disadvantage is that a muddy filtrate is produced.

The optimum filter aid should have maximum pore size and ensure a prespecified filtrate clarity. Desirable properties characteristics for the optimum filter aid include:

1. The additive should provide a thin layer of solids having high porosity (0.85 to 0.90) over the filter medium's external surface. Suspension particles will ideally form a layered cake over the filter aid cake layer. The high porosity of the filter aid layer will ensure a high filtration rate. Porosity is not determined by pore size alone. High porosity is still possible with small size pores.
2. Filter aids should have low specific surface, since hydraulic resistance results from frictional losses incurred as liquid flows past particle surfaces. Specific surface is inversely proportional to particle size. The rate

of particle dispersity and the subsequent difference in specific surface determines the deviations in filter aid quality from one material to another. For example, most of the diatomite species have approximately the same porosity; however, the coarser materials experience a smaller hydraulic resistance and have much less specific surface than the finer particle sizes.

3. Filter aids should have a narrow fractional composition. Fine particles increase the hydraulic resistance of the filter aid, whereas coarse particles exhibit poor separation. Desired particle-size distributions are normally prepared by air classification, in which the finer size fractions are removed.
4. In applications where the filter aid layer is to be formed on open-weave synthetic fabric or wire screens, wider size distributions may have to be prepared during operation. Filter aids should have the flexibility to be doped with amounts of coarser sizes. This provides rapid particle bridging and settling of the filter aid layer. For example, diatomite having an average particle size of 8 μm may be readily applied to a screen with a mesh size of 175 μm by simply adding a small quantity of filter aid with sizes that are on the same order but less in size than the mesh openings. Particle sizes typically around 100 μm will readily form bridges over the screen openings and prevent the loss of filter aid in this example.
5. The filter aid should be chemically inert to the liquid phase of the suspension and not decompose or disintegrate in it.

The ability of an admix to be retained on the filter medium depends on both the suspension's concentration and the filtration rate during this initial precoat stage. The same relationships for porosity and the specific resistance of the cake as functions of suspension concentration and filtration rate apply equally to filter aid applications.

FILTER AID APPLICATIONS

Filter aids are added in amounts needed for a suspension to acquire desirable filtering properties and to prepare a homogeneous suspension before actual filtration. Filter aids increase the concentration of solids in the feed suspension. This promotes particle bridging and creates a rigid lattice structure for the cake. In addition, they decrease the flow deformation tendency. Irregular or angular particles tend to have better bridging characteristics than spherical particles. As a rule, the weight of aid added to the suspension should equal the particle weight in suspension. Typical filter aid additions are in the range of 0.01 percent to 4 percent by weight of suspension; however, the exact amount can only be determined from experiments. Excess amounts of filter aid will decrease the filter rate. Operations based on the addition of admixes to their suspensions may be described by the general equations of filtration with cake formation. A plot of filtration time versus filtrate volume on rectangular coordinates results in a nearly parabolic curve passing through the origin. The same plot on logarithmic coordinates, assuming that the medium resistance may be neglected, results in a straight line. This convenient linear relationship allows results obtained from short-time filtration tests to be extrapolated to long-term operating performance (i.e., for several hours of operation). This reduces the need to make frequent, lengthy tests and saves time in the filter selection process.

In precoating, the prime objective is to prevent the filter medium from fouling. The volume of initial precoat normally applied should be 25 to 50 times greater than that necessary to fill the filter and connecting lines. This amounts to about 5–10 lb/100 ft^2 of filter area, which typically results in a $\frac{1}{16}$-in. to $\frac{1}{8}$-in. precoat layer over the outer surface of the filter medium. An exception to this rule is in the precoating of continuous rotary drum filters where a 2-in. to 4-in. cake is deposited before filtration. The recommended application method is to mix the precoat material with clear liquor (which may consist of a portion of the filtrate). This mixture should be recycled until all the precoat has been deposited onto the filter medium. The unfiltered liquor follows through immediately without draining off excess filter aid liquor. This operation continues until a predetermined head loss develops, when the filter is shut down for cleaning and a new cycle.

In precoating, regardless of whether the objective is to prevent filter medium clogging or to hold back fines from passing through the medium to contaminate the filtrate, the mechanical function of the precoat is to behave as the actual filter medium. Since it is composed of incompressible, irregularly shaped particles, a high-porosity layer is formed within itself, unless it is impregnated during operation with foreign compressible materials. Ideally, a uniform layer of precoat should be formed on the surface of the filter medium. However, a nonuniform layer of precoat often occurs due to uneven medium resistance or fluctuations in the feed rate of filter aid suspension. Cracks can form on the precoat layer that will allow suspension particles to penetrate into the medium. To prevent cracking, the filter aid may be applied as a compact layer. On a rotating drum filter, for example, this may be accomplished by applying a low concentration of filter aid (2 percent to 4 percent) at the maximum drum rpm. In other filter systems, maintaining a low-pressure difference during the initial stages of precoating and then gradually increasing it with increasing layer thickness until the start of filtration will help to minimize cake cracking. Also, with some filter aids (such as diatomite or perlite), the addition of small amounts of fibrous material will produce a more compact precoat cake. At low-suspension concentrations (typically 0.01 percent), filter aids serve as a medium under conditions of gradual pore blocking. In this case the amount of precoat is 10 to 25 N/m^2 of the medium and its thickness is typically 3 to 10 mm. In such cases, the filter aid chosen should have sufficient pore size to allow suspension particle penetration and retention within the precoat layer.

PROPERTIES OF TYPICAL FILTER AIDS

Diatomite

The most important filter aids from a volume standpoint are the diatomaceous silica type (90 percent or better silica). These are manufactured from the siliceous fossil remains of tiny marine plants known as diatoms.

Diatomaceous filter aids are available in various grades. This is possible because the natural product can be modified by calcining and processing, and because filter aids in different size ranges and size distributions have different properties. The filter aids may be classified on the basis of cake permeability to water and water flow rate. Finer grades are the slower-filtering products; however, they provide better clarification than do faster-filtering grades. Thus a fast-filtering aid may not provide the required clarification. However, by changing the physical character of the impurities (e.g., by proper coagulation), the same clarity may be obtained by using the fast-filtering grades.

Calcinated diatomaceous additives are characterized by their high retention ability with relatively low hydraulic resistance. Calcining dramatically affects the physical and chemical properties of diatomite, making it heat resistant and practically insoluble in strong acids. Further information is given in the literature.

Perlite

Perlite is glass-like volcanic rock, called *volcanic glass*, consisting of small particles with cracks that retain 2 percent to 4 percent water and gas. Natural perlite is transformed to a filter aid by heating it to its melting temperature (about 1,000°C), where it acquires plastic properties and expands due to the emission of steam and gas. Under these conditions its volume increases by a factor of 20. Beads of the material containing a large number of cells are formed. The processed material is then crushed and classified to provide different grades.

The porosity of perlite is 0.85 to 0.9 and its volumetric weight is 500 to 1,000 N/m^3. Compared to diatomite, perlite has a smaller specific weight and compatible filter applications typically require 30 percent less additive. Perlite is used for filtering glucose solutions, sugar, pharmaceutical substances, natural oils, petroleum products, industrial waters, and beverages. The principal advantage of perlite over diatomite is its relative purity. There is a danger that diatomite may foul filtering liquids with dissolved salts and colloidal clays. Further discussions are given in the literature.

Asbestos

The primary use of asbestos is in covering metallic cloths to improve separation characteristics. Asbestos fibers form a highly compressible, adsorbent cake. The cost of asbestos as a filter aid is much higher than that of diatomite. Due to its carcinogenic properties, its use as a filter aid has decreased if not disappeared.

Cellulose

As with asbestos, cellulose fiber is applied to cover metallic cloths. The fibers form a highly compressed cake with good permeability for liquids, but a smaller retention ability for solid particles than that of diatomite or perlite. The use of cellulose is recommended only in cakes where its specific properties are required. These properties include a lack of ashes and good resistance to alkalies. The cost of cellulose is higher than those of diatomite and perlite.

Sawdust

This filter aid may be employed in cases where the suspension particles consist of a valuable product that may be roasted. For example, titanium dioxide is manufactured by calcining a mixture of sawdust and metal titanium acid. The mixture is obtained as a filter cake after separating the corresponding suspension with a layer of filter aid.

Charcoal

Charcoal is not only employed in activated form for decoloring and adsorbing dissolved admixtures but also in its unactivated form as a filter aid. It can be used in suspensions consisting of aggressive liquids (e.g., strong acids and alkalies). As with sawdust, it can be used to separate solids that may be roasted. On combustion, the charcoal leaves a residue of roughly 2 percent ash. Particles of charcoal are porous and form cakes of high density but that have a lesser retention ability than does diatomite.

Fly Ash

This material has a number of industrial filtering applications but primarily is applied to dewatering sewage sludge. The precoat is built up to 2-in. thick from a 60 percent solid slurry. On untreated sludges, filtration rates of 25 lb/ft^2-hr are obtainable. This rate can be doubled with treated sludges. The sludge is reduced from a liquid to a semidry state. Fly ash may also be used as a precoat in the treatment of papermill sludge.

EXPERIMENTAL METHODS FOR FILTER AID SELECTION

Filter aid selection must be based on planned laboratory tests. Guidelines for selection may only be applied in the broadest sense, since there is almost an infinite number of combinations of filter media, filter aids, and suspensions that will produce varying degrees of separation. The hydrodynamics of any filtration process are highly complex; filtration is essentially a multiphase system in which interaction takes place between solids from the suspension, filter aid, and filter medium, and a liquid phase. Experiments are mandatory in most operations not only in proper filter aid selection but in defining the method of application. Some general guidelines can be applied to such studies:

1. The filter aid must have the minimum hydraulic resistance and provide the desired rate of separation.
2. An insufficient amount of filter aid leads to a reduction in filtrate quality. Excess amounts result in losses is filtration rate.
3. It is necessary to account for the method of application and characteristics of filter aids. Some of the experimental methods for filter aid selection follow.

Precoat applications to rotary filters have been studied extensively. Work has been aimed at determining the relative usefulness of various filter aids in rotary precoat filtration as well as providing a basis for comparing filter aids. The following variables are important.

1. type and grade of filter aid used as the precoat
2. drum speed
3. knife advance rate
4. vacuum pressure
5. filter medium
6. temperature of feed
7. filter medium submergence
8. manner of precoat application
9. sharpness of knife
10. knife angle and bevel

It can be observed that the manner in which filter aid is removed depends on its properties and the drum speed. In some cases the filter aid is removed as flakes, while other tests reveal the remaining portion of the cake acquires a rough surface. High drum speed and filter aid dispersity are found to enhance the quality of the precoat layer. Figure 7.1 illustrates the progressive loss of filtering capacity as a function of rate of knife advance.

As shown in Figure 7.1, for a low rate of knife advance (1 mm/min), the resistance of the filter aid decreases in spite of the decrease in cake thickness. This is explained by the fact that for a single rotation of the drum, the particles of the suspension penetrate into the pores of filter at depths greater than 1.5 mm and blockage occurs. These systems are characterized not only by their rate, but by the ratio of filtrate volume to the weight of filter aid cut. The following observations typify rotary filter precoat applications:

1. The greater the permeability of the filter aid, the greater the amounts of filtrate collected.
2. A gradual decrease in filtration rate is observed; however, at some point during the operation a large reduction in rate can be observed.
3. In the operating regime characterized by a sharp drop in the filtration rate, the permeability of the filter aid is not zero, but has a definite value.
4. Two operating periods lead to reduction in permeability. During the first period, filtration primarily proceeds with pore blocking on the outer surface of the filter aid. During the second period, the filter aid pores undergo blocking within the filter aid cake.

Figure 7.2 schematically shows the operation of a precoat filter with cake in place. Point *A* represents the entrance of the cake into the unfiltered suspension. Point *B* represents the point of emergence and *C* is the point at which the cake, together with its accumulated suspension particles, reaches minimum permeability. Maximum filtration economy can be achieved by proper combination of submergence depth and rate of rotation so that points

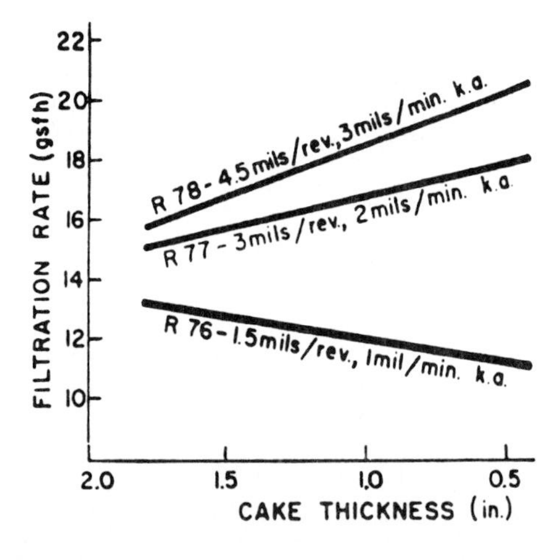

Figure 7.1 Performance of a typical perlite filter and relative to knife advance and cake thickness. Filtration of 0.5% $A1(0H)_3$ at 2/3 rpm.

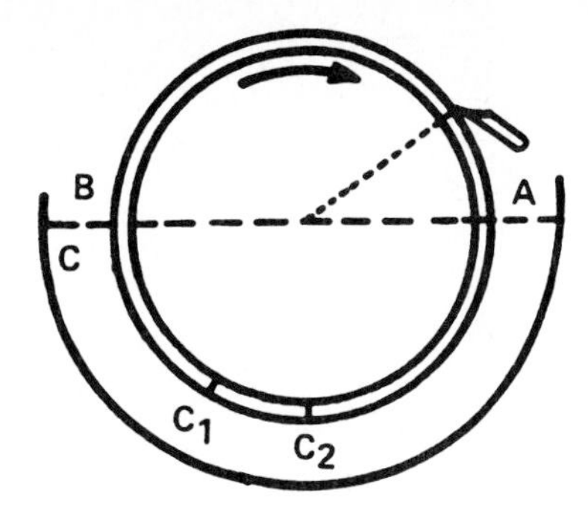

Figure 7.2 Cross-sectional view of a precoat filter.

B and *C* coincide. If minimum permeability is not achieved before point *C* reaches location *B*, the drum is rotating too rapidly or the submergence is too low, resulting in a portion of the filter aid capacity being wasted. Conversely, if minimum permeability is reached before *C* reaches *B*, either the submergence level could be lowered or the drums rotated at a faster rate. Choice of decreasing the submergence level or increasing the rate of rotation would be dictated by the time requirement for the drainage of accumulated filtrate from the filter cake. In accordance with this thinking, if points *B* and *C* coincide with 50 percent submergence and 10-rpm drum speed, a reduction in drum speed to 0.67 rpm would move point *C* back to C_1; a further reduction in drum speed to 0.5 rpm would locate it at C_2. If the blinding medium has zero effective permeability, filtrate collection under these circumstances should have a linear relationship with respect to drum speed. However, if the binding medium has appreciable permeability, additional filtrate will be collected in proportion to that permeability through the distances BC_1 and BC_2 for 0.67-rpm and 0.5-rpm speeds, respectively.

A graphical method is available for determining filter aid efficiency. Figure 7.3 shows a plot of filtrate rate versus cake thickness. Curve *AB* corresponds to the filtration of clear liquid through a gradually decreasing layer of filter aid when the resistance of the filter medium is negligible. The curve is based on the basic law of filtration using a special permeability for the filter aid. The line *EF* corresponds to suspension separation through a gradually decreasing layer of filter aid and filter medium of known resistance. Areas *ABCD* and *EFCD* are proportional to the filtrate volumes in an ideal and actual process, respectively. Since the areas under the curves are proportional to total output of the cycle, the observed output (*CDEF*) is much lower than that theoretically obtainable with clear liquid (*ABCD*). If we arbitrarily call the filter aid 100 percent rate efficient when operating with solids-free liquid under ideal conditions, then the rate efficiency with the slurry involved may be calculated as follows:

$$\eta_r = \frac{\text{area } CDEF}{\text{area } ABCD} \times 100 = \frac{CDEF}{ABCD} \times 100 \qquad (7.1)$$

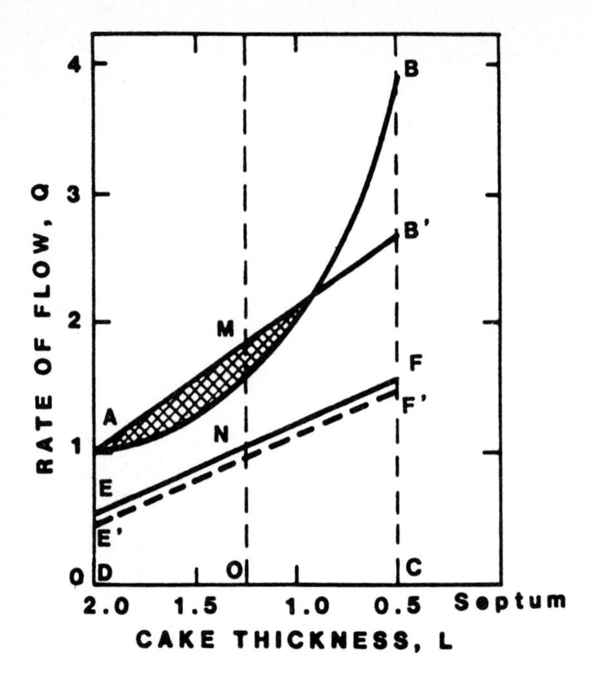

Figure 7.3 Relationship between blinding losses and theoretical efficiency.

The expression for curve *AB* in terms of *a* and *b* is:

$$Q = a\left(\frac{1}{L}\right) - b \tag{7.2}$$

Hence, area *ABCD* may be computed by integration of the curve's formula. An alternative approach is to draw the equivalent line *AB'*, so that the cross-hatched areas are equal, from whence the rate efficiency is:

$$\eta_r = \frac{\overline{CDEF}}{AB'CD} \times 100 \tag{7.3}$$

The following approximate procedure may also be used. Note in Figure 7.3 that the center of the cake is 1.25 in. Areas *AB, CD,* and *CDEF* can be computed from the product of the intercept of the diagonal line *AB'* or *EF* and base *CD*.

$$\eta_r\ (\%) = \frac{NO \times CD}{MO \times CD} \times 100 = \frac{NO}{MO} \times 100 \tag{7.4}$$

The location of line *AB* can be determined for any filter aid from its permeability.

During the period when the knife performs a full cut, the filtration rate is at a maximum. During the inactive periods of the knife, the filtration rate is reduced to a level representative of the permeability of the blinding layer. In Figure 7.3, this results in a filtration rate curve approximately parallel to the theoretical curve and displaced downward. As shown, these curves are quite linear. This supports the assumption that the resistance of the filter aid cake does not significantly restrict filtrate flow through the filter medium. Note that the total area *ABCD* theoretically represents 100 percent filter aid rate efficiency and the area under observed curve *EF* is the performance attained. Then, the area above the observed curve denotes the loss of rate efficiency due to blinding.

Blinding and penetration are rarely separate effects. Figure 7.4 illustrates a graphic evaluation of the rate efficiency together with applicable permeability corrections for the blinding medium. The total area under theoretical curve *AB′* represents 100 percent rate efficiency. That area under the actual line *EF* represents the filtration rate efficiency of the filter aid alone. The middle region denotes the loss of filter aid capacity due to blinding (i.e., pore blocking). The top triangular region denotes the loss of capacity of the filter aid due to penetration of solids into the cake beyond the cutting depth of the knife.

The average filtration rate and filtrate volume produced per unit weight of filter aid spent may be determined from Figure 7.4. The combined influence of both factors on the filtration efficiency is different depending on actual operating conditions.

Studies on the separation of $Al(OH)_3$ and bentonite reveal that the penetration of fine solids through the filter aid results in larger pores and consequently in smaller specific resistances, which are less applicable than those with pores of smaller size where the fine particles do not penetrate. This can

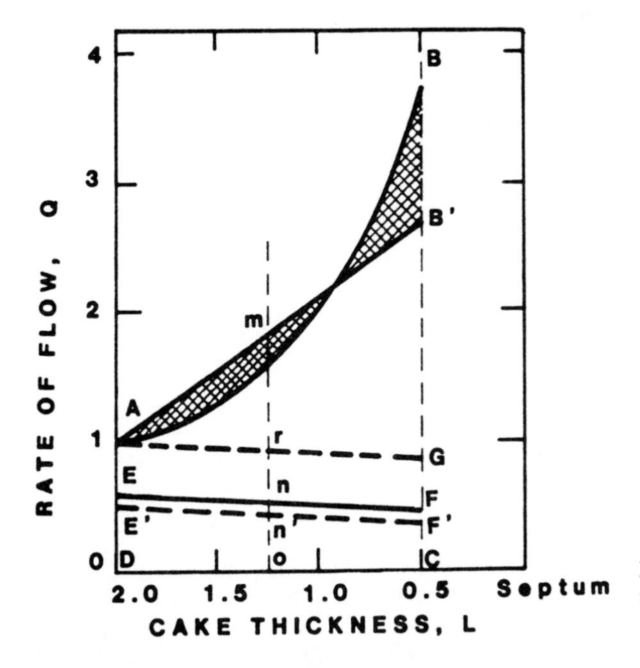

Figure 7.4 Graphical evaluation of filter aid efficiency and blinding and penetration losses.

be explained by the fact that the specific resistance of the first filter aid where particles penetrated becomes higher than that of the second filter aid.

The efficiency of a rotary drum filter is expressed by the product of the average filtration rate and the ratio of filtrate volume to weight of filter aid spent. Efficiency reaches a maximum at different rates of knife motion and varies according to the properties of the specific filter aid. An increase in the thickness of a cake layer cut results in a ratio of filtrate volume to weight of filter aid spent. One should account for the volume weight of different filter aids since this parameter varies considerably even among different grades of the same material.

To maintain an acceptable filtration rate, the following general guidelines apply:

1. If suspension particles are observed to intensely penetrate the filter aid cake, another grade with better solids retention properties should be selected.
2. If blinding of the outer surface of the filter aid is observed, then the rate of rotation should be increased.

After preliminary selection of the filter aid type, the optimum amount of addition to a suspension and the preferred rate of filtration must be established. Final selection should be based on comparison of the suspension separation efficiency for different filter aids and filter aid grades.

Figure 7.5 shows particle-size distributions for different grades of diatomaceous earth. As shown, grade B was the most finely dispersed of the material tested.

Figure 7.6 shows the relationships between filtration rate and rate of filtrate turbidity versus filter aid content in the suspension. Note that grade

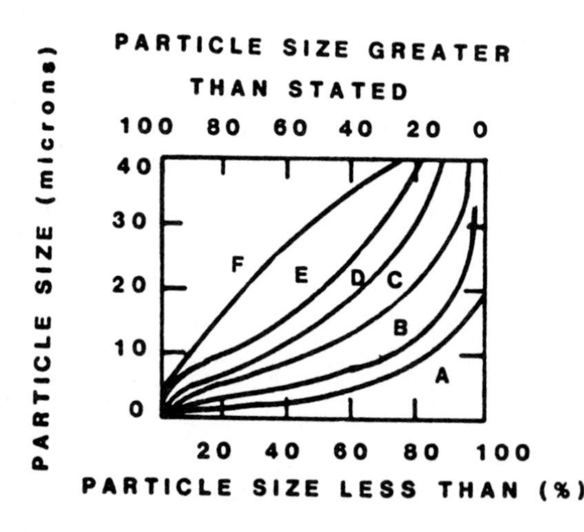

Figure 7.5 Particle size distribution for different grades of diatomite.

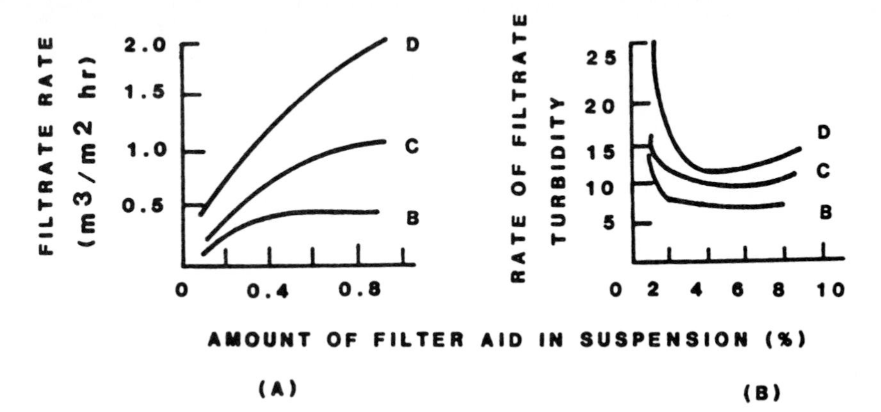

Figure 7.6 Influence of filter aid addition on filtrate turbidity.

B, having the greatest dispersity, provided the least filtrate turbidity at a reduced rate of filtration, and grade D, which has the smallest dispersity, produced a filtrate of increased turbidity with increasing filtration rate.

To explain the mode of operation of a filter aid in the separation of a suspension and to obtain optimum filtration performance, let's consider the porous structure of the unblocked filter aid with no filter-impeding impurities present. For any cross section through the cake, the total area open for flow A_o is $\epsilon_a A$ for the pure filter aid, where ϵ_a is the bulk porosity of the filter aid in the cake and A is the filter area. The cake of thickness L may be considered to be composed of a large number of layers or thickness Q, where Q is a characteristic dimension of the particles of the filter aid (e.g., diameter). The cake is therefore treated as a series of slabs with thickness of one particle diameter.

The volume of each of these thin slabs is QA: thus, the volume of slurry V_l required to deposit each layer is given by:

$$V_l = \frac{Al\,\rho_a\,(1 - \epsilon_a)}{a} \tag{7.5}$$

where

ρ_a = particle density
a = filter aid concentration

Thin sections of cake can be considered as a filter septum. Then the resistance R_s of a slab will be given by the special form of the Fanning equation for laminar flow:

$$\frac{1}{A}\frac{dV}{dt} = \frac{1}{\mu}\frac{\Delta P_s}{R_{sa}}$$

where

V = volume of filtrate
t = time
μ = viscosity
ΔP_s = pressure differential across the slab
R_{sa} = total filtration resistance of the slab of pure filter aid

Summing over the slabs:

$$\frac{1}{A}\frac{dV}{dt} = \frac{1}{\mu}\frac{\Delta P}{R_a} \tag{7.6}$$

where ΔP and R_a pertain to the cake of thickness L.

If filtration-impeding impurities are introduced to the system and some blocking occurs in the cake, there will be a drop in the area of cross section open for flow. For a partially blocked slab:

$$\frac{1}{A}\frac{dV}{dt} = \frac{1}{\mu}\frac{\Delta P_s}{R_{sb}} \tag{7.7}$$

The subscript b refers to the filter aid being partially blocked. Similarly:

$$\frac{1}{A}\frac{dV}{dt} = \frac{1}{\mu}\frac{\Delta P}{R_b} \tag{7.8}$$

Treating each slab as a filter septum we obtain:

$$\frac{R_{sa}}{R_{sb}} = \frac{A_{ob}}{A_o}$$

and for constant-pressure filtration with a compressible cake, and under all pressure conditions for an incompressible cake:

$$\frac{R_a}{R_b} = \frac{A_{ob}}{A_o}$$

Similarly:

$$\frac{r_a}{r_b} = \frac{A_{ob}}{\epsilon_a A} \tag{7.9}$$

where

r = specific cake resistance
$A_o = \epsilon_a A$

This last expression assumes that the flow channels in the porous mass are of approximately the same diameter. We may then directly relate the flow to the open area in the cross section as more pores become blocked.

Consider the manner in which the value of A_{ob} varies as filter-blocking impurities are added to the system. Assume that the impurities pack within the interstices of the bed of filter aid, blocking a certain number of the flow passages, but not contributing to the overall volume of the cake. Each of the large number of thin slabs (of which the cake is considered to be composed) is like a filter septum.

The filter aid impurities are intimately mixed before filtration. The blocking material will be distributed randomly over the cross section. Fresh filter septum with a fixed proportion of impurities is being continuously deposited; however, multiple blocking of an opening is possible. More than one impurity particle may be deposited in any particular pore, since the blocking material is of somewhat smaller size than the slab thickness. Multiple blocking may continue until most of the void space in the cake is filled, but beyond this the presence of impurities will start to increase the overall volume of cake.

To calculate the effect of the concentration of filtration-impeding impurities c on the unblocked area A_{ob}, c must be increased in small stages.

We add to the system a very small concentration c of impurity. This will block a small fraction p of the open area of the slab given by:

$$P = \frac{kV_1\delta c}{A_o} \tag{7.10}$$

where k is a constant for the system and depends on the geometry of the filter aid, and the particle size, density, and shape of the impurities. Parameter k is a measure of the blocking power of the impurities in a certain system.

Combining Equations 7.5 and 7.10 and substituting for A_o we obtain:

$$\delta c = \frac{pa\,\epsilon_a}{kl\rho_a\,(1 - \epsilon_a)}$$

Then the number of equal increments N required to reach concentration c will be:

$$N = \frac{c}{\delta c} = kl\,\rho_a\,(1 - \epsilon_a)\,\frac{c}{ap\epsilon_a} \tag{7.11}$$

After the first increment in concentration in the open area in the thin slab A_{o1} will be given by:

$$A_{o1} = (1 - p)\,A_o$$

Since the filter aid and impurities are intimately mixed it can be assumed that with a second increment δc, the blocking will again be randomly distributed over the cross section and multiple blocking may occur. The open area after the second increment in concentration A_{o2} will be given by:

$$A_{o2} = (1 - p)^2\,A_o$$

Similarly, after N stages, when the concentration of impurities is c, the open area A_{oN} is:

$$A_{oN} = A_{ob} = (1 - p)^N\,A_o \tag{7.12}$$

The increments must be small so that multiple blocking may be neglected during the first stage. Also, the value of c must be sufficiently small so that the impurities may be accommodated within the filter aid structure.

Combining the preceding expressions we obtain:

$$\frac{r_b}{r_a} = (1 - p)^{-k(c/a)} \tag{7.13}$$

where

$$K = \frac{kl\ \rho_a\ (1 - \epsilon_a)}{P\ \epsilon_a}$$

or

$$\log \frac{r_b}{r_a} = \frac{0.4343\ kl\ \rho_a\ (1 - \epsilon_a)}{\epsilon_a)} \frac{c}{a} \tag{7.14}$$

in the limit of $p \rightarrow 0$.

This model predicts a log-linear relationship between the specific cake resistance of the partially blocked filter aid, and the ratio of the concentrations of the impurities and the filter aid body feed. The slope of the linear form of the plot is related to the properties of both the filter aid and the impurities.

The filtration rate of an impure solution is then measured simply by the volume of filtrate (δV) collected under the test conditions and expressed as a percentage of the standard value (ΔV_{st}) for a pure substance measured with unit addition of filter aid:

$$f = \frac{f\%}{100} = \frac{\Delta V}{\Delta V_{st}} \tag{7.15}$$

This filtration rate must be maximized by optimizing the filter aid concentration. Integration of the filtration equation under the standardized constant-pressure conditions of the experiments gives:

$$\Delta V = \left(\frac{2A^2\ \Delta p\ \Delta t}{\mu r \upsilon}\right)^{1/2}$$

where υ is the volume of cake deposited by unit volume of filtrate.

Then, combining with Equation 7.15, since A, ΔP, Δt and μ are constant:

$$f = \left(\frac{r_a\ \upsilon_a}{r_p\ \upsilon_b}\right)^{1/2} \tag{7.16}$$

where υ_a refers to unit concentration of filter aid.

The volume of cake deposited by unit volume of slurry is:

$$\upsilon_b = \frac{a}{\rho_a\ (1 - \epsilon_a)} \tag{7.17}$$

Combining Equations 7.13, 7.16, and 7.17 we obtain:

$$f = \left[\frac{1}{a}(1 - p)^{k(c/a)}\right]^{1/2} \tag{7.18}$$

To optimize f we differentiate with respect to a and set $df/da = 0$. This gives:

$$a_{opt} = \frac{k\,\rho_a\,(1 - \epsilon_a)lc}{\epsilon_a} \tag{7.19}$$

The value of a_{opt} may be substituted into Equation 7.18 to give the limit for small values of p:

$$f\,(a_{opt})^{1/2} = 1/e^{1/2} = 0.607 \tag{7.20}$$

where e = base of Naperian logarithms.

The equation is independent of the concentration of the impurities c, and so may be used to relate the optimum conditions for different suspensions.

Figures 7.7 and 7.8 show plots of specific cake resistance versus the impurity to filter aid ratio for two different filter aids using a sugar suspension. The data are replotted in Figure 7.9 in the form of f versus a, for fixed percentages of impure sugar solution.

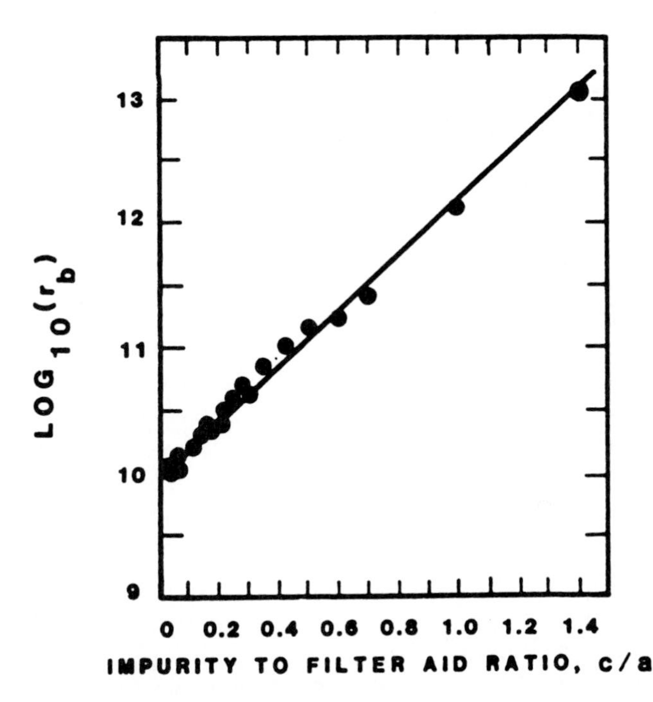

Figure 7.7 Specific cake resistance in Celite 505/C sugar system.

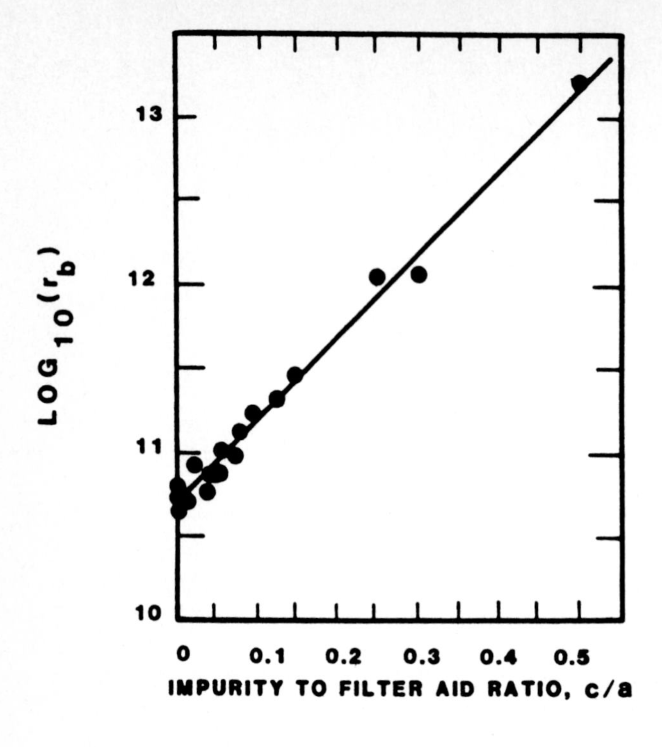

Figure 7.8 Specific cake resistance in asbestos/C sugar system.

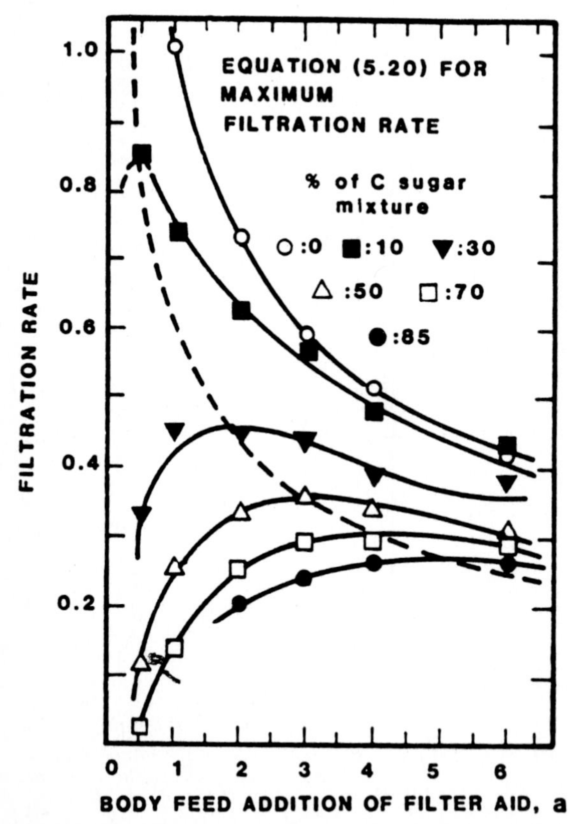

Figure 7.9 Optimum addition of filter aid Celite 505 to sugar suspension separation analysis.

Comparison of Equations 7.14 and 7.19 shows that measurement of the slopes of Figures 7.7 and 7.8 leads directly to the optimum filter aid addition. This indicates the principal application of the simple theory in predicting filter aid operating conditions. For a system that obeys the exponential law, a minimum of two measurements is sufficient to determine the optimum filter aid addition. A plot of log r_b versus c/a as in Figure 7.7 and 7.8 gives a straight line with gradient *C*, which is related to a_{opt} by the simple relation:

$$\frac{a_{opt}}{c} = \frac{G}{0.4343}$$

a_{opt} is the optimum of theoretical interest. In practice, however, the economic optimum has the greatest importance. This is of a more complex nature, and the filterability term is only one aspect of it. The simplest form of the total filtering cost *F* could be compounded for continuous filter operation under the conditions of the filterability test and become:

$$F = ak' + k''/f$$

where k' and k'' are cost factors related to the costs of filter aid and filter process operation, respectively. This simple form of the cost equation does not yield a ready analytical solution for the optimum and would require numerical treatment.

If the filtration is difficult, k'' will be the dominant factor and the economic optimum will not be far removed from a_{opt}. If, however, k' is the important cost parameter and it is uneconomical to use much filter aid, the preceding analysis is unlikely to be applicable. In the latter case the ratio c/a will generally be too high for the simple blocking model to be obeyed. Hence, the application of the model is most useful in cases of very difficult slurries when large additions of filter aid must be tolerated.

RULES FOR PRECOAT FILTER AIDS

Various investigators have developed guidelines for filter aid precoating as applied to batch filters with a flat filter medium operating under pressure. Examples of specific filter types that fall into this category are filter presses and leaf filters.

It is generally recommended that a suspension of filter aid be prepared in a pure liquid by agitation before filtration. Delivery to the filter medium can usually be done with a centrifugal pump. Filtrate can be recycled to the filter aid suspension feed until the finished precoat is formed. Depending on the ratio of particle size of the filter aid to the pore diameters of the medium, as well as the rate of suspension feed and its concentration, two processes arise on the filter. First, the medium retains only some particles and the remainder pass through with filtrate. Gradually, the medium's retention ability increases and the amount of solid particles in the filtrate decreases. A short time after starting the process, the retention ability of the medium becomes so pronounced (for example, due to bridging) that the pure filtrate passes through.

In both cases, the suspension concentration in the filter body decreases gradually. The suspension of filter aid is usually delivered on the filter not filled with liquid; however, in some cases the filter is filled with pure liquid during the precoat stage. In such cases, the filter and suspension entering the unit is diluted. Its concentration first increases, reaching a maximum value; then as the particles settle onto the filter, it decreases.

In most cases we may assume that the filtrate volume is equal to the suspension volume (the concentration of filter aid in suspension is typically 0.1 percent to 0.5 percent). The cake may be assumed incompressible. Its specific resistance is independent of concentration. Also, within the filter body, the suspension can be assumed to be perfectly mixed, with filtration at a constant rate.

The following equations were derived from theory for the case where the suspension of filter aid is delivered to a filter not filled with liquid. For filter aid particles completely retained by the medium:

$$c = c_o e^{-n} \tag{7.21}$$

For the case where the retention ability of the medium increases gradually:

$$c = c_o \left[cn\left(\hat{K} \frac{\lambda}{n} \tau \right) \right]^{n/\hat{k}} \tag{7.22}$$

where

C = variable concentration of suspension in the filter body (m^3/m^3)
C_o = initial concentration of suspension in the mixer (m^3/m^3)
n = V/V_s = degree of circulation of suspension in the system
V = filtrate volume (m^3)
V_s = initial volume of suspension in the mixer (m^3)
$\hat{K}$ = constant
λ = Q/V_s = coefficient of dissolution (sec^{-1})
Q = amount of filtered liquid (m^3/sec)

It must be noted that the filter medium is the workhorse of any filtration process. Its selection is often the most important consideration for assuring efficient suspension separation. A good filter medium should have the following characteristics in varying degrees, depending on the specific problem:

1. It should be able to retain a wide size distribution of solid particles from the suspension.
2. It should offer minimum hydraulic resistance to the filtrate flow.
3. It should allow easy discharge of cake.
4. It should be resistant to chemical attack.

5. It should not undergo swelling when in contact with filtrate and washing liquid.
6. It should be heat resistant within the ranges of filtration.
7. It should have sufficient strength to withstand filtering pressure and mechanical wear.
8. It must be capable of avoiding wedging of particles into its pores.

There are many filter media from which to choose; however, the optimum type often depends on the properties of the suspension and specific process conditions. Filter media may be classified into several groups. Flood et al. [1] classified filter media into surface type and depth-media type.

Surface-type filter media are distinguished by the fact that the solid particles of suspension on separation are mostly retained on the medium's surface. That is, particles do not penetrate into the pores. Examples are filter paper, filter cloths, and wire mesh.

Depth-type filter media, mostly used for liquid clarification, are characterized by the fact that the solid particles penetrate into the pores, where they are retained. The pores of such media are considerably larger than the particles of suspension. The suspension's concentration is generally not high enough to promote particle bridging inside the pores. Particles are retained inside the pores by adsorption, settling, and sticking. As a rule, the depth-type filter media cannot retain all suspended particles. Their retaining capacity is typically 90 percent to 99 percent. Sand and filter aids, for example, fall into this category.

Some filter media may act as either surface type or depth type, depending on the pore size and suspension properties (e.g., particle size, solids concentration, and suspension viscosity). Filter media may be classified by the materials of construction. Examples are cotton, wool, linen, glass fiber, porous carbon, metals, and rayons. Such a classification is convenient for selection purposes, especially when resistance to aggressive suspensions is a consideration. Dicky [2] classifies filter media according to structure. Two groups are defined: rigid and flexible media (Table 7.1).

TABLE 7.1 FLEXIBLE AND RIGID MEDIA

Flexible Media
Woven Fabrics
Nonmetallic, e.g., cotton, synthetics, and glass
Metallic, e.g., wire cloth
Combinations of nonmetallic and metallic (asbestos and wire)
Nonwoven
Preformed, e.g., paper and felts
Nonpreformed, e.g., asbestos and cellulose pulps
Rigid Media
Loose, e.g., sand and diatomaceous earth
Fixed, e.g., aluminum oxide plates and perforated steel plates

FLEXIBLE FILTER MEDIA

Flexible nonmetallic materials are widely used as filter media. They are manufactured as fabrics or as preformed unwoven materials, but rarely in the form of perforated plates.

Fabric filter media are characterized by mesh count, mesh opening, yarn size, and type of weave. The mesh count or thread count of a fabric is the actual number of threads per inch. Thread counts in both warp and weft directions are the same and are indicated by one number (Figure 7.10). Warp threads run lengthwise in a fabric and are parallel to the salvage edge. Weft or filling threads run across the width of a fabric at right angles to the warp.

The space between threads is the mesh opening (refer to Figure 7.10). It is measured in units of micrometers or inches. Different yarn sizes are normally specified as a measurement of diameter in micrometers or mils (thousandths of an inch). Yarn sizes in the warp and weft directions are normally the same, and are indicated as one number.

Fabrics are available in differing mesh openings and varying thread diameters. The thread diameter affects the amount of open area in a particular cloth, which in turn determines flow rate or throughput.

A plain weave is the most basic weave, with a weft thread alternately going over one warp thread and then under one warp thread. A twill weave produces a diagonal or twill line across the fabric face. These diagonals are caused by moving the yarn intersections one weft thread higher on successive warp yarns. A twill weave is designated 2/1, 2/2, or 3/1 depending on how many weft threads the warp threads go over and under. A satin weave has a smooth surface caused by carrying the warp (or the weft) on the fabric surface over many weft (or warp) yarns. Intersections between warp and weft are kept to a minimum, just sufficient to hold the fabric firmly together and still provide a smooth fabric surface. The percentage of open area in a textile filter indicates the proportion of total fabric area that is open. This calculation

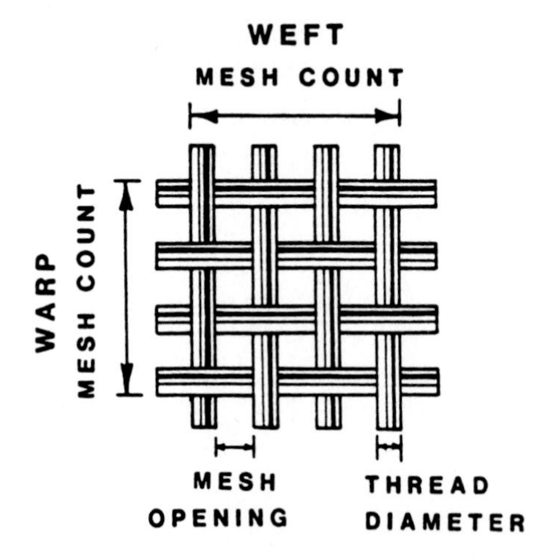

Figure 7.10 Construction parameters that determine the characteristics of a fiber-based fabric.

is determined by:

$$\% \text{ open area} = \frac{(\text{mesh opening})^2}{(\text{mesh opening} + \text{thread diameter})^2} \times 100 \quad (7.23)$$

Asbestos Cloths

Asbestos cloths have high heat resistance as well as chemical resistance to acids. However, the use of asbestos is normally limited to cases where there is no mechanical stress or high pressure.

The combination of metallic wires and asbestos tends to produce a non-uniform cloth. Cotton blends well with asbestos; however, the mixture is soon destroyed by corrosive liquor or high temperatures. Karacharov and Vorontsova [3] provide further discussions.

Glass Cloths

Glass cloths are manufactured from glass yarns. They have high thermal resistance, high corrosion resistance, high tensile strength, and are easily handled; the composition and diameter of the fibers can be altered as desired. The disadvantages of glass cloth are the lack of flexibility of individual fibers, causing splits and fractures, and its low resistance to abrasion. However, backing glass cloth with a lead plate, rubber mats, or other rigid materials provides 100 percent greater life. Backing with cotton or rubber provides about 50 percent greater life than in cases where no backing is used.

Cotton Cloths

Cotton filter cloths are perhaps the most widely used filter media. They have a limited tendency to swell in many liquids and are used for separation of neutral suspensions at temperatures up to 100°C, as well as suspensions containing acids up to 3 percent or alkalies with concentrations up to 10 percent at 15 to 20°C. Hydrochloric acid at 90° to 100°C destroys cotton fabric in about 1 hour, even at concentrations as low as 1.5 percent. Nitric acid has the same effect at concentrations of 2.5 percent, and sulfuric acid at 5 percent. Phosphoric acid (70 percent) destroys the cloth in about six days. Water and water solutions of aluminum sulfate cause cotton fabrics to undergo shrinkage. Further discussions are given in the literature cited at the end of this chapter.

Woven Cotton Cloths

Woven cotton filter cloths comprise ducks, twills, chain weaves, canton flannel, and unbleached muslins. Cotton duck is a fabric weave that is a plain cloth with equal-thickness threads and texture in the "over one and under one" of the warp and woof. The twill weave is over two and under two with the next filling splitting the warp strands and giving a diagonal rib at 45° if the number of warp and filling threads are equal. Canton flannel is a twill weave in which one surface has been brushed up to give a nap finish. A muslin cloth is a very thin duck weave, which is unbleached for filtering. In chain weave one filling goes over two warp threads and under two, the next reversing this; the third is a true twill sequence, and the next repeats the cycle. Certain characteristics of each weave may be generalized on a comparative basis as shown in Table 7.2.

A duck may be preferable to a twill of higher porosity because the hard surface of the duck permits freer cake discharge. Under high increasing pressure a strong, durable cloth (duck) is required, since the first resistance is small as compared with that during cake building. Certain types of filters, such as drum filters, cannot stand uneven shrinkage and, in some cases, cloths must be preshrunk to ensure fitting during the life of the cloth.

Nitro-filter (nitrated cotton cloth) cloths are about the same thickness and texture as ordinary cotton filtration cloths but are distinguished by a harder surface. It is claimed that the cake is easily detached and that clogging is rare. Their tensile strength is 0 percent to 80 percent of that of the specially manufactured cotton cloths from which they are prepared. They are resistant to the corrosive action of sulfuric, nitric, mixed nitration, and hydrochloric acids. They are recommended for filtering sulfuric acid solutions to 40 percent and at temperatures as high as 90°C, with the advantage of removing finely divided amorphous particles, which would quickly clog most ceramic media.

Nitro-filter cloths are composed of cellulose nitrate, which is an ester of cellulose. Any chemical compound that will saponify the ester will destroy the cloth. Caustic soda or potash in strengths of 2 percent at 70°C or over; alkali sulfides, polysulfides, and sulfohydrates; or mixtures of ethyl alcohol and ether, ethyl, amyl and butyl acetates, pyridine, ferrous sulfates, and other reducing agents are detrimental to the cloth.

Cellulose nitrate is inflammable and explosive when dry, but when soaked in water it is considered entirely safe if reasonable care is taken in handling. For this reason it is colored red and packed in special containers. Users are cautioned to keep the cloths wet and to handle them carefully.

TABLE 7.2 CHARACTERISTICS OF DIFFERENT COTTON WEAVES

Weave	Tensile strength	Filtrate clarity	Flow rate	Cake discharge	Resistance to binding	Shrinkage
Canton Flannel	Fair	High	Fair	Poor	Good	Even
Chain	Low	High	High	Fair	Fair	Uneven
Duck	High	Fair	Poor	Good	Poor	Even
Muslin	Low	Poor	High	Good	Good	Even
Twill	Fair	Fair	High	Fair	Fair	Uneven

Wool Cloths

Wool cloths are used in handling acid solutions with concentrations up to 5 percent to 6 percent. Wool cloth has a life comparable to that of cotton in neutral liquors. Wool is woven in the duck-like square cloth weave, or with a raised nap; or it may be formed as a felt. Originally the smooth cloth weave was used for filtering electrolytic slimes and similar slurries. The hairlike fibers, as in cotton cloth, ensure good filtrate clarity. Long-nap wool cloth has found wide application in sewage sludge dewatering and in cases where only ferric chloride is used for conditioning. The wool has a long life and it does not clog easily. Wool cloths are sold by weight, usually ranging 10 to 22 oz/yd^2 with the majority at 12 oz/yd^2. The clarity through wool cloths is considerably less than through cotton cloths, and their high cost (four to five times that of cotton) limits their use.

Paper Pulp and Fiber Cloths

Paper pulp and fiber cloths are excellent materials for precoats and filter aids. Paper pulp was extensively used at one time in sugar-juice clarification, although now it has been largely replaced by diatomaceous earth. Paper pulp gives a high rate of flow, is easily discharged and shows little tendency to clog.

Paper pulp's disadvantage lies in its preparation. Soda or sulfate pulp,most commonly used, must be disintegrated and kept in suspension by agitation before precoating. This requires considerable auxiliary equipment. Diatomaceous earths, while they should be kept in suspension, are very light, easy to handle, and do not undergo disintegration.

Paper pulp compressed into pads is used in pressure filters for beverage clarification. After becoming dirty, as evidenced by decrease in the rate of flow, the paper may be repulped, water-washed, and reformed into pads. Although this involves considerable work, excellent clarity and high flow rates are obtained. The impurities do not form a cake as such, but penetrate into the pad and can only be removed by repulping and washing the pad.

Pads of a mixture of paper pulp and asbestos fiber are used in bacteriological filtrations. In sheet form it is employed in the laboratory for all kinds of filtration. Filter papers are made in many grades of porosity for use in porcelain and glass funnels. Industrially, paper in the form of sheets is used directly or as a precoat in filter presses.

Used directly in lubricating oil clarification in a "blotter press," it acts in much the same manner as the paper pads, but it is much thinner and is not reused. As a precoat, paper protects in the filter medium from slimy fines; it may be peeled off and discarded after clogging, leaving the medium underneath clean.

Rubber Media

Rubber media appear as porous flexible rubber sheets and microporous hard rubber sheets. Commercial rubber media have 1,100 to 6,400 holes/in.^{2} with pore diameters of 0.012 in. to 0.004 in. They are manufactured out of soft rubber, hard rubber, flexible hard rubber, and soft neoprene.

The medium is prepared on a master form, consisting of a heavy fabric belt, surfaced on one side with a layer of rubber filled with small round pits uniformly spaced. These pits are 0.020 in. deep, and the number per unit area and their surface diameter determine the porosity of the sheet. A thin layer of latex is fed to the moving belt by a spreader bar so that the latex completely covers the pits, yet does not run into them. This process traps air in each pit. The application of heat to the under-surface of the blanket by a steam plate causes the air to expand, blowing little bubbles in the film. When the bubbles burst, small holes are left, corresponding to the pits. The blown rubber film, after drying, is cooled and the process repeated until the desired thickness of sheet is obtained. The sheet is then stripped off of the master blanket and vulcanized.

Approximately 95 percent of the pits are reproduced as holes in the rubber sheet. The holes are not exactly cylindrical in shape but are reinforced by slight constrictions which contribute to strength and tear resistance. This type is referred to as *plain*, and can be made with fabric backing on one or both sides to control stretching characteristics. If the unvulcanized material is first stretched, and then vulcanized while stretched, it is called *expanded*. Resulting holes are oval and have a higher porosity (sometimes up to 30 percent). Special compounds have been developed for resistance to specific chemicals under high concentrations at elevated temperatures, such as 25 percent sulfuric acid at 180°F.

The smooth surface allows the removal of thinner cakes than is possible with cotton or wool fabrics. Rubber does not show progressive binding and it can be readily cleaned and used at temperatures up to 180°F. On the other hand, because a clear filtrate is difficult to obtain when filtering finely divided solids, a precoat often becomes necessary.

Synthetic Fiber Cloths

Cloths from synthetic fibers are superior to many of the natural cloths thus far considered. They do not swell as do natural fibers, are inert in many acid, alkaline, and solvent solutions, and are resistant to various fungus and bacterial growths (the degree depending on the particular fiber and use). Several synthetic fibers resist relatively high temperatures and have a smooth surface for easy cleaning and good solids discharge. Some of the most widely used synthetic filter media are nylon, Saran, Dacron, Dynel, Vinyon, Orlon, and Acrilan. Table 7.3 shows comparable physical properties of several synthetic fiber filter media.

TABLE 7.3 PROPERTIES OF WOVEN FILTER CLOTH FIBERS

Fibers	Acids	Alkalies	Solvents	Fibert tensile strength	Temperature limit (°F)
Acrilan	Good	Good	Good	High	275
Asbestos	Poor	Poor	Poor	Low	750
Cotton	Poor	Fair	Good	High	300
Dacron	Fair	Fair	Fair	High	350
Dynel	Good	Good	Good	Fair	200
Glass	High	Fair	Fair	High	600
Nylon	Fair	Good	Good	High	300
Orlon	Good	Fair	Good	High	275
Saran	Good	Good	Good	High	240
Teflon	High	High	High	Fair	180
Wool	Fair	Poor	Fair	Low	300

Tightly woven, monofilament (single-strand) yarns consist of small-diameter filaments. They tend to lose their tensile strength because their small diameters reduce their permeability; thus, multifilament yarns are normally used. Monofilament yarns in loose weaves provide high flow rates, good solids discharge, easy washing, and high resistance to blinding, but the turbidity of the filtrate is high and recirculation is usually necessary, initially at least. Table 7.4 provides technical data on various synthetic filter fabrics. Excellent results have been obtained with high-twist multifilament synthetic yarn fabrics handling materials with particles in the submicrometer range.

Flexible Metallic Media

Flexible metallic media are especially suitable for handling corrosive liquors and for high-temperature filtration. They have good durability and are inert to physical changes. Metallic media are fabricated in the form of screens, wire windings, or woven fabrics of steel, copper, bronze, nickel, and different alloys.

Perforated sheets and screens are used for coarse separation as supports for filter cloths or as filter aids. Metallic cloths are characterized by the method of wire weaves as well as by the size and form of holes and by the wire thickness. Metallic cloths may be manufactured with more than 50,000 holes/cm^2 and with hole sizes less than 20 μm.

Metallic/Nonmetallic Cloths

Combination metallic and nonmetallic cloths consist of metallic wires and weak asbestos threads. There are some difficulties in weaving when attempting to maintain uniformity between wires and the asbestos, and considerable

TABLE 7.4 TECHNICAL DATA FOR VARIOUS SYNTHETIC FABRIC FILTERS

Style no.	Weave	Weight (oz/yd²)	Threads/in., warp × weft	Thread diam., warp × weft (μm)	Mesh opening (μm)	Air permeability (ft³/min)	Thickness (μm)
Nylon 6, 6.6							
Warp and Weft Monofilament							
111-020	Plain	4.5	22 × 22	305 × 305	850 × 850	NA[a]	570
111-110	Plain	4.6	50 × 50	200 × 200	300 × 300	NA	350
111-150	Plain	3.2	62 × 62	150 × 150	250 × 250	NA	270
111-160	Plain	2.4	107 × 76	100 × 100	250 × 230	NA	220
111-170	Plain	4.6	29 × 29	250 × 250	600 × 600	NA	450
111-180	Plain	2.3	66 × 66	130 × 130	210 × 210	NA	270
111-190	Twill	5.5	38 × 38	250 × 250	420 × 420	NA	530
111-206	Plain	5.3	183 × 43	150 × 150		170–210	410
111-220	Plain	2.9	80 × 80	125 × 125	175 × 175	NA	240
111-230	Plain	5.7	40 × 40	250 × 250	420 × 420	NA	450
111-056	Satin	7.2	109 × 42	205 × 300		350 × 400	450
Warp and Weft Multifilament							
1053	Plain	2.1	147 × 97			150–200	160
1093	Twill	3.5	297 × 122			75–100	250
1103	Twill	3.7	297 × 135			40–70	250
1123	Twill	3.2	195 × 140			15–25	190
1153	Satin	3.2	236 × 99			15–25	210
1193	Satin	5.3	300 × 99			45–70	300
1203	Satin	6.8	152 × 76			20–30	390
1212	Satin	6.5	178 × 97			50–80	710
1283	Plain	1.8	112 × 97			170–220	130
1338	Leno	5.6	7 × 5			NA	630
1353	Plain	3.2	64 × 48			50–100	255
1363	Plain	11.5	69 × 28			1–3	660
1393	Twill	9.7	236 × 53			5–10	560
122-053	Twill	4.7	80 × 117			50–80	410
122-073	Oxford	15.5	72 × 21			0.5–2	720
Warp Monofilament, Weft Spun							
1233	Satin	5.0	320 × 71			60–100	410
Warp and Weft Multifilament and Metal Spun							
9165	Twill	4.1	297 × 132			25–40	300
Nylon							
Warp and Weft Monofilament							
1656	Satin	8.2	99 × 53	180 × 290		200–300	520
1666	Satin	9.0	111 × 53	180 × 290		125–200	530

1686	Satin	7.1	107 × 91	180 × 180		125–200	395
111-096	Satin	9.5	99 × 53	205 × 300		150–300	490
Nomex							
Warp Multifilament, Weft Spun							
1513	Plain	7.7	107 × 66			50–80	510
Polyester							
Warp and Weft Monofilament							
1713	Plain	4.1	345 × 84			80–120	190
1716	Plain	4.1	350 × 79			10–20	180
1733	Plain	8.8	175 × 36			300–400	380
9656 B	Satin	8.3	104 × 53	200 × 300		300–400	450
9813 B	Satin	11.8	112 × 53	200 × 300		200–250	510
9884 B	Satin	12.7	145 × 53	200 × 250		80–150	440
311-010 HS[b]	Plain	6.9	38 × 38	250 × 250	420 × 420	NA	400
311-020	Plain	6.9	38 × 38	250 × 250	420 × 420	NA	400
Warp and Weft Multifilament							
1813	Plain	10.3	43 × 30			1–3	500
1853	Leno	10.0	7 × 5			NA	800
1893	Twill	4.1	295 × 112			40–75	250
1896	Twill	4.1	295 × 112			5–10	200
1933	Satin	3.2	290 × 99			35–55	190
1943	Twill	6.5	257 × 79			10–20	330
1953	Plain	10.0	36 × 30			2–4	420
1956	Plain	10.0	36 × 30			1–2	380
1973	Plain	3.5	94 × 81			20–40	195
1976	Plain	3.5	99 × 84			5–15	140
9881	Plain	2.7	152 × 81			40–60	270
322-013	Twill	3.6	219 × 147			5–10	180
322-033	Plain	7.1	81 × 56			3–7	300
322-036	Plain	7.1	81 × 56			1–3	280
322-043	Broken Twill	5.3	216 × 100			30–50	250
322-070	Plain	11.5	52 × 32			1–2.5	600
322-073	Plain	13.0	54 × 36			0.5–1.5	580
322-123	Twill	5.3	216 × 100			40–60	270
Warp Multifilament, Weft Spun							
323-013	Twill	13.2	76 × 53			4–6	580
Warp and Weft Spun							
1823	Twill	10.9	51 × 25			30–40	770
1873	Plain	10.6	53 × 30			7–15	610
1923	Plain	5.6	64 × 53			70–100	360
333-020	Plain	9.6	47 × 29			20–30	570

(continued)

TABLE 7.4 TECHNICAL DATA FOR VARIOUS SYNTHETIC FABRIC FILTERS *(continued)*

Style no.	Weave	Weight (oz/yd^2)	Threads/in., warp × weft	Thread diam., warp × weft (μm)	Mesh opening (μm)	Air permeability (ft^3/min)	Thickness (μm)
Acrylic							
Warp and Weft Multifilament							
2013	Plain	7.1	81 × 33			3–7	480
2023	Twill	4.4	102 × 69			15–20	250
Warp Multifilament, Weft Spun							
2033	Twill	6.5	102 × 61			25–40	480
Warp and Weft Spun							
2063	Twill	9.4	69 × 41			10–20	700
9856	Twill	7.1	51 × 64			40–60	580
Polyvinylchloride							
2973	Twill	6.2	66 × 51			60–90	325
Polypropylene							
Warp and Weft Monofilament							
2773	Satin	9.7	86 × 38	300 × 250		70–100	620
2776	Satin	9.7	86 × 38	320 × 290		45–90	680
2793	Twill	10.9	61 × 25	300 × 500		300–400	700
511-013	Satin	8.0	86 × 46	250 × 250		150–300	550
511-015	Satin	8.2	89 × 46	250 × 250		40–60	500
511-016	Satin	8.2	89 × 46	250 × 250		100–150	580
511-050	Plain	3.5	50 × 50	200 × 200	300 × 300	NA	370
511-066	Basket	5.9	110 × 45	200 × 200		50–100	450
511-070 HS	Twill	7.2	28 × 28	375 × 375	530 × 530	NA	650
511-086	Plain	5.3	108 × 32	200 × 200		100–150	320
511-090	Plain	3.7	30 × 30	250 × 250	600 × 600	NA	490
511-110 HS	Twill	5.4	38 × 33	305 × 305		NA	600
511-126	Twill	8.8	76 × 28	305 × 305		100–200	590
511-136	Satin	6.5	110 × 60	200 × 200		20–300	450
511-150	Plain	3.5	20 × 20	305 × 305	950 × 950	NA	510
Warp Monofilament, Weft Multifilament							
512-013	Satin	9.1	86 × 30			100–150	740
512-015	Satin	9.9	86 × 30			10–30	510
512-016	Satin	9.7	86 × 30			40–80	600
Warp Monofilament, Weft Spun							
513-016	Satin	11.4	93 × 36			5–10	680
Warp and Weft Multifilament							
2723	Plain	6.5	33 × 20			5–10	470
2726	Plain	6.5	33 × 20			3–8	425
2843	Twill	10.0	241 × 43			10–15	640

2853	Twill	16.8	58 × 30			4–8	1,200
2856	Twill	16.8	58 × 30			2–4	1,100
9870	Twill	19.8	69 × 30			1–2	1,250
522-013	Plain	10.0	43 × 24			3–6	690
522-053	Twill	18.8	70 × 30			4–9	1,160
522-060	Plain	14.3	76 × 21			1–2.5	960
522-063	Plain	15.1	77 × 21			0.5–2	950
522-070	Twill	8.3	58 × 39			20–40	630
Warp Multifilament, Weft Spun							
2873	Twill	20.6	175 × 33			5–10	1,500
523-040	Twill	10.3	70 × 38			4–7	840
523-043	Twill	11.0	71 × 40			3–5	800
523-053	Twill	18.6	72 × 30			2–5	1,200
Warp amd Weft Spun							
2783	Twill	7.7	53 × 33			50–80	800
2833	Twill	10.6	53 × 30			8–13	1,000
2893	Twill	13.9	61 × 23			6–8	1,200
9811	Twill	19.2	53 × 23			10–15	1,550
533-013	Plain	11.7	50 × 23			1–2	850
533-033	Twill	15.8	58 × 26			3–6	1,200
533-070	Twill	14.0	48 × 31			4–6	1,050
Polyethylene							
Warp and Weft Monofilament							
2473	Satin	5.3	155 × 58	150 × 150	25 × 300	200–300	425
2483	Satin	8.3	89 × 48	250 × 250	40 × 300	250–350	665
2486	Satin	8.3	89 × 48	280 × 270	15 × 190	100–150	500
2503	Leno	5.3	6 × 5	500 × 500	4,200 × 3,000	NA	1,500
411-010	Plain	7.2	105 × 42	220 × 220		200–300	580
411-013	Plain	9.0	111 × 44	220 × 220		100–175	650
411-015	Plain	10.1	112 × 50	220 × 220		50–80	650
Warp Monofilament, Weft Multifilament							
2573	Satin	6.3	150 × 36			130–200	540
Saran							
Warp and Weft Monofilament							
2333	Satin	11.2	66 × 43	250 × 250	65 × 340	600–700	700
2336	Satin	11.2	66 × 43	280 × 370	25 × 180	200–300	370
2363	Plain	7.7	15 × 15	320 × 380	1,050 × 1,050	NA	900
611-010	Twill	6.6	50 × 50	200 × 200	300 × 300	NA	500
611-020	Plain	7.5	30 × 30	250 × 250	600 × 600	NA	490
611-040	Plain	6.5	20 × 20	305 × 305	950 × 950	NA	550

[a] NA = not available.

[b] HS = high shrink.

(*Courtesy of SciTech Publishers, Morganville, N.J.*)

TABLE 7.5 PHYSICAL PROPERTIES AND CHEMICAL RESISTANCES OF FIBERS

Type of fiber	Specific gravity	Moisture absorption at 20°C (68°F), 65% RH	Moisture expansion (%)	Ironing temperature (°C)	Maximum working temperature (°C)	Melting point (°C)[a]	Resistance[b] to light and weather
Cotton	1.5	7–8.5	45	200	80–100	D > 200	S
Silk	1.37	9–9.5	40	140–160	90	B = 200–400 D = 170	C
Wool	1.3	13–15	42	140–160	80	B = 200–400 D = 130	C
Glass	2.54	0	0	300	250–300	S = 815 M = 845–1150	R
Steel Fibers (Brunsmet®)	7.9	0	0	300	500	M = 1440–1455	R
Polyamides							
PA 6 (Perlon®)	1.14	4.5	10–14	120	100	S = 170 M = 215	C
PA 6.6 (Nylon)	1.14	4.5	10–14	150	100	S = 235 M = 250–255	C
PA 11 (Rislan®)	1.04	1.2		120	100	S = 160 M = 183–186	C
PA 12 (Vestamid®)	1.02	0.95			90 Stab. 120	S = 168 M = 175	C
PA Nomex®	1.38	5		230	200–300	M = 375 D = 371	C
Polyester	1.38	0.4	3–5	150	140–160	S = 230–249 M = 256	G
Polyacrylonitrile	1.15	1–1.5	4.5–6	160	145	S = 235–250 D = 300	R
Polyvinylchloride	1.38	0.1	0	60–80	60	S = 70 D = 180–190	R
Polyvinylidenechloride (Saran®)	1.7	≤0.1	0	80	70	S = 115–138 M = 150–170	G
Polyolefins							
Polyethylene							
High-Pressure	0.92	≤0.1	0	70	60	S = 107–110 M = 110–120	U
Low-Pressure	0.95	≤0.1	0	80	70	S = 115 M = 124–138	U

Polypropylene	0.9	≤0.1	0	100	90	S = 150–155 M = 163–175	C
Polytetrafluoroethylene (Teflon®)	2.1	0	0	250	200	S = 327 D > 275	R

Type of fiber	Insect proof	Resistance to aging	Resistance against					
			Acid	Alkali	Chloro-carbonic hydride	Ketone	Phenol	Benzene
Cotton	Medium	Low	Unstable	Low resistance, swelling	Resistant	Resistant	Resistant	Resistant
Silk	Medium	Low	Low resistance	Unstable	Resistant	Resistant	Resistant	Resistant
Wool	Bad	Low	Low resistance	Unstable	Resistant	Resistant	Resistant	Resistant
Glass	Good	Good	Low resistance	Unstable	Resistant	Resistant	Resistant	Resistant
Steel Fibers (Brunsmet®)	Good	Good	Low resistance	Resistant	Resistant	Resistant	Resistant	Resistant
Polyamides								
PA 6 (Perlon®)	Good	Good	Unstable	Resistant	Resistant	Resistant	Unstable	Resistant
PA 6.6 (Nylon)	Good	Good	Unstable	Resistant	Resistant	Resistant	Unstable	Resistant
PA 11 (Rislan®)	Good	Good	Low resistance	Resistant	Resistant	Resistant	Unstable	Resistant
PA 12 (Vestamid®)	Good	Good	Low resistance	Resistant	Swelling	Swelling	Unstable	Swelling
PA Nomex®	Good	Good	Low resistance	Resistant	Resistant	Resistant	Unstable	Resistant
Polyester	Good	Good	Resistant	Low resistance	Resistant	Resistant	Unstable	Resistant
Polyacrylonitrile	Good	Good	Resistant	Low resistance	Resistant	Resistant	Resistant	Resistant
Polyvinylchloride	Good	Good	Resistant	Resistant	Resistant	Unstable	Unstable	Unstable
Polyvinylidenechloride (Saran®)	Good	Good	Resistant	Resistant except NH_4OH	Resistant	Resistant	Unstable	Resistant
Polyolefins								
Polyethylene								
High-Pressure	Good	Good	Resistant	Resistant	Swelling	Resistant	Resistant	Resistant
Low-Pressure	Good	Good	Resistant	Resistant	Swelling	Resistant	Resistant	Resistant
Polypropylene	Good	Good	Resistant	Resistant	Resistant	Resistant	Resistant	Resistant
Polytetrafluoroethylene (Teflon®)	Good	Good	Resistant	Resistant	Resistant	Resistant	Resistant	

[a] B = burning point; D = disintegration; M = melting point; S = softening point.

[b] R = recommended; G = good; S = satisfactory; C = conditional; U = unsatisfactory.

(*Courtesy of SciTech Publishers, Morganville, N.J.*)

dissatisfaction has been experienced with such construction. While cotton weaves well with the asbestos, the cotton fibers destroy the fabrics' resistance to heat and corrosion. Its use is therefore quite limited, despite its resistance to high temperatures, acids, and mildew.

Cotton cloths are sometimes treated with metallic salts (copper sulfate) to improve their corrosion-resistant qualities. Such cloths are in the usual cotton filter cloth grades; and while they are not equivalent to metallic cloths, the treatment does materially prolong the life of the cotton fiber.

Nonwoven Media

Nonwoven media are fabricated in the form of belts or sheets from cotton, wool, synthetic, and asbestos fibers or their mixtures, as well as from paper mass. They may be used in filters of different designs, for example, in filter presses, filters with horizontal disks, and rotary drum vacuum filters for liquid clarification. Most of these applications handle low-suspension concentrations; examples are milk, beverages, lacquers, and lubricating oils. Individual fibers in nonwoven media are usually connected among them as a result of mechanical treatment. A less common approach is the addition of binding substances. Sometimes the media are protected from both sides by loosely woven cloth. Nonwoven media of various materials and weights, and in several grades of retentiveness per unit weight can be formed, in either absorbent or nonabsorbent material. These filter media retain less dispersed particles (more than 100 per unit) on their surface or close to it and more dispersed particles within the depths of the media.

Nonwoven filter media are mostly used for filter medium filtration with pore clogging. Because of the relatively low cost of this medium, it is often replaced after pore clogging [2, 19]. In some cases, nonwoven media are used for cake filtration. In this case cake removal is so difficult that it must be removed altogether from the filter medium. Nonwoven filter media can be prepared so that pore sizes decrease in the direction from the surface of the filter media contacting the suspension to the surface contacting the supporting device [20]. This decreases the hydraulic resistance of filtration and provides retention of relatively large particles of suspension over the outer layer of the nonwoven medium. Nonwoven filter media of synthetic, mechanically pressed fibers are manufactured by puncturing the fiber layer with needles (about 160 punctures/cm^2), and subsequent high-temperature treatment with liquid which causes fiber contraction. Such filter media are distinguished by sufficient mechanical strength and low hydraulic resistance, as well as uniform fiber distribution. Filter media from fibers connected by a binder are manufactured by pressing at 70 N/cm^2 and 150°C. These media have sufficient mechanical strength, low porosity, and are corrosion resistant.

Filter media may be manufactured by lining a very thin layer of heat-resistant metal (e.g., nickel 360) over a fiber surface of inorganic or organic material. Such filter media may withstand temperatures of 200°C and higher.

Of the flexible filter media described, the synthetic fabrics are perhaps the most widely relied on in industrial applications. Each filtration process must meet certain requirements in relation to flow-rate clarity of filtrate, moisture of filter cake, cake release, and nonbinding characteristics. The ability of a filter fabric to help meet these criteria and to resist chemical and physical

Table 7.6 PHYSICAL PROPERTIES AND CHEMICAL RESISTANCES OF POLYESTER FIBERS USED FOR BELT FILTERS

Specific Gravity	1.38
Moisture Regain	
At 65% RH and 68°F (20°C) (%)	0.4
Water Retension Power (%)	3–5
Tensile Strength	
cN/dtex	7–9.5
Wet in % of dry	95–100
Elongation at Break	
%	10–20
Wet in % of dry	100–105
Ultraviolet Light Resistance	R[a]
Resistance to Fungus, Rot and Mildew	R
Resistance to Dry Heat	
Continuous	
°F	302
°C	150
Short-Term Exposure	
°F	392
°C	200
Chemical Resistance to	
Acids	C[b]
Acetic Acid Concentration	R
Sulfuric Acid 20%	R
Nitric Acid 10%	C
Hydrochloric Acid 25%	C
Alkalies	C
Saturated Sodium Carbonate	R
Chlorine Bleach Concentration	R
Caustic Soda 25%	U[c]
Ammonia Concentration	U
Potassium Permanganate 50%	R
Formaldehyde Concentration	R
Chlorinated Hydrocarbons	R
Benzene	R
Phenol	C
Ketones, Acetone	R

[a] R = recommended.

[b] C = conditional.

[c] U = unsatisfactory.

(Courtesy of SciTech Publishers, Morganville, N.J.)

TABLE 7.7 STANDARD FIBERS AND MICROMETER RATINGS FOR BAG FILTERS

Construction		Available micrometer ratings																	
		1	3	5	10	15	25	50	75	100	125	150	175	200	250	300	400	600	800
Felts	Polyester	X	X	X	X	X	X	X	X	X				X					
	Polypropylene	X	X	X	X		X	X		X									
Multifilament Meshes	Polyester								X	X	X	X		X	X	X	X		X
	Nylon (heavy)																		X
Monofilament Meshes	Nylon							X	X	X	X	X	X	X	X	X	X	X	X
	Polypropylene															X		X	

Compatibility and temperature limits for standard bag materials

Fiber	Compatibility with							Temperature limits (°F)
	Organic solvents	Animal, vegetable and petro oils	Microorganisms	Alkalies	Organic acids	Oxidizing agents	Mineral acids	
Polyester	Excellent	Excellent	Excellent	Good	Good	Good	Good	300
Polypropylene	Excellent	Excellent	Excellent	Excellent	Excellent	Good	Good	225
Nylon	Excellent	Excellent	Excellent	Good	Fair	Poor	Poor	325

Bag capacities and dimensions

Bag size no.	Fits Rosedale model no.	Bag surface area (ft^2)	Bag volume (gal)	Bag dimensions	
				Length (in.)	Diameter (in.)
1	8 15	2.0	2.1	16.5	7
2	8 30	4.4	4.6	32.0	7
1 (inner)	8 15	1.6	1.7	14.5	5.75
2 (inner)	8 30	3.6	3.8	30.0	5.75

(Courtesy of SciTech Publishers, Inc., Morganville, N.J.)

attack depend on such characteristics as fiber type, yarn size thread count, type of weave, fabric finish, and yarn type (monofilament, multifilament, or spun).

Monofilament yarns consist of a single, continuous filament with a relatively smooth surface. The different sizes are specified by a measurement of weight known as denier. These yarns are generally used for filter fabrics which require a smooth surface and relatively tight weave. Spun yarns are made from filaments which are chopped in short lengths and then spun or

TABLE 7.8 COMPARATIVE PARTICLE SIZES

U.S. Mesh	in.	μm
3	0.265	6,730
$3\frac{1}{2}$	0.223	5,660
4	0.187	4,760
5	0.157	4,000
6	0.132	3,360
7	0.111	2,830
8	0.0937	2,380
10	0.0787	2,000
12	0.0661	1,680
14	0.0555	1,410
16	0.0469	1,190
18	0.0394	1,000
20	0.0331	841
25	0.0280	707
30	0.0232	595
35	0.0197	500
40	0.0165	420
45	0.0138	354
50	0.0117	297
60	0.0098	250
70	0.0083	210
80	0.0070	177
100	0.0059	149
120	0.0049	125
140	0.0041	105
170	0.0035	88
200	0.0029	74
230	0.0024	63
270	0.0021	53
325	0.0017	44
400	0.0015	37

twisted together. Spun yarns are made into filter fabrics with a hairy, dense surface very suitable for filtration of very fine particles.

It is necessary to select the type of fiber that will offer the most resistance to breakdown normally caused by chemical, temperature, and mechanical conditions of the filter process. Tables 7.5 through 7.7 can serve as rough guides to proper media selection. Table 7.8 provides linear conversion units between mesh size, inches, and micrometers.

RIGID FILTER MEDIA

Fixed Rigid Media

Fixed rigid media are manufactured in the forms of disks, pads, and cartridges. They are composed of firm, rigid particles set in permanent contact with one another. The media formed have excellent void uniformity, resistance to wear, and ease in handling as piece units. Depending on the particle size forming the filter media, temperature, pressure, and time for caking, it is possible to manufacture media with different porosities. The higher the pore uniformity, the more uniform the shape of the particles. These media are distinguished by long life, high corrosion resistance, and easy cake removal. However, the particles that penetrate inside the pores are very difficult to extract.

Metallic Media

Metallic filter media are widely used throughout the chemical and process industries in the form of perforated or slotted plates of steel, bronze, or other materials. These designs provide for easy removal of coarse particles and for supporting loose rigid media.

Powdered metal is a porous medium. The physical characteristics, chemical composition, structure, porosity, strength, ductility, shape, and size can be varied to meet special requirements. The porosity ranges up to 50 percent void by volume, tensile strength up to 10,000 psi varying inversely with porosity, and ductility of 3 percent to 5 percent in tension, and higher in compression. Powdered metal cannot be readily ground or machined. It is made in disks, sheets, cones, or special shapes for filtering fuel oil, refrigerants, solvents, and so on.

The smooth surface associated with a perforated plate permits brush cleaning or scrubbing in addition to the naturally easier discharge from such surfaces. The hard metallic material has a long life, not being subject to abrasion or flexing. The size of particles filtered on such plates must be relatively large. Normally, plates are confined to free-filtering materials where there is little danger of clogging.

Metallic filter media may be used either for cake filtration or depth filtration, i.e., pore clogging. Regeneration of media may be achieved by dissolving solid particles inside the pores or by back thrust of filtrate.

Ceramic Media

Ceramic filter media are manufactured from crushed and screened quartz or chamotte, which is then thoroughly mixed with a binder (for example, with silicate glass) and sintered. Quartz media are resistant to concentrated mineral acids but not resistant to low-concentration alkalies or neutral water solutions of salts. Chamotte media are resistant to dilute and concentrated mineral acids and water solutions of their salts, but have poor resistance to alkali liquids.

The rough surface of ceramic filter media promotes adsorption of particles and bridging. Sintering of chamotte with a binder results in large blocks from which filter media of any desired shape can be obtained. Using synthetic polymers as binders, ceramic filter media that do not contain plugged pores are obtained.

Diatomaceous Media

Diatomaceous media are available in various shapes. Their relatively uniform particle size establishes high efficiency in retaining solid particles of sizes less than 1 μm as well as certain types of bacteria. Media in the form of plates and cartridges are manufactured by sintering a mixture of diatomite with a binder.

Coal Media

Coal media are manufactured by mixing a fraction of crushed coke with an anthracene fraction of coal tar and subsequent forming under pressure, drying, and heating in the presence of a reducing flame. These media of high mechanical strength are good for use in acids and alkalies.

Ebonite Media

Ebonite media are manufactured from partially vulcanized rubber, which is crushed, pressed, and vulcanized. These media are resistant to acids, salt solutions, and alkalies. They may be used for filtration at temperatures ranging from $-10°C$ to $+110°C$.

Foam Plastic Media

Foam plastic media are manufactured from polyvinylchloride, polyurethane, polyethylene, polypropylene, and the other polymer materials. The foam plastic media are economical, since the material and fabrication are not expensive.

Loose Rigid Media

Filter media may also be composed of particles that are rigid in structure but are applied in bulk loose form. That is, individual particles merely contact each other. This form has the advantage of being cheap and easy to keep clean by rearrangement of the particles. When the proper size and shape of particles are selected, the section of passage may be regulated over extremely wide limits. The disadvantages of a rigid medium in simple contact are that it can be used conveniently only in a horizontal position and that it does not allow removal of thick deposits or surface cleaning except by backwashing, without disturbing the filter bed.

Coal and Coke

Coal (hard) and coke are used in water filtration, primarily for the removal of coarse suspensions, care being taken to prevent them from scouring or washing away, because of their relative lightness and fine division. Coal is principally composed of carbon and is inert to acids and alkalies. Its irregular shapes are advantageous at times over silica sand. Though inert to acids, sand is affected by alkalies, and its spherical particle shape allows deeper solids penetration and quicker clogging than does coal. With the lighter weight of coal (normally 50 lb/ft^3, compared with 100 lb/ft^3 for sand), a greater surface area is exposed for solids entrapment.

Charcoal

Charcoal, whether animal or vegetable, when used as a filter medium, is required to perform the dual services of decoloring or adsorbing and filtering. The char filters used in the sugar industry are largely decoloring agents and the activated carbons used in water clarification are for deodorizing and removal of taste. There are many types of charcoal in use as filter media, ranging from ordinary wood char to specially prepared carbons.

Diatomaceous Earth

Diatomaceous earths may resemble the forms of the charcoals. The earths are primarily filter aids, precoats, or adsorbents, the function of the filter medium being secondary. Fuller's earth and clays are for decoloring; diatomaceous earths are for clarification.

Precipitates and Salts

Precipitates or salts are used when corrosive liquor must be filtered, and where there is no available medium of sufficient fineness that is corrosion resistant and will not contaminate the cake. In these cases, precipitates or salts are used on porous supports. In the filtration of caustic liquors, ordinary salt (sodium chloride) is used as the filter medium in the form of a precoat over metallic cloth. This procedure has the advantage that the salt medium will not be detrimental to either the cake or the filtrate if inadvertently mixed with it.

Sand and Gravel

Sand and gravel are the most widely used of the rigid media in simple contact. Most of the sand used this way is for the clarification of water for drinking or industrial uses. Washed, screened silica sand is sold in standard grades for this work and is used in depths ranging from a few inches to several feet, depending on the type of filter and clarification requirements. Heavy irregular grains, such as magnetite, give high rates of flow and low penetration by the solid particles and are easily cleaned. They are, however, considerably more expensive than silica sand, so their use is limited. Sand beds are often gravel supported, but gravel alone is seldom used as the filter medium.

Crushed Stone or Brick

Crushed stone or brick is used for coarse filtration of particularly corrosive liquors. Their use, however, is extremely limited and they are not considered important filter media.

Due to the wide variety of filter media, filter designs, suspension properties, conditions for separation, and cost, selection of the optimum filter medium is complex. Filter media selection should be guided by the following rule: A filter medium must incorporate a maximum size of pores while at the same time providing a sufficiently pure filtrate. Fulfillment of this rule invokes difficulties because the increase or decrease in pore size acts in opposite ways on the filtration rate and solids retention capacity.

The difficulty becomes accentuated by several other requirements that cannot be achieved through the selection of a single filter medium. Therefore, selection is often reduced to determining the most reasonable compromise between different, mutually contradictory requirements as applied to the filter medium at a specified set of filtration conditions. Because of this, some problems should be solved before final medium selection. For example, should attempts be made to increase filtration rate or filtrate purity? Is cost or medium life more important? In some cases a relatively more expensive filter medium, such as a synthetic cloth, is only suitable under certain filtration conditions, which practically eliminates any cost consideration in the selection process.

Thus, the choice may only be made after consideration of all requirements. It is, however, not practical to analyze and compare each requirement with the hope of logically deducing the best choice. There is, unfortunately, no generalized formula for selection that is independent of the intended application. Each cake requires study of the specific considerations, which are determined by the details of the separation process.

It is possible to outline a general approach for medium selection along with a test sequence applicable to a large group of filter media of the same type. There are three methods of filter media tests: laboratory-scale or bench-scale tests, pilot-unit tests, and plant tests. The laboratory-scale test is especially rapid and economical, but the results obtained are often not entirely reliable and should only be considered preliminary. Pilot-unit tests provide results that approach plant data. The most reliable results are obtained from plant tests.

Different filter media, regardless of the specific application, are distinguished by a number of properties. The principal properties of interest are the permeability of the medium relative to a pure liquid, its retention capacity relative to solid particles of known size, and the pore-size distribution. These properties are examined in a laboratory environment and are critical for comparing different filter media.

The permeability relative to a pure liquid, usually water, may be determined with the help of different devices that operate on the principle of measurement of filtrate volume obtained over a definite time interval at known pressure drop and filtration area. The permeability is usually expressed in terms of the hydraulic resistance of the filter medium. This value is found from:

$$V = \frac{\Delta PS}{\mu(r_o h_c + R_f)} \tau \tag{7.23A}$$

When cake thickness is 0, we may write the equation as:

$$R_f = \frac{\Delta PS\tau}{V\mu} \tag{7.23B}$$

Note that:

$$\Delta P = \Delta P_t - \Delta P_f \tag{7.24}$$

where

ΔP_t = pressure difference accounting for the hydrostatic pressure of a liquid column at its flow through the filter medium, supporting structure and device channels

ΔP_f = same pressure drop when the flow of liquid is through the supporting structure and device channels

Analytical determination of the hydraulic resistance of the medium is difficult. However, for the simplest filter medium structures, certain empirical relationships are available to estimate hydraulic resistance. The relationship of hydraulic resistance of a cloth of monofilament fiber versus fiber diameter and cloth porosity is given by Rushton et al. [34]. On the basis of a fixed-bed model, they obtained a relationship for the hydraulic resistance of metallic cloth.

In evaluation and selection of a filter medium, one should account for the fact that hydraulic resistance increases gradually with time. In particular, the relationship between cloth resistance and the number of filter cycles is defined by:

$$R = R_{in}e^{KN} \tag{7.25}$$

The retentivity relative to solid particles (e.g., spherical particles of polystyrene of definite size) is found from experiments determining the amount of these particles in the suspension to be filtered before and after the filter media. The retentivity K is determined as follows:

$$K = \frac{g' - g''}{g''} \tag{7.26}$$

where

g', g'' = amounts of solid particles in liquid sample before and after the medium, respectively

The pore distribution by sizes, as well as the average pore size, is determined by the bubble method. The filter medium to be investigated is located over a supporting device under a liquid surface that completely wets the medium material. Air is introduced to the lower surface of the medium. Its pressure is gradually increased, resulting in the formation of single chains of bubbles. This corresponds to air passages through the largest-diameter pores. As pressure is increased, the number of bubble chains increases due to air passing through the smaller pores. In many cases a critical pressure is achieved where the liquid begins to "boil." This means that a filter medium under investigation is characterized by sufficiently uniform pores. If there is no boiling, the filter medium has pores of widely different sizes. The pore size through which air passes is calculated from known relations. For those pores whose cross section may be assumed to be close to a triangle, the determining size should be the diameter of a circle that may be inscribed inside the triangle.

For orientation in cloth selection for a given process, the following information is essential: filtration objectives (obtaining cake, filtrate or both), and complete data (if possible) on the properties of solid particles (size, shape, and density), liquid (acid, alkali or neutral, temperature, viscosity, and density), suspension (ratio of solids to liquid, particle aggregation, and viscosity), and cake (specific resistance, compressibility, crystalline, friable, plastic, sticky, or slimy). Also, the required capacity must be known as well as what constitutes the driving force for the process (e.g., gravity force, vacuum, or pressure). Based on such information, an appropriate cloth that is resistant to chemical, thermal, and mechanical aggression may be selected. In selecting a cloth based on specific mechanical properties, the process driving force and filter type must be accounted for. The filter design may determine one or more of the following characteristics of the filter cloth: tensile strength, stability in bending, stability in abrasion, and the ability of taking the form of a filter-supporting structure. Tensile strength is important, for example, in belt filters. Bending stability is important in applications of metallic woven cloths or synthetic monofilament cloths. If the cloth is subjected to abrasion, then glass cloth cannot be used even though it has good tensile strength.

From the viewpoint of accommodation to the filter-supporting structure, some cloths cannot be used, even though the filtering characteristics are excellent. For rotary drum filters, for example, the cloth is pressed onto the drum by the caulking method, which uses cords that pass over the drum. In this case, the closely woven cloths manufactured from monofilament polyethylene or polypropylene fiber are less desirable than more flexible cloths of polyfilament fibers or staple cloths.

Depending on the type of filter device, additional requirements may be made of the cloth. For example, in a plate-and-frame press, the sealing properties of cloths are very important. In this case, synthetic cloths are more applicable staple cloths, followed by polyfilament and monofilament cloths. In leaf filters operating under vacuum and pressure, the cloth is pulled up onto rigid frames. Since the size of a cloth changes when in contact with the suspension, it should be pretreated to minimize shrinkage.

In selecting cloths made from synthetic materials, one must account for the fact that staple cloths provide a good retentivity of solid particles due to the short hairs on their surface. However, cake removal is often difficult from these cloths—more than from cloths of polyfilament and, especially, monofilament fibers. The type of fiber weave and pore size determine the degree of retentivity and permeability. The objective of the process and the properties of particles, suspension, and cake should be accounted for. The cloth selected in this manner should be confirmed or corrected by laboratory tests. Such tests can be performed on a single filter. These tests, however, provide no information on progressive pore plugging and cloth wear. However, they do provide indications of expected filtrate pureness, capacity, and final cake wetness.

A single-plate filter consists of a hollow flat plate, one side of which is covered by cloth. The unit is connected to a vacuum source and submerged into the suspension (filtration), then suspended in air to remove filtrate, or irrigated by a dispersed liquid (washing). The filter cloth is directed downward or upward or located vertically, depending on the type of filter that is being modeled in the study.

The following is a recommended sequence of tests that can assist in cloth selection for ten types of continuous vacuum filters:

1. drum filter with a device for cake removal with cords
2. drum filter with cake removal with a knife
3. drum filter with cake removal using a roller
4. drum filter with falling cloth

5. drum filter with a layer of filter aid
6. drum filter with internal surface of filtration
7. disk filter with a knife for cake removal
8. disk filter with a roller for cake removal
9. table filter with a screw conveyor for cake removal
10. a rotary drum filter

If the cycle consists of only two operations (filtration and dewatering), tests should be conducted to determine the suspension weight concentration after 60 sec of filtration and 120 sec of dewatering. The cake thickness should be measured and the cake should be removed to determine the weight of wet cake and the amount of liquid in it. The weight of filtrate and its purity are also determined. If the cake is poorly removed by the mentioned devices, it is advisable to increase the dewatering time, vacuum, or both. If the cake is poorly removed after an operating regime change, it should be tested with another cloth. If the cake is removed satisfactorily, filtration time should be decreased under increased or decreased vacuum. Note that the compressible cakes sometimes plug pores faster at higher vacuum. After the filtration test for a certain filter cycle (which is based on the type of the filter being modeled), the suspension's properties should be examined. Based on the assumed cycle, a new filtration test should be conducted and values characteristics of the process noted. Capacity (N/m^2-hr), filtration rate (m^3/m^2-hr), and cake wetness can then be evaluated. Also, if possible, the air rate and dewatering time should be computed. The results of the first two or three tests should not be taken into consideration because they cannot exactly characterize the properties of the cloth. A minimum of four or five tests is generally needed to achieve reproducible results of the filtration rate and cake wetness to within 3 percent to 5 percent.

When the cycle consists of filtration, washing, and dewatering, the tests are considered principally in the same manner. The economic aspects of cloth selection should be considered after complete determination of cloth characteristics. Donat and Vagina [20] and Begiov [37] describe the method of selection for synthetic, cotton, wool and glass cloths from the point of view of their permeability and retentivity.

REFERENCES

1. Flood, J. E., H. F. Porter, and F. W. Rennie, *Chem. Eng.*, 73(13) (1966), 163.
2. Dicky, G. D., *Filtration*. New York: Reinhold Publishing Co., 1961.
3. Karacharov, E., and E. I. Vorontsova, *Gigiena Sanitarya*, (6) (1952), p. 29.
4. Smucker, C. A., and W. C. Marlow, *Ind. Eng. Chem.* 46(1) (1954), 176–178.
5. Wente, V. A., and R. T. Lucas, *Ind. Eng. Chem.*, 48(2) (1956), 219–222.

6. Stein, L., *Chem. Apparatur.*, 13 (1926), 37–39, SS-S7, 81–83.
7. Stein, L., *Chemikerzeitung*, 50 (1926), 97–98, 110, 12S–127.
8. Walla, G. E., *Silicat Tech.*, 3 (1952), p. 161.
9. Grune, A., *Oesterr. Chemiker Ztg.*, 64(9) (1963), 261–266.
10. Kufferath, A., *Filtration Filter*. Berlin: 1953–1954.
11. Farguhar, B. S., *Prod. Eng.*, 25(12) (1954), 190–193.
12. Hall, A. J., *Fibers*, 18 (1957), pp. 239–242.
13. Nauder, W., *Paper Mill News*, 84(37) (1961), 17–22.
14. Turnball, A. D., *Corr. Prev. Control*, 9(7) (1962), 49–52.
15. Duyster, H., *Chem.-Ing.-Tech.*, 36(3) (1964), 241–246.
16. Boyer, C. W., and C. E. Avery, *Chem. Eng.*, 67(12) (1960), 226–230.
17. "Titanium Wire Cloth," *Steel*, 139(10) (1956), 106.
18. "Stainless Steel Micronic Cloth," *Chem. Age*, (London) 87(2225) (1962), 362.
19. Wrotnowski, A. C., *Chem. Eng. Prog.*, 58(12) (1962), 61.
20. Donat, E. V., and A. M. Vagina., *Trudy Unikhim*, 10 (1963), p. 105.
21. Wrotnowski, A. C., *Chem. Eng. Prog.*, 53(7) (1957), 313.
22. Frehn, F., W. Hotop, and G. Stempel, "Filtern aus Sintermetallen, Hufbau Werkstoffeigenschaften and Einsatzmoglichkeiten," *Dechema Monog.*, 28 (1956), pp. 363–391.
23. Agte, C., and K. Ocetec, *Metallfilter*. Berlin: Akademisches Verlag, 1957.
24. McBride, D., "Stainless Steel Filters," *Purdue Eng.*, 51(5) (1956), 38.
25. Strobel, E., and R. Siegel, *Chem. Technik*, 13(11) (1961), 646.
26. Bates, R., *Filtration*, 1(2) (1964), 107.
27. Morgan, V. T., *Filtration*, 1(4) (1964), 215.
28. Poulter, J.E., *Instrum. Prod. Eng.*, 32 (1953), pp. 443–448.
29. Singer, F., *Chem. Age*, (London) 70 (1954), pp. 141–146.
30. Norden, R. B., *Chem. Eng.*, 66(16) (1969), 158, 160, 162.
31. Mounsey, R., *Am. City*, 68(5) (1953), 83–85.
32. Tarrant, J., *Waterworks Sewerage*, 92 (1945), p. 218.
33. Piskarev, I.V., Filtrovalnye Tkani, An. (1963).
34. Rushton, A., D. J. Green, and H. E. Khoo, *Filrra. Sep.*, 5(3) (1968), 213.
35. Gross, R. J., *Hydraul. Pneumat.*, 16(8) (1963), 74.
36. Wheeler, H. L., *Hydraul. Pneumat.*, 16(4) (1963), 102.
37. Beglov, V. M., *Uzb. Khim. J.*, (1) (1966), 50.

8

Filtration Practices for Wastewater Treatment

STRAINING OPERATIONS

Physical straining processes are defined as those processes which remove solids by virtue of physical restrictions on a medium which has no appreciable thickness in the direction of liquid flow.

Physical straining devices may be grouped according to the nature of their straining action. Refer to Table 8.1. The first of these devices to describe is the wedge-wire screen. Among this group is the inclined screen device. The screen consists of three sections with successively flatter slopes on the lower sections (Figure 8.1). The screen wires are triangular in cross section as shown in Figure 8.2, and usually spaced 0.06 in. apart for raw wastewater screening applications. In some units, these wires bend in the plane of the screen, as illustrated in Figure 8.3.

Above the screen and running across its width is a headbox; Figure 8.1 shows two possible inlet designs. A lightweight hinged baffle at the top portion of the screen can be used to reduce turbulence. To collect the solids coming off the end of the screen, several arrangements can be used, including a trough with a screw conveyor.

Inclined screening units are generally constructed entirely of stainless steel. Lighter units with a fiberglass house and frame costing about 25 percent

TABLE 8.1 PHYSICAL STRAINING PROCESSES

Principal applications	Device	Hydraulic capacity	Straining surface	Waste solids composition	Percent SS removals
Pretreatment & Primary Treatment	Inclined wedge-wire stainless steel screens	High flow rates 4–16 gpm/in of screen width	Coarse .01–.06 in. (250–1,500 microns)	10–15% solids by weight	5–25
	Rotary stainless steel wedge-wire screens	16–112 gpm/sq ft	Coarse .01–.06 in. (250–1,500 microns)	16–25% solids by weight	5–25
	Centrifugal screens	40–100 gpm/sq ft	Medium 105	0.05–0.1%	60–70
Secondary and Tertiary SS Removal	Microscreens	Medium flow rates 3–10 gpm/sq ft	Medium 15–60 microns	250–700 mg/l (app. 0.05%)	50–80
	Diatomite filters	Medium flow rates 0.5–1.0 gpm/sq ft	N/A		up to 90
	Ultrafilters	Low flow rates 5–50 gpd/sq ft	Fine[a] 10^{-3}–15 microns	1,500 mg/l (1.5%)	>99

[a] These values typify the range of solids filtered by the media. Removals are a function of media thickness and not media opening sizes.

less may also be obtained. Dimensions and capacities for hydrasieve units are given in Table 8.2.

Influent wastewater enters and overflows the headbox onto the upper portion of the screen. On the screen's upper slope most of the fluid is removed from the influent. The solids mass on the following slope because it is flatter, and additional drainage occurs. On the screen's final slope the solids stop momentarily, simple drainage occurs, and the solids are displaced from the screen by oncoming solids.

In test studies and actual installations, hydrasieves have been operated satisfactorily at loading capacities of 4 to 16 gpm per in. of screen width. This hydraulic capacity is a function of the viscosity (which is a function of the temperature of the fluid), the solids loading, and the spacing of the individual slots. Slot width is selected by actual tests using sample screens. Once the slot opening has been chosen, the screen's capacity per foot of width can be determined from empirical relationships. Since work to date has not been sufficiently extended to actual municipal wastewater conditions, pilot studies should be the prime basis for design.

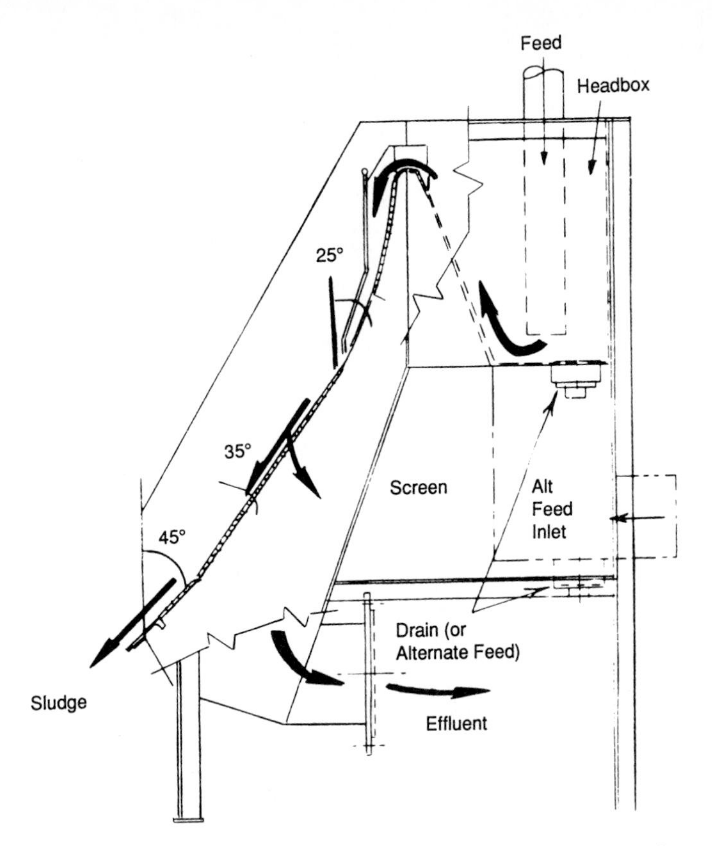

Figure 8.1 Schematic of sieve type filter for wastewater treatment.

Little quantitative work has been done on the solids loading capacity of a hydrasieve but generally speaking, for good performance, the influent should be dilute enough for smooth flow over the weir. Unit sizes designed to accommodate more than 1 mgd are available; however, for pilot studies a 6-in.-wide by 22-in.-long screen can be used provided flow rates are limited to 5 to 10 gpm.

Although inclined screens cannot remove suspended solids (SS) to the same extent as a sedimentation tank, they have been favorably received by operators because they do an excellent job of removing trashy materials which may foul subsequent treatment of sludge-handling units. Their ability to remove fine grit is limited by size openings. Separate grit removal equipment, if needed, should be installed after the inclined screens.

The functional design of a microscreen unit involves:

1. Characterization of suspended solids in feed as to concentration and degree of flocculation, as these factors have been shown to affect microscreen capacity, performance, and backwashing requirements.

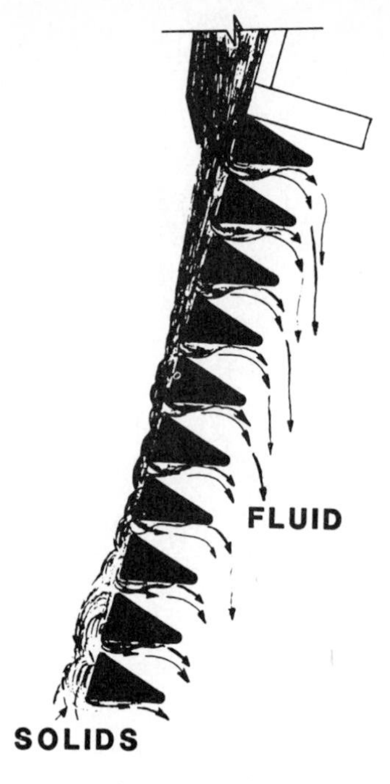

Figure 8.2 Shows details of screen.

2. Selection of unit design parameters which will assure sufficient capacity to meet maximum hydraulic loadings with critical solids characteristics, and provide the required performance over the expected range of hydraulic loadings and solids characteristics.
3. Provision of backwash and supplemental cleaning facilities to maintain the design capacity.

Table 8.3 shows typical values for microscreen and backwash design parameters for solids removal from secondary effluents. Similar values would apply to direct microscreening of good-quality effluent from fixed-film bio-

Figure 8.3 Shows curved screen bars.

TABLE 8.2 SPECIFICATIONS OF HYDRASIEVES

Overall dimensions				
Width (ft)	Depth (ft)	Height (ft)	Weight (lb)	Capacity (mgd)
2	3.5	5	350	0.2
3.5	4	5	550	0.4
4.5	5	7	650	0.9
5.5	5	7	800	1.2
6.5	5	7	1,000	1.5
7	9.5	7.3	1,800	2.9
14	9.5	7.3	3,600	5.8
21	9.5	7.3	5,400	8.7
28	9.5	7.3	7,200	11.6
35	9.5	7.3	9,000	14.5

logical reactors such as trickling filters or rotating biological contractors, where the microscreens replace secondary settling tanks. This application is not widely practiced, however.

Microscreening has been used for the removal of algae from uncoagulated lagoon effluents. However, many classes of algae (e.g., chlorella) are too small to be removed, even on fine screens (23 microns) and excessive loadings (up to 2×10^6 algae per ml) make this application a limited one.

TABLE 8.3 MICROSCREEN DESIGN PARAMETERS

Item	Typical value	Remarks
Screen Mesh	20–25 microns	Range 15–60 microns
Submergence	75% of height 66% of area	
Hydraulic Loading	5–10 gpm/sq ft of submerged drum surface area	
Head Loss (HL) through Screen	3–6 in.	Maximum under extreme condition: 12–18 in. Typical designs provide for overflow weirs to bypass part of flow when head exceeds 6–8 in.
Peripheral Drum Speed	15 fpm at 3 in. (HL) 125–150 fpm at 6 in. (HL)	Speed varied to control extreme maximum speed 150 fpm.
Typical Diameter of Drum	10 ft	Use of wider drums increases backwash requirements.
Backwash Flow and Pressure	2% of throughput at 50 psi 5% of throughput at 15 psi	

Note: Among these parameters peripheral speed, hydraulic loading, and major variations in mesh size also affect performance on a given feed flow. In addition, drum speed and diameter affect the wastewater flows and pressures needed to effect proper cleaning of the screen.

The parameters of mesh size, submergence, allowable head loss, and drum speed [rpm = peripheral speed/ $\frac{\pi}{4}$ (diameter)] are sufficient to determine the flow capacity of a microscreen with given suspended solids characteristics.

The *filterability index* quantifies the effect of the feed solids characteristics on the flow capacity of a particular fabric. This relation assumes that at any constant laminar flow rate the head loss, ΔP in ft, through any given strainer fabric would increase exponentially with the volume passed per unit area (V in. cu ft/sq ft):

$$\frac{\Delta P}{AP_O} = e^{IV}$$

In the preceding relation, the filterability index is the exponential rate constant I (in l/ft).

From the filterability index concept, hydraulic capacity relations for continuous operation of a rotating drum microscreen can be expressed as follows:

$$\mu = \frac{Q}{A} = \frac{1n\left[\left(\frac{\Delta P}{C_F}\right)\left(\frac{I\phi}{R}\right) + 1\right]}{\left(\frac{I\phi}{R}\right)}$$

where

μ = mean flow velocity through submerged screen area (fps)
Q = total flow through microscreen (cfs)
A = submerged screen area (sq ft)
P = pressure drop across screen (ft)
C_F = fabric resistance coefficient (ft/fps or sec) (clean fabric head loss at 1 fps approach velocity)
I = filterability index (1/ft)
ϕ = decimal fraction of screen area submerged
R = drum rotational speed (rpm)

The expression $\Delta P/C_F$ represents the initial flow velocity through the clean screen as it enters submergence. C_F is a particular characteristic of the screen fabric, varying inversely with mesh opening size as follows:

Mesh size (mu)	Fabric resistance C_F (ft/fps)
15	3.6
23	1.8
35	1.0
60	0.8

Limits on ΔP reflect screen fabric mechanical strength and expected operating conditions for the unit. A typical value is 0.5 ft at normally expected maximum flow.

The relation of parameters in the expression $(I\phi/R)$ shows that the effect of a higher index or faster buildup of head loss on the screen may be offset by maintaining a higher drum rotational speed.

Figure 8.4 is a graphical representation of the preceding relation, obtained by plotting Q/A against $\Delta P/C_F$ for various values of the parameter $I\phi/R$.

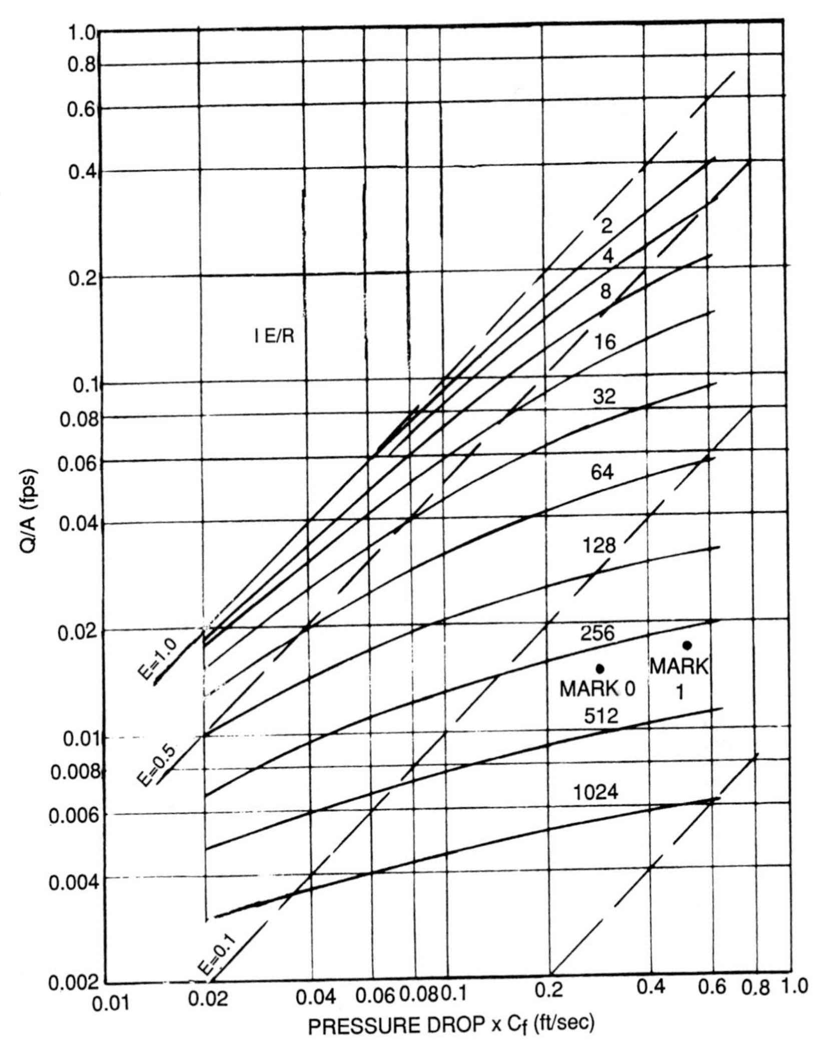

Figure 8.4 Typical microscreen capacity chart.

The graph shows lines of constant value for the ratio

$$E = \frac{Q/A}{\Delta P/C_F}$$

which is the ratio of the mean velocity through the screen to the initial velocity when the screen enters submergence. Recognizing the effect of drum speed on performance, it is suggested selecting $I\phi/R$ to keep the ratio E below 0.5. Above this limit he assumes that insufficient opportunity is given for a mat to form on the drum and solids removal efficiency is likely to suffer.

Suitable relationships have not been developed for quantitative predictions of microscreen performance from knowledge of influent characteristics and key design parameters. Where performance must be predicted closely, pilot studies should be made. Where close prediction is less critical, performance data from other locations with generally similar conditions may serve as a guide.

Some general conclusions can be made about the microscreen as a device for removing SS from secondary effluents:

1. Under best operating conditions microscreen units can reduce solids to as low as 5 mg/1.
2. Although the SS removal pattern is irregular, performance tends to be better at lower hydraulic loadings.
3. Increases in influent-suspended solids are reflected in the effluent but with noticeable damping of peaks.
4. Microscreens are applicable in place of clarifiers to polish effluent from low-rate trickling filters, if the solids are generally low in concentration and well flocculated.

Under peripheral speed and hydraulic loading limits set by manufacturers and regulatory authorities, the ratio E for most microscreen designs actually falls below 0.1.

The basic screen-support structure is a drum-shaped, suitably stiffened rigid frame supported on bearings to allow rotation. Designs using water-lubricated axial bearings or greased bearings located on the upper inside surface of the rotating drum allow submergence well above the central axis.

Both plastic (polyester) and stainless steel are used for the microscreen media itself. Greater mechanical strength, especially at higher temperatures, is the prime advantage stated for stainless steel. Greater economy and chemical resistance are pointed to as advantages of plastic.

Depending on manufacturer, screen fabric is supplied either in small sections (8 in. × 12 in.) supported by and fastened directly to the drum frame or in larger (18 in. × 24 in.) panels consisting of fabric integrally bonded to a grid-like supporting mesh of stainless steel. These panels are attached to the drum frame.

In the past cast bronze and cast carbon steel were used as drum and frame construction materials. The present practice is to use fabricated carbon steel. Generally, smaller units are factory assembled in steel tanks, while large units are placed in concrete structures.

Table 8.4 illustrates the approximate size and power requirements for various microscreen units.

Microscreen fabrics normally are woven of stainless steel or plastic (polyester with polypropylene-supporting grid) with openings in the range of 15 to 60 microns. Plastic fabric is less subject to chemical attack by strong chlorine or acid-cleaning solutions. Stainless steel can better withstand temperatures encountered in steam cleaning.

Suggested operating head-loss limits for microscreens are based on observation of the effect of differential head on screen life. Standard design calls for a 3-in. head loss at average flows and 6 in. for normally expected maximum flows. For occasional peaks (less than 3 percent of the time) head losses up to 24 in. can be tolerated. Stainless steel screens operated under the preceding conditions would have a life of ten years: if operated continuously at a 24-in. head loss, the same screen might only last six months.

Hydraulic control of microscreening units is effected by varying drum speed in proportion to the differential head across the screen. The controller is commonly set to give a peripheral drum speed of 15 fpm at 3 in. differential and 125 to 150 fpm at 6 inches. In addition, backwash flow rate and pressure may be increased when the differential reaches a given level.

The operating drum submergence is related to the effluent water level and head loss through the fabric. The minimum drum submergence value for a given installation is the level of liquid inside the drum when there is no flow over the effluent weir. The maximum drum submergence is fixed by a bypass weir which permits flows in excess of unit capacity to be bypassed; at maximum submergence the maximum drum differential should never exceed 15 inches.

TABLE 8.4 TYPICAL MICROSCREEN POWER AND SPACE REQUIREMENTS

	Drum sizes		Floor space		Motors		
Source code	Diam. (ft)	Length (ft)	Width (ft)	Length (ft)	Drive (BHP)	Wash pump (BHP)	Approx. ranges of capacity (mgd)
A	5.0	1.0	8	6	0.50	1.0	0.07–0.15
A	5.0	3.0	9	14	0.75	3.0	0.2–0.4
A	7.5	5.0	11	16	2.00	5.0	0.5–1.0
A	10.0	10.0	14	22	5.00	7.5	1.5–3.0
B	4.0	4.0	7	15	0.75	1.0	0.2–0.4
B	6.0	6.0	10	17	2.00	1.5	0.5–1.0
B	10.0	10.0	14	22	5.00	5.0	1.5–3.0

Effluent and bypass weirs should be designed as follows:

1. Select drum submergence level (70 percent to 75 percent of drum diameter) for no flow over the effluent weir.
2. Locate top of effluent weir at selected submergence level.
3. Determine maximum flow rate.
4. Size effluent weir to limit liquid depth in effluent chamber above the weir to 3 in. at the maximum flow rate.
5. Position the bypass weir 9 to 11 in. above effluent weir. (3 in. head on effluent weir maximum flow plus 6 to 8 in. differential on drum at maximum drum speed and maximum flow).
6. Size bypass weir length to prevent the level above effluent weir flow exceeding 12 to 18 in. at peak maximum flow or overflowing the top of the backwash collection hopper.

Backwash jets are directed against the outside of the microscreen drum as it passes the highest point in its rotation. About half the flow penetrates the fabric, dislodging the mat of solids formed on the inside. A hopper inside the drum receives the flushed-off solids. The hopper is positioned to compensate for the trajectory that the solids follow at normal drum peripheral velocities.

Microscreen effluent is usually used for backwashing. Straining is required to avoid clogging of backwash nozzles. The in-line strainers used for this purpose will require periodic cleaning; the frequency of cleaning will be determined by the quality of the backwash water.

The backwash system can employ two header pipes; one operates continuously at 20 psi, while the other operates at 40 psi when the unit receives a high solids loading. Also one can use two sets of jets but both operate continuously at pressures from 15 to 55 psi. Under normal operating conditions these jets operate at 35 psi. Once a day they can be operated at 50 psi for one-half hour to keep the jets free of slime buildup. Should this procedure fail to keep the jets clean, the pressure is raised to 55 psi. At this pressure the spring-loaded jet mouth widens to allow for more effective cleaning.

Over a period of time screen fabrics may become clogged with algal and slime growths, oil, and grease. To prevent clogging, cleaning methods in addition to backwashing are necessary. To reduce clogging from algae and slime growth, use of ultraviolet lamps placed in close proximity to the screening fabric and monthly removal of units from service to permit screen cleaning with a mild chlorine solution is recommended.

Where oil and grease are present, hot water and/or steam treatment can be used to remove these materials from the microscreens. Plastic screens with grease problems are cleaned monthly with hot water at 120°F to prevent damage to the screen material. Downtime for cleaning may be up to 8 hours.

In starting a microscreening unit care should be taken to limit differential water levels across the fabric to normal design ranges of 2 to 3 inches. For example, while the drum is being filled it should be kept rotating and the backwash water should be turned on as soon as possible. This is done to limit the formation of excessive differential heads across the screen which would stress the fabric during tank fill-up.

Leaving the drum standing in dirty water should be avoided because suspended matter on the inside screen face which is above the water level may dry and prove difficult to remove. For this reason introducing unscreened waters, such as plant overloads, into the microscreen effluent compartment should also be avoided.

If the unit is to be left standing for any length of time, the tank should be drained and the fabric cleaned to prevent clogging from drying solids.

Conventional mesh screens have not been used with success in municipal wastewater treatment. Recently, however, a centrifugal screen, the Sweco Concentrator, has demonstrated its effectiveness. In this unit, influent is directed against the inside of a rotating cylindrical screen cage. It is claimed that the rotational speed (centrifugal force of 3 to 6 gravities) increases hydraulic capacity and, together with the impingement angle, permits separation of solids finer than the screen openings (150 to 165).

Separated solids and the rejected portion of the liquid flow are removed from the bottom of the unit while effluent is taken off at the periphery. Screen blinding is cleared by timer-actuated spray cleaning systems which direct water jets against both the inner and outer screen surfaces.

DIATOMACEOUS EARTH FILTERS

Diatomaceous earth (DE) filters have been applied to the clarification of secondary effluents at pilot scale. No full-scale installations have been characterize in the literature. They produce a high-quality effluent but appear unable to handle the solids loadings normally expected in this application.

DE filtration utilizes a thin layer of precoat formed around a porous septum to strain out the suspended solids in the feed water which passes through the filter cake and septum. The driving force can be imposed by vacuum from the product side or pressure from the feed side. As filtration proceeds, head loss through the cake increases due to solids deposition until a maximum is reached. The cake and associated solids are then removed by flow reversal and the process is repeated. In the cases where secondary effluents have been treated by this process, a considerable amount of diatomaceous earth (body feed) has been required for continuous feeding with the influent in order to prevent rapid buildup of head losses. Generally, the DE filtration process is capable of excellent removal of suspended solids but not colloidal matter.

A wide variety of diatomaceous earth (diatomite) grades are available

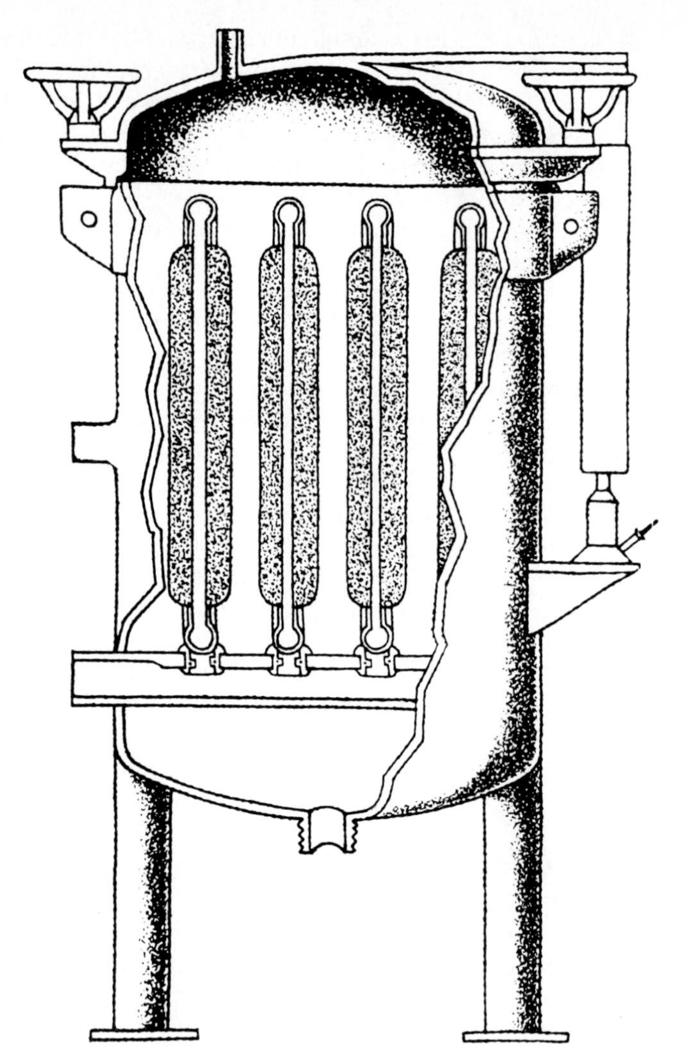

Figure 8.5 Shows a vertical leaf pressure filter arranged in a vertical vessel.

for use. As might be expected, the coarser grades have greater permeability and solids-holding capacities than do the finer grades which will generally produce a better effluent. Some grades of diatomite are pretreated to change their characteristics for improved performance. A number of vessel configurations are available, with open-basin vacuum and vertical pressure designs most common. Refer to Figure 8.5.

The filtration cycle can be divided into two phases, run time and down time. Down time includes the periods when the dirty cake is dislodged from the septum and removed from the filter and when the new precoat is formed. Run time commences when the feed is introduced to the filter and ends when a limiting head loss is reached. The single most important factor in secondary effluent filtration by DE filters is the amount of body feed required during the filtration or run time. The body feed rate is the largest operating cost factor and strongly effects the operating economics of the process. Similarly, it is related to cycle time between backwashing which determines the installed filtering area, hence the capital cost economics.

ULTRAFILTRATION

Ultrafiltration (UF) is the title given to a form of membrane separation which employs relatively coarse membrane separation at relatively low pressures. The process should be differentiated from reverse osmosis, which is a similar process used for dissolved solids separation using fine membranes and high pressures. Ultrafiltration, using a thin semipermeable polymeric membrane, is reported most successful in separating suspended solids as well as large-molecule colloidal solids (0.002 to 10.0 μ) from wastewater. Fluid transport and solids retention are achieved by regulating pore-size openings. Thus, the ultrafiltration process is a physical screening through molecular-sized openings rather than one controlled by molecular diffusion.

A system employing high-MLSS aeration followed by UF operated at Pikes Peak since 1970 has proven capable of removing virtually 100 percent of the suspended matter and 93 percent to 100 percent of the associated BOD, COD, and TOC from aerated mixed liquors.

Several installations have proven the ability of the activated sludge-ultrafiltration process to remove all SS and almost all bacteria and BOD. These systems are typified by Figure 8.6. Because UF installations have all produced zero SS effluents, other parameters are given to illustrate process capability.

The major drawback of ultrafiltration is the high capital and operating costs. Phosphate and color removal are both negligible, but they may not be necessary in many places. The high cost may be offset by compactness where space is a critical factor, such as on a ship or a mountain top.

The most important design considerations are:

1. Membrane area
2. Membrane configuration
3. Membrane material
4. Membrane life
5. Driving force

Membrane area is a function of flux which is determined by membrane construction and the fouling characteristics of the wastewater. Membrane flux tends to decrease with time due to surface fouling. It has been found that physical elimination of foulants, mostly organic acids and polarized materials,

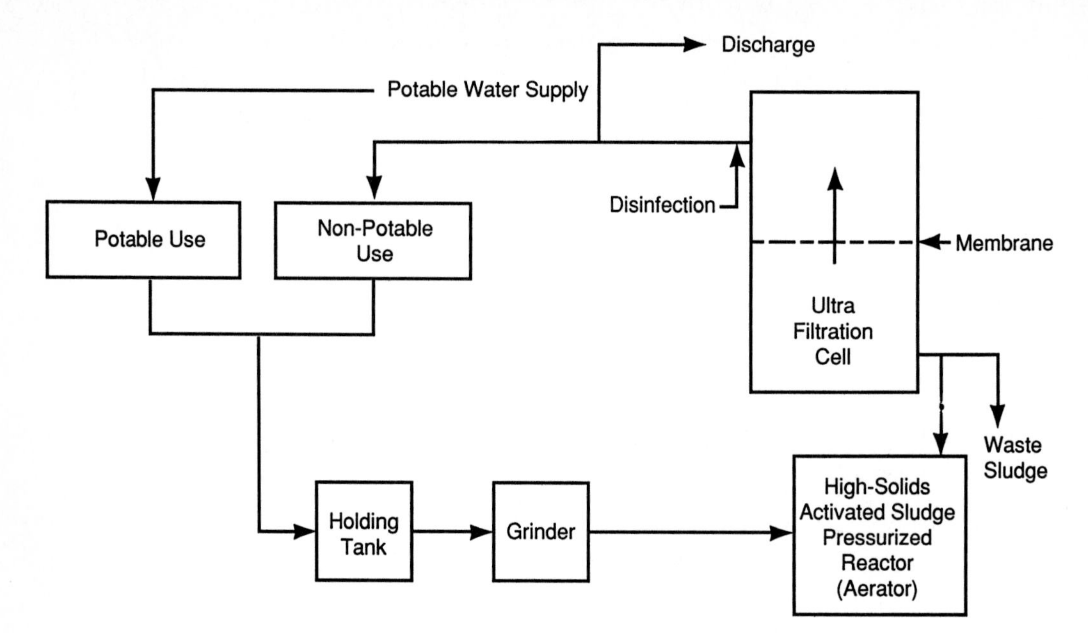

Figure 8.6 Flow diagram of a treatment and reuse system.

lessens their flux-reducing effect. By operating the process at liquid velocities of 3 to 10 fps parallel to the membrane surface, scouring of contaminants can be accomplished and a more stable flux achieved. At such velocities, with normal membrane fluxes, single-pass design would require impractically large membrane area. Therefore, the wastewater is recirculated as shown in Figure 8.6. Some blowdown of concentrated waste results to prevent excessive solids buildup. The blowdown can be intermittent, at a rate sufficient to keep the MLSS within acceptable ranges, usually 4,000 to 15,000 mg/l.

Membrane configuration concerns the amount of membrane surface area which can be incorporated into a module. Because of low membrane fluxes, it is imperative to design the module to maximize membrane surface area. One configuration adopted solely for ultrafiltration is the storage battery configuration, as shown in Figure 8.7. The membrane is cast on both external faces of a hollow plate. A number of these plates are arranged in a parallel array. The edges of these plates face the incoming stream of solids and act as a coarse screen which can be backwashed by reversing the direction of the approaching flow. Other designs include tubular support elements over which a membrane is wound helically or in which the membrane is enclosed in a continuous spiral.

The membrane itself is made up of two basic layers:

1. Surface—an extremely thin homogenous polymer of 0.1 to 10.00 microns (typically 5.0 microns).
2. Surface support—an open cell of 5 to 10 mil thickness.

TABLE 8.5 TYPICAL MEMBRANE SPECIFICATIONS

Material	Most organic polymers
Water Permeability	7–290 gpd per sq ft at 30 psi
Molecular weight	340–45,000
Retentivity	60–100 percent
Maximum Operating Temperature	50–120°C

The membrane, in turn, is supported on a porous sheet (paper) for added mechanical support.

The thin surface layer controls the transport and rejection properties of the membrane. Numerous means and types are available and can be tailored to the particular application. Typical membrane specification ranges are listed in Table 8.5.

Water permeability is used to characterize the porosity of the membrane, but does not represent the stabilized, long-term flux on a process fluid. In the waste treatment field, fluxes of 7 to 10 gallons per square foot of membrane surface per day are typical. Given the current state of membrane manufacturing technology, almost any set of a clean water performance characteristics, without consideration for fouling can be produced.

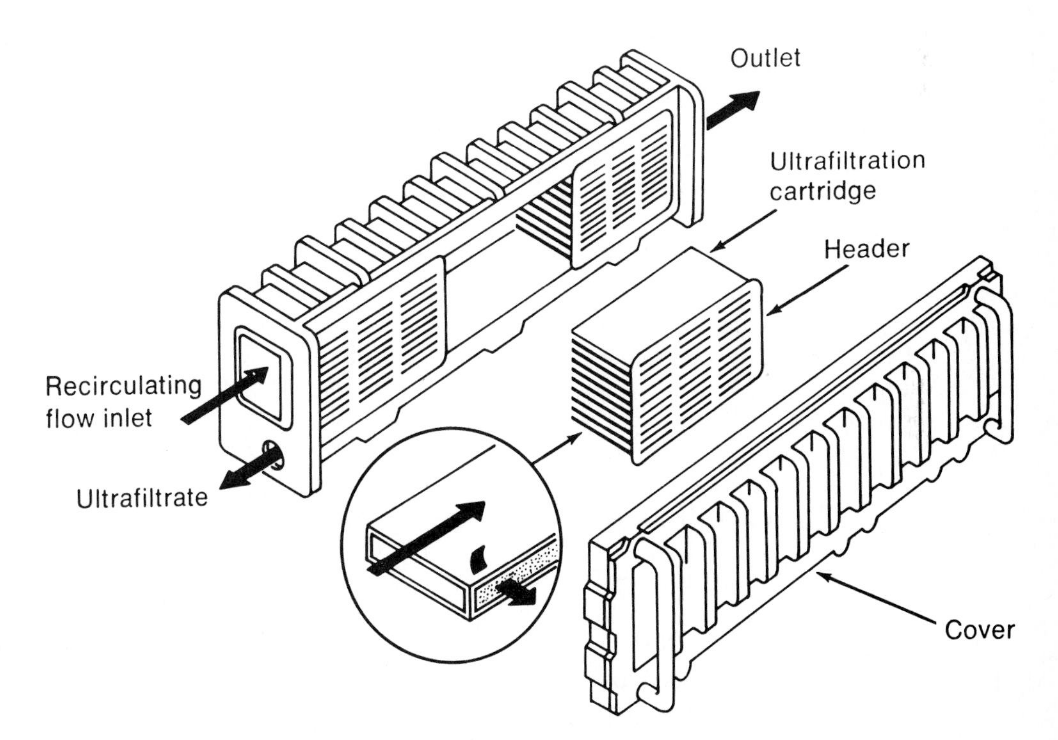

Figure 8.7 Shows "storage battery" membrane modules.

Membrane life is a function of fouling and required flux rates. A membrane may be considered acceptable for a life span of six months in continuous operation with an initial flux of 18 and a final flux of 8 gpd/sq ft. A plant must be designed for the lower figure and membrane replacement made when the design figure is reached.

The driving force for transport of water through the membrane is pressure. Operation is achieved at pressure gradients of approximately 25 psi. Total system pressures do not exceed 50 psi.

GRANULAR MEDIA FILTRATION

This process has long been applied in the treatment of municipal and industrial water supplies. Next to gravity, sedimentation is the most widely used process for separation of wastewater solids. The following specific applications are noted:

1. Removal of residual biological floc in settled effluents from secondary treatment by trickling filters or activated sludge processes.
2. Removal of residual chemical-biological floc after alum, iron, or lime precipitation of phosphates in secondary settling tanks of biological treatment processes.
3. Removal of solids remaining after the chemical coagulation of wastewaters in tertiary or independent physical-chemical waste treatment.

In these applications filtration may serve both as an intermediate process to prepare wastewater for further treatment (such as carbon adsorption, clinoptilolite ammonia exchange columns, or reverse osmosis) and as a final polishing step following other processes.

Granular media filtration involves passage of water through a bed of granular material with resulting deposition of solids. Eventually the pressure drop across the bed becomes excessive or the ability of the bed to remove suspended solids is impaired. Cleaning is then necessary to restore operating head and effluent quality to acceptable levels. Most filters operate on a batch basis, the entire unit being removed from service for periodic cleaning.

The time in service between cleanings is termed the *run length*. The head loss at which filtration is interrupted for cleaning is called the *terminal head loss*.

Filter design involves selection of the following filter characteristics:

1. Filter configuration
2. Media sizes and depths and materials
3. Filtration rate (gpm/sq ft)
4. Terminal head loss (ft of water)

5. Method of flow control
6. Backwashing design features

The major goal in design is to achieve effluent quality objectives at low capital and operating costs. The most important characteristic in determining capital costs is the filtration rate, which fixes the filter size. Operating costs are affected primarily by filtration rate, terminal head loss, media characteristics, and backwash design. The first three filter characteristics determine the cost of power for operating head and the production of the filter per run. The backwash design determines the cost per cleaning of operator attention, washwater pumping, air scouring (compressor operation), and treatment of dirty washwater. The cost of cleaning per unit volume treated (cost per cleaning divided by production per run) depends on all four factors.

Wherever possible, designs should be based on pilot filtration studies of the actual waste. Such studies are the only way to assure:

1. Meaningful cost comparisons between different filter designs capable of *equivalent performance*, that is, producing the same output quantity and quality over the same time period.
2. Most economical selection of filter rate, terminal head loss, and run length for a given media application.
3. Definite effluent quality performance for a given media application.

Pilot studies are also useful for determining effects of pretreatment variations or for characterizing, filterability in terms of performance attainable with a specific filter design. Where there is no opportunity for pilot studies, parameters for workable designs can still be determined from the discussions of wastewater and filter characteristics in the following sections. The parameters will necessarily be conservative and will tend to give more costly designs and less assurance of effluent quality than parameters based on testing. Facilities designed without pilot testing are likely to be small ones, for which the design should provide long filter runs and minimize required operating attention.

Another approach to obtaining economical facilities is to prepare a functional specification which will permit competitive bidding between suppliers of alternative filter systems. Functional requirements should include:

1. Guarantees of specified performance; both capacity and effluent quality.
2. Guarantees of proposed values of all factors which affect operating costs such as head or power requirements and backwash volume to be recycled.

Bids should be evaluated based on total present worth including operating costs, which should be calculated by a predetermined formula using factors in the guarantee.

This approach will work best when bidders are supplied test results characterizing the filterability of the waste flow. In any case, bidders should be given full information on the wastewater and treatment to be provided ahead of filtration plus the opportunity of testing effluents from such treatment where already in operation.

Filter units generally consist of a containing vessel, the granular media, structures to support or retain the media, distribution and collection devices for influent, effluent and washwater flows, supplemental cleaning devices, and necessary controls for flows, water levels, or pressures. Some of the more significant alternatives in filter layout follow.

Most filter designs employ a static bed with vertical flow either downward or upward through the bed. The downflow designs traditionally used in potable water treatment (Figures 8.8 and 8.9A, D, and E) are most common but recently a number of installations have been designed for upward flow (Figures 8.10 and 8.9B). The European biflow design (Figure 8.9C) employs both flow directions with the effluent withdrawn from the interior of the bed. Upflow washing is used regardless of the operating flow direction. Two special filter designs employ horizontal radial flow through an annular bed. Media is cycled downward through the bed, withdrawn at the bottom, externally washed, and returned to the top.

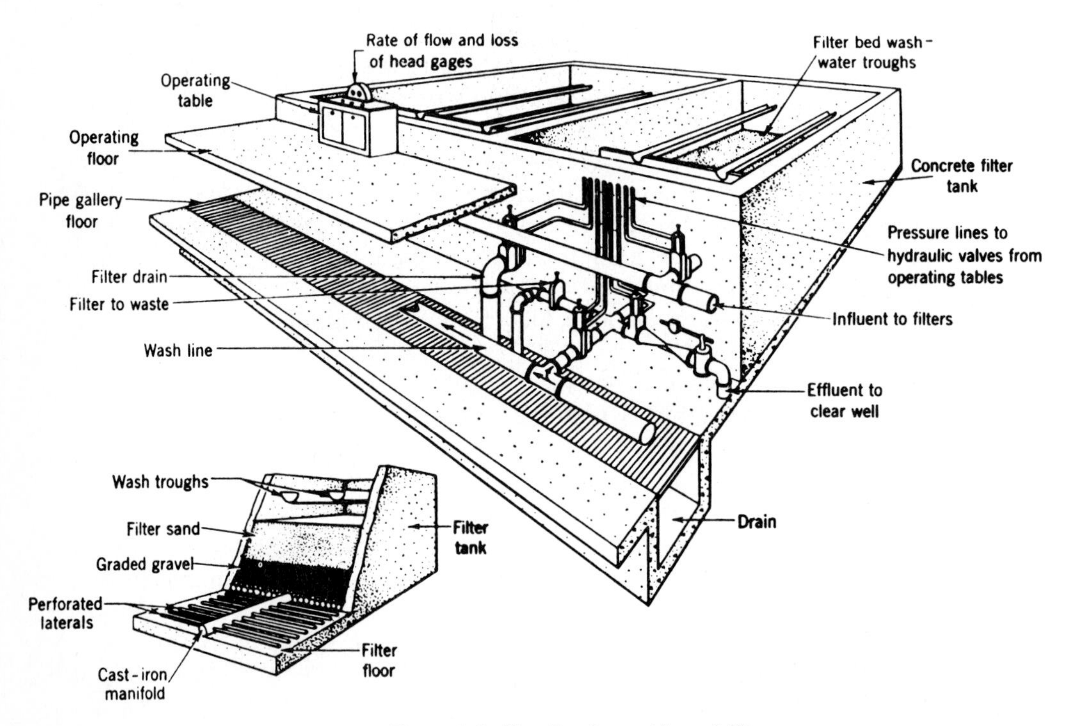

Figure 8.8 Details of a rapid sand filter.

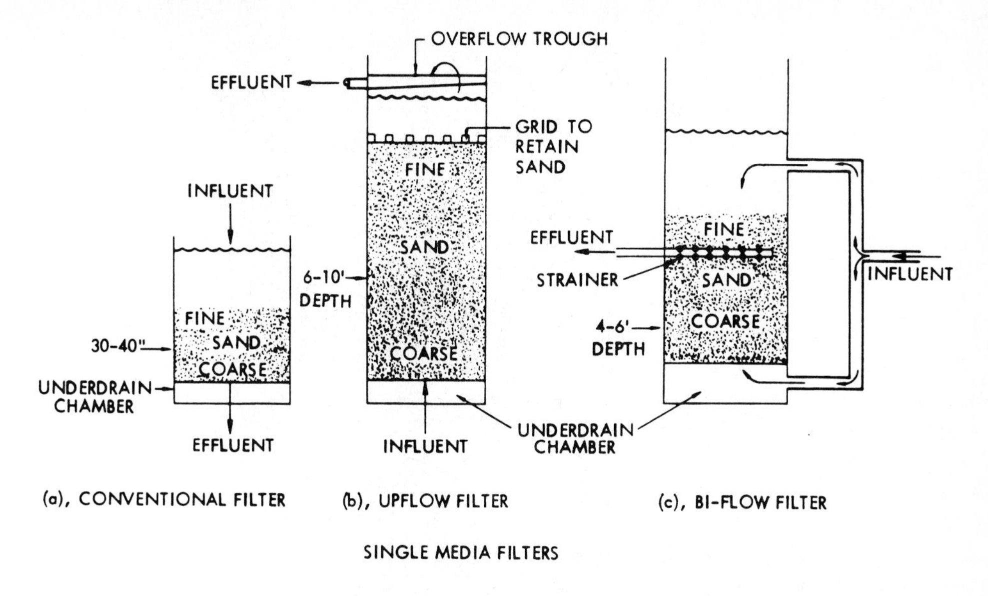

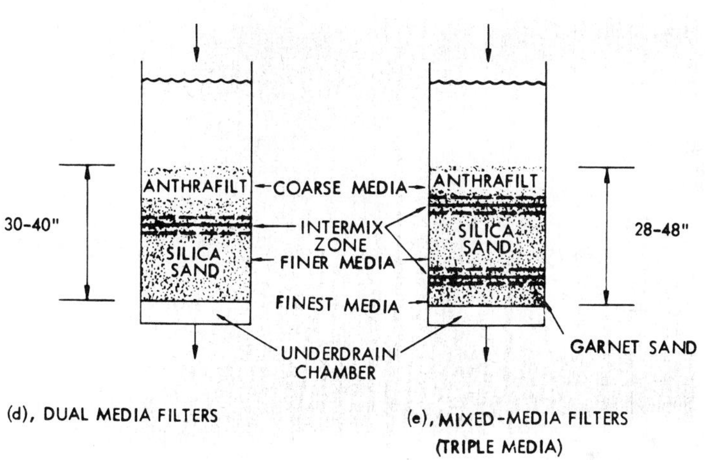

Figure 8.9 Typical filter configurations

Filters may be designed with closed vessels permitting influent pressures above atmospheric (Figure 8.11) or with open vessels where only the hydrostatic pressure over the bed is available to overcome filter head losses. Pressure units are generally preferable where high terminal head losses are expected or where the additional head will permit flow to pass through downstream units without repumping. They are most commonly used in small-to-medium-sized treatment plants where steel-shell package units are economical.

Figure 8.9 shows schematically a number of different filter configurations using fixed-bed media. The beds shown are all graded during upflow

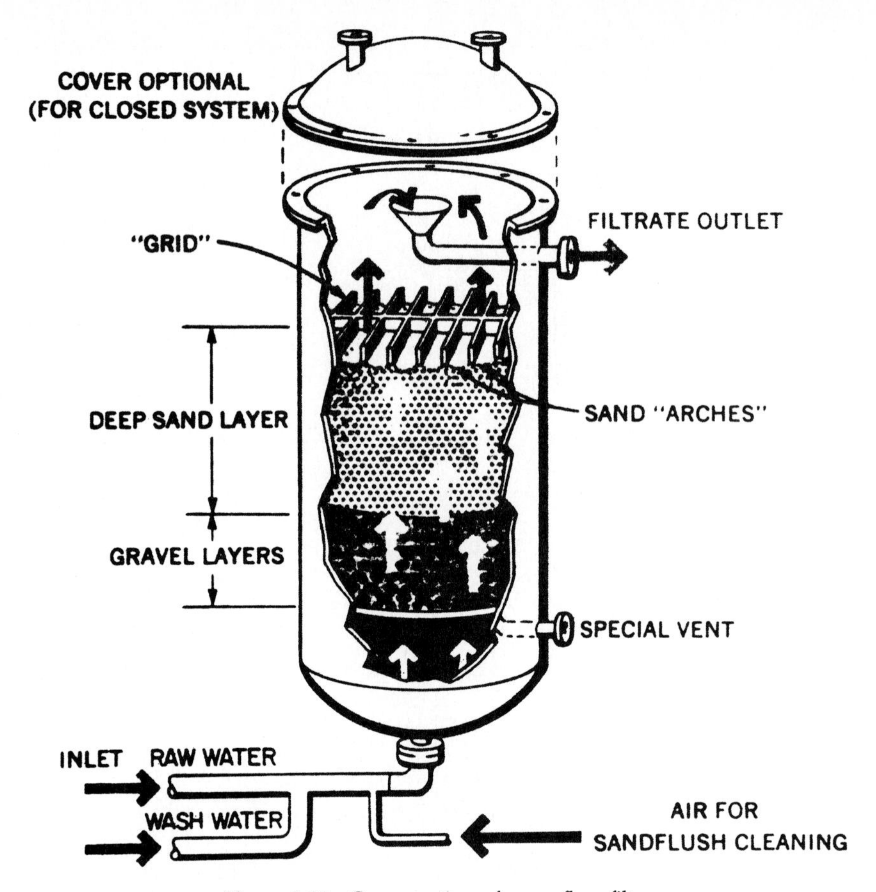

Figure 8.10 Cross-section of an upflow filter.

washing so that the finer material of a given specific gravity is on top. It should be noted that the conventional single-media filter used in potable water treatment (Figure 8.9A) is generally unsatisfactory for wastewater treatment because the wastewater solids cause a high head-loss buildup at the fine surface layer.

In upflow designs, flow passes first through the coarser media which for a given head-loss buildup has greater capacity for retaining filtered solids. This is advantageous in lengthening filter runs and increasing output. Dual and multi-media (Figure 8.9D and E) obtain the same effect under downflow operation by placing coarser layers of lighter material over finer denser material. An alternative downflow single-media configuration not shown attempts to get the same advantage from use of beds of uniform-sized coarse media with depths of 60 in. or more. The effects of significant media characteristics such as size gradation, specific gravity, and depth on filter performance are discussed later.

In filters using external wash, the media is not vertically graded; particle-size distribution tends to be the same throughout the bed.

It is normal practice to design filters to operate on a batch basis with entire units taken out of service for cleaning according to schedule or as required. Several special designs, however, provide more or less continuous cleaning, either externally with media cycled through the bed, or in place with techniques such as traveling backwash or air pulsing of the bed and air mixing of the liquid above it.

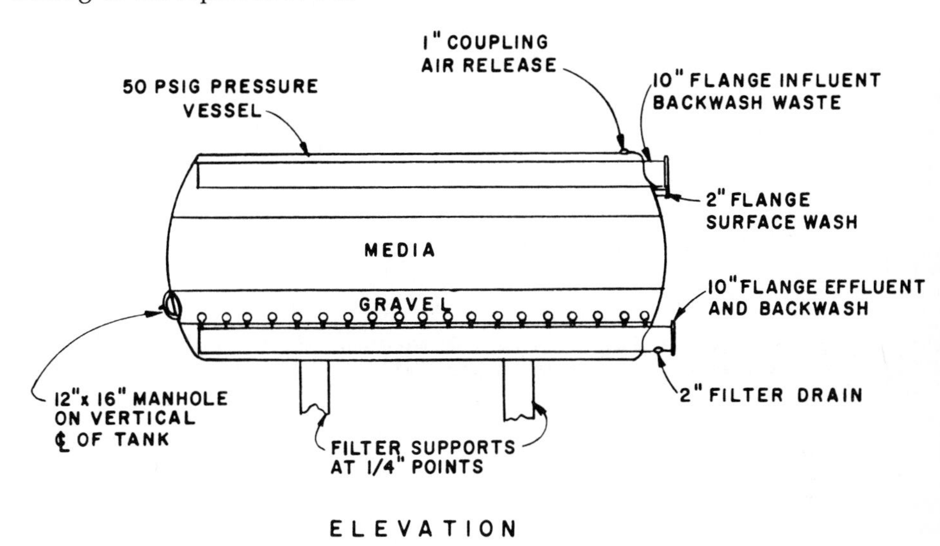

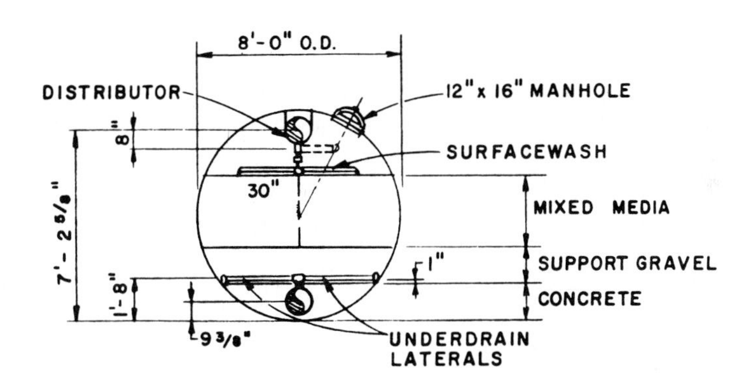

Figure 8.11 Shows a typical pressure filter.

Most of this section relates to fixed-bed systems with intermittent upflow washing. Both upflow and downflow designs will be included as normal design.

The measures of filter performance are output quality and quantity. The variables which determine or limit performance fall into two major groups: influent characteristics and the physical characteristics of the filter. The latter include media characteristics, filtration rate, available and applied operating head, and the design and operating parameters of the filter cleaning system.

In determining the fundamental limits on quality performance, the characteristics of prime importance are those of the influent solids: concentration, strength, size, and the physical-chemical properties governing adhesion of particles to each other or to the media surfaces. Commonly a number of filters with different physical characteristics can come close to the limiting quality performance for a given influent. In contrast, quality performance of given filters can vary widely for different solids characteristics.

In determining output quantity from filters, the influent solids characteristics—especially floc strength and solids concentration—are again very important, but the physical characteristics of the filter become significant too.

At run lengths of 24 hours or more, output depends almost totally on filter rate. As run lengths become shorter, however, the effects of downtime and washwater recycle during cleaning become increasingly important. The washwater recycle volume depends on the backwash flow rates and the wash cycle duration needed for adequate cleaning. Factors governing backwash system design include:

1. Size distribution, depth, and specific gravity of media
2. Nature of solids removed, principally their adhesion to the media and their tendency to compact in a dense layer at the medium surface
3. Type of supplementary cleaning provided

Run length may be limited either by available head or by deterioration of effluent quality as the filter bed becomes filled with solids (breakthrough). Which factor governs depends on the interaction of several variables including:

1. Influent solids characteristics (all those which affect quality performance)
2. Flow rate
3. Temperature and viscosity of the wastewater
4. Media characteristics
5. The amount of head available

Head loss in a clean bed varies directly with filter rate and inversely with grain size. In determining head-loss buildup, the most significant media characteristic is the grain or pore size at the influent surface of the bed (or in some cases within finer denser layers of multimedia filters). In downflow filtration through a graded bed, influent solids particles larger than about 7 percent of the minimum grain size will be removed by straining, provided their strength is sufficient to withstand the shear at the surface. Shear varies with filter rate and liquid viscosity.

In surface straining, head loss increases exponentially with time or solids accumulation. Where significant solids loads are removed predominantly by surface straining, head-loss buildup will be rapid, filter runs short, and backwash frequency high. In addition the solids removed at the surface tend to be compressed into a dense mat which is difficult to remove in backwashing.

Removal of solids within the bed rather than just at the surface is termed *depth filtration*. Both surface and depth filtration are usually involved to some degree in any given application.

In depth filtration head loss tends to build up linearly with time or with solids accumulation. Compression of the solids removed is limited by the granular structure of the bed. For downflow filtration within a single media, the farther solids penetrate into the bed, the slower will be the rate of head-loss buildup, but the sooner solids will breakthrough into the effluent.

The factors which determine breakthrough for a single media are the media size and depth, the flow rate, and the resistance of deposited materials to shear within the bed. One approach to characterizing the resistance of solids to breakthrough is by an index, K, calculated from the physical characteristics of the filter and the head loss at which breakthrough occurs. The expression for the index is:

$$K = Vd^3H/L$$

where

V = filtration rate (gpm/sq ft)
H = head loss at breakthrough (ft)
d = effective size of media (mm)
L = bed depth (ft)

The influent characteristics of prime importance in determining filter performance are those of the solids to be removed. The only significant characteristic of the wastewater liquid—as opposed to the solids—is viscosity which varies with temperature. Its effects on development of filter head loss are generally small, however, in comparison to the effects of solids accumulations or filter rates.

The characteristics of wastewater solids which govern or limit filter performance are determined by the treatment processes ahead of filtration. In direct filtration of secondary biological effluent, the residual solids applied to the filter are predominantly biological floc growth in the treatment process. In filtration of effluent following tertiary coagulation for phosphate removal, the residual solids are predominantly chemical flocs. In filtration of chemically precipitated raw wastewater or primary effluent, the solids consist of inorganic chemical floc with varying quantities of precipitated organics.

The only influent solids characteristic included in routine filter testing is the concentration, perhaps because it is the only one that is easily measured. A few special studies have attempted to take into account other characteristics such as floc strength, particle-size distribution (concentration versus size) and properties governing adhesion of particles to each other or to the media. Some other studies have tried to distinguish differences in filter performance according to parameters of the treatment prior to filtration.

Biological flocs tend to be significantly stronger or more resistant to shear than chemical flocs, at least those from alum or iron coagulants. Consequently, in filtering biological flocs, surface straining is generally significant and runs are almost always terminated by excessive head loss. Breakthrough is rarely observed. In one study head losses as high as 30 ft were applied without deterioration of effluent quality. This contrasts with alum and iron (hydroxide) flocks which have been shown to penetrate readily into filters and to breakthrough at relatively low heads ranging from 3 to 6 feet. In contrast to flocs from other common coagulants, calcium carbonate precipitates are strongly removed at the filter surface where they may form a dense compressed layer hard to remove during washing. Comparative data are lacking on the strength of flocs from precipitation of phosphates in wastewater using alum, iron, or lime. It is reasonable to assume, however, that they are similar to aluminum or ferric hydroxide floc.

Polymer filter aids may be added to the filter influent to strengthen weak chemical flocs, thereby permitting operation at higher rates without breakthrough. Doses of 0.1 mg/1 or less are often adequate. Polymers added as coagulant aids in upstream settling or flocculating units may similarly strengthen the residual floc applied to the filters. Ample head loss must be available to meet losses due to the tougher floc, and doses must be kept as low as possible to avoid excessive head loss.

Floc particle sizes in settled biological effluent tend to be bimodally distributed. Mean sizes for the two modes in one study were 3 to 5 microns and 80 to 90 microns. About half of the weight was in each mode. Theoretical work suggests that particles in the lower size range are much less effectively removed by filtration than those in the higher range. Hence, for the best quality performance from filtration, the proportion of smaller-sized particles must be reduced to a minimum by proper flocculation.

The filterability of residual solids from secondary settling varies with solids retention time and with liquid contact time in the biological process. For biological systems with higher solids retention times and longer liquid contact times, filtered effluents tend to have lower suspended solids. Expected performance of multimedia filters for plain filtration in secondary effluents are shown in Table 8.6.

It is significant that the solids in extended aeration effluents filter particularly well, in as much as they often settle poorly, leaving high concen-

TABLE 8.6 EXPECTED EFFLUENT-SUSPENDED SOLIDS FROM MULTIMEDIA FILTRATION OF SECONDARY EFFLUENT

Effluent type	Effluent SS (mg/l)
High-Rate Trickling Filter	10–20
2-Stage Trickling Filter	6–15
Contact Stabilization	6–15
Conventional Activated Sludge	3–10
Extended Aeration	1–5

trations in the secondary effluent. Sludges with high solids retention times lose their tendency to agglomerate into larger easily settleable particles, but increase in strength so that fewer are broken up into particles of a size not readily filtered.

While effluent quality reflects the solids which pass through the filters, head-loss development reflects the amount and location of solids which deposit in the bed. Both solids loading (solids concentration times flow rate) and filter efficiency are important in determining the buildup of head loss with increasing solids capture. Various studies relating head-loss buildup to solids capture show widely different results. This would be expected in view of the wide range of solids characteristics, media characteristics, and filter rates, and the very different head-loss patterns that result from surface and depth filtration. It is reasonable to expect the highest values of specific capture where the filter and influent solids characteristics permit depth filtration and extremely low values where they promote a high degree of surface straining.

Most wastewater filter designs employ media configurations and loadings which minimize surface straining and promote depth filtration. A few special designs with fine media are intended to remove solids primarily by surface filtration or straining. These designs include provisions for overcoming the adverse effects of rapid head-loss buildup. Where surface filtration predominates, the media characteristics have little effect on quality performance or head loss. In addition, removal of solids is quite independent of filter rate or influent solids concentration. Hence, the effects of physical characteristics of filters are discussed later only in relation to depth filtration, not surface straining.

The most important media characteristic in determining performance is size. Studies using uni-size media have clearly demonstrated that finer media have greater removal efficiency. Various investigators have related percent removal to powers of diameter ranging from -1 to -3. In finer media head loss per unit of removal (lb/cu in. of media) is also higher.

In a media graded from fine to coarse in the direction of flow, the highest solids concentration is applied to the layers with the greatest removal efficiency. As a result, removal is concentrated in a small depth with accompanying high head losses.

In media graded from coarse to fine in the direction of flow, substantial penetration occurs but most of the solids are removed in the coarser media where less head-loss buildup results. The finer layers, protected from heavy solids loadings, are available for polishing and to prevent breakthrough as the coarser layers become filled with solids.

Media depth is most significant in coarse uniform beds. Because of the uniformity, the efficiency of removal (as a percent of the solids applied to each depth) is nearly constant for all layers of the filter. Penetration is substantial and extra depth is relied upon for polishing and to retard breakthrough. Size and specific gravity of media together are significant in determining expansion during backwash and the degree of intermixing in multimedia beds.

The effect of filter rates on quality performance can vary widely depending on application. In filtering biological floc at reasonably low influent solids concentration, the effect on effluent quality of rates up to 10 gpm/sq ft is not very significant. With weaker chemical flocs or with high influent concentrations of biological solids (usually indicating poorly functioning biological treatment), filter effluent quality tends to degrade at filter rates above about 5 gpm/sq ft. Sudden changes in filter rates may affect effluent quality more adversely than sustained higher rates.

Higher filter rates tend to increase solids penetration. In cases where this significantly reduces surface removal, head-loss buildup per unit volume filtered may actually be less at higher rates. Existence of an optimum rate has been suggested as typical of combined surface and depth filtration. It has also been suggested that the advantages of using a coarse top media layer may be lost if the filter rate is not high enough to force solids into the bed and limit surface straining.

In addition to upflow washing, some form of auxiliary scouring of the media appears essential to adequately clean wastewater filters. If cleaning is inadequate, two serious problems will develop: filter bed cracking and mud ball formation. Cracks open in filter beds because of compression of excessively thick coatings on the filter grains. The resulting localized heavy penetration of solids may both lower effluent quality and contribute to mud ball formation.

Mud balls are compressed masses of filtered solids large and dense enough to remain in the bed during backwashing. If conditions favoring their formation persist, mud balls tend to increase in size and to sink deeper in the bed. Their presence increases head loss and may lead to loss of effluent quality.

Both air scrubbing and surface or internal water jets have been used for auxiliary scouring of the media. Air injected below the media produces shear as the bubbles rise through the bed. Water jets, positioned at the top of the expanded bed, produce high shear around the surface media, which is the most heavily loaded with solids. In multi-media beds, jets may be similarly provided at the expanded height of the media interface.

The main upflow wash and the auxiliary scouring systems should be controlled independently to permit use together or separately. The key parameters for design of the cleaning system are the upflow wash rate capacity and the air scour rate of surface wash rate capacity. Typically, upflow wash rates are about 20 gpm/sq ft. The maximum capacity is selected to provide the desired degree of fluidization and expansion of the media under critical high temperatures. Capacities for auxiliary scouring are generally established empirically. Air scour rates typically range from 3 to 5 scfm/sq ft, and surface wash rates from 1 to 3 gpm/sq ft.

Given adequate information on performance, the filter rate and terminal head loss for a particular media design should be selected by making economic trade-offs between filter size, operating head requirements, and run length, all within the limits dictated by effluent quality requirements. This section outlines procedures for such trade-offs and provides an alternative basis for selection where specific performance information is lacking.

Adequate information for making economic trade-offs can be obtained only from pilot studies of the specific media application. Pilot studies should indicate the buildup of head loss with time for various filter rates and for average and peak influent solids concentrations. Results may be indicated in a form similar to Figure 8.12. With this information it is possible to estimate the filter run length, the net production, and the capital and operating costs of the filter for the given influent solids concentrations and for different combinations of filter rate and terminal head loss.

In determining net production, allowances must be made for downtime during cleaning and for recycle of washwater through the treatment plant. The downtime effects are calculated from the cleaning frequency, cleaning cycle duration, and the number of individual filters. Washwater recycle effects are calculated from the cleaning frequency and the backwash rate and duration. Washwater recycle has no effect on net production if filter influent is used for washing. Net production may be expressed as volume (filter rate $\times$ run length) or as an average rate (gpm/sq ft) over one filter cycle (run length plus cleaning time). The net production rate is almost the same as the filter rate for runs of 24 hours or more. For run lengths below 10 to 12 hours the differences become significant; below 6 to 8 hours the effect on production may be critical.

Short-term peak loadings due to downtime or recycle during backwash need not be considered directly in economic trade-offs. After the design filtration rate and terminal head loss are determined, however, the design should be checked to assure that it can accommodate these peaks within the available head loss and effluent quality limits. If not, peak effects should be reduced or eliminated by increasing the number of filter units or by providing equalizing storage for the backwash and wastewater flows.

The design should also be checked for its ability to handle the sustained peak loads imposed when a unit is taken out of service for repairs. If the resultant shorter run lengths do not provide enough capacity, the design may

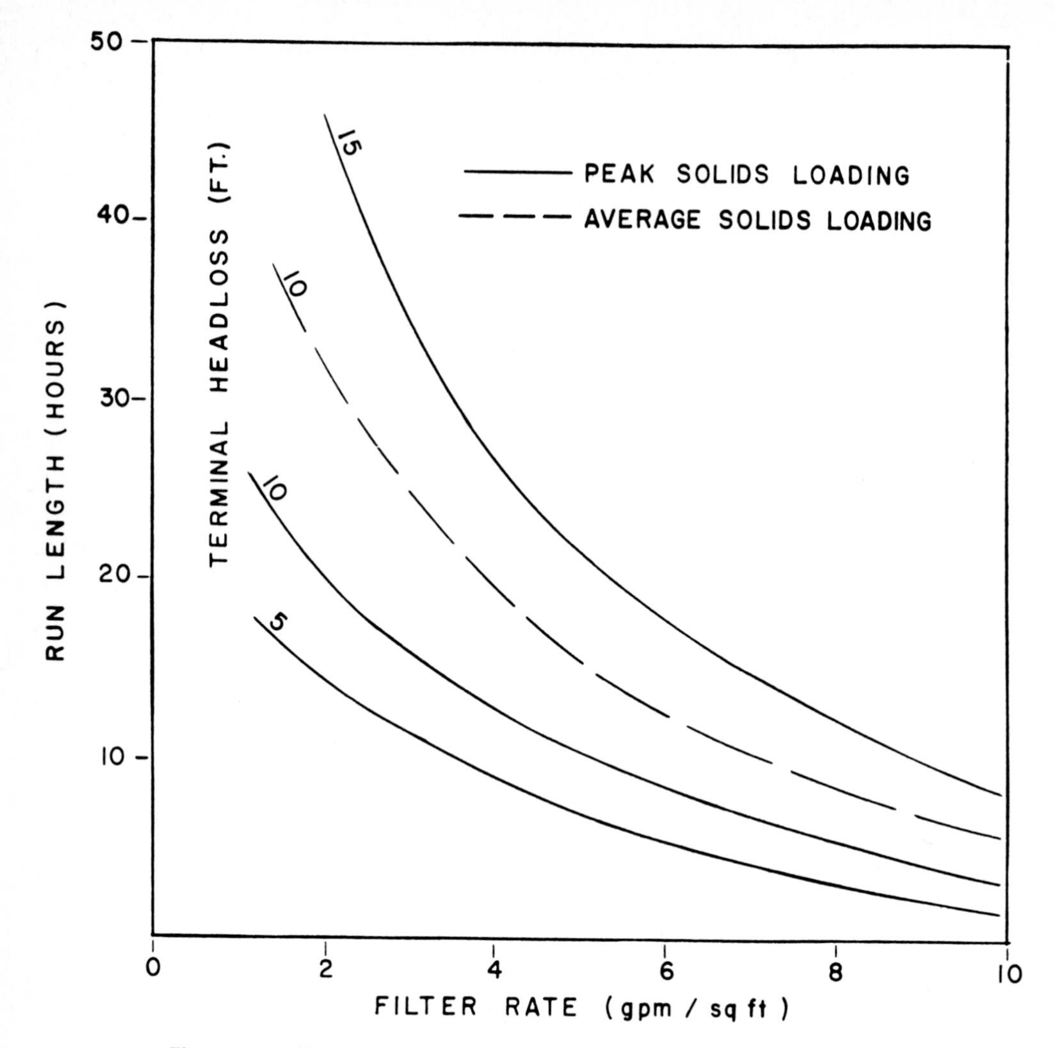

Figure 8.12 Illustrates run length vs filter rate for different terminal head losses.

be revised as follows: peak hydraulic loadings should first be reduced by increasing the number of filters keeping the total area the same. If this reduction is not sufficient, the area should be increased beyond that determined in the original design.

Before cost trade-offs are made, the following must be defined:

1. Maximum flows and solids loadings for various durations up to 24 hours. A tentative decision is required on the use of equalization to limit maximum wastewater flows.
2. Run-length limits. The lower limit should be 6 to 8 hours to maintain

reasonable net production. The upper limit should be 36 to 48 hours to avoid anaerobic decomposition of solids in the filter.

3. Head-loss limits. For gravity filters allowable head losses generally are blow 10 feet. Use of heads much above this commits the design to pressure filters. Use of pressure filters would be favored where pressurized discharge to following facilities is needed. Gravity filters would be favored where the extra head for pressure filters would require intermediate pumping, but head for the gravity units is available without such pumping.
4. Backwash design and expected cost per cycle. Labor costs should reflect whether the operation is to be automated. Backwashing costs should include costs of treating recycled backwash in units ahead of the filters, and the recycled flow should be deducted in determining the net production of the filters.
5. Space limitations. These may force use of higher filter rates.
6. Number of filter units. This should be tentatively selected to facilitate cost estimates, but may be varied with little effect on the trade-off calculations provided labor is not a major factor in the operating cost per backwash. For reliability and economy, a minimum of four to six units should generally be provided, with at least two in even the smallest installations. Above these minimums, the number of filter units depends on the actual size of individual units. The practical maximum size of gravity filters is about 800 sq ft.

In addition to limits indicated previously, pilot testing may reveal: (1) upper limits on head loss or rate required to avoid solids breakthrough and effluent quality deterioration, and (2) an optimum filter rate for minimizing head-loss buildup. No filter rates lower than the optimum should be considered in the trade-offs.

The following procedures are suggested for determining the most cost-effective filter sizing, design terminal head loss, and run length. Figure 8.13, relating net production to filter rate and run length, has been prepared to facilitate the analysis. The figure should be modified before application if backwashing conditions differ significantly from those assumed in its development.

1. From filter test data for average and maximum design influent solids concentration, prepare a head-loss development plot (see example plot Figure 8.12).
2. Assume initial trial values for terminal head loss and filter run length. (See Step 12).
3. For the assumed terminal head loss and run length determine the filter rate from the head-loss development plot for maximum solids concentration.

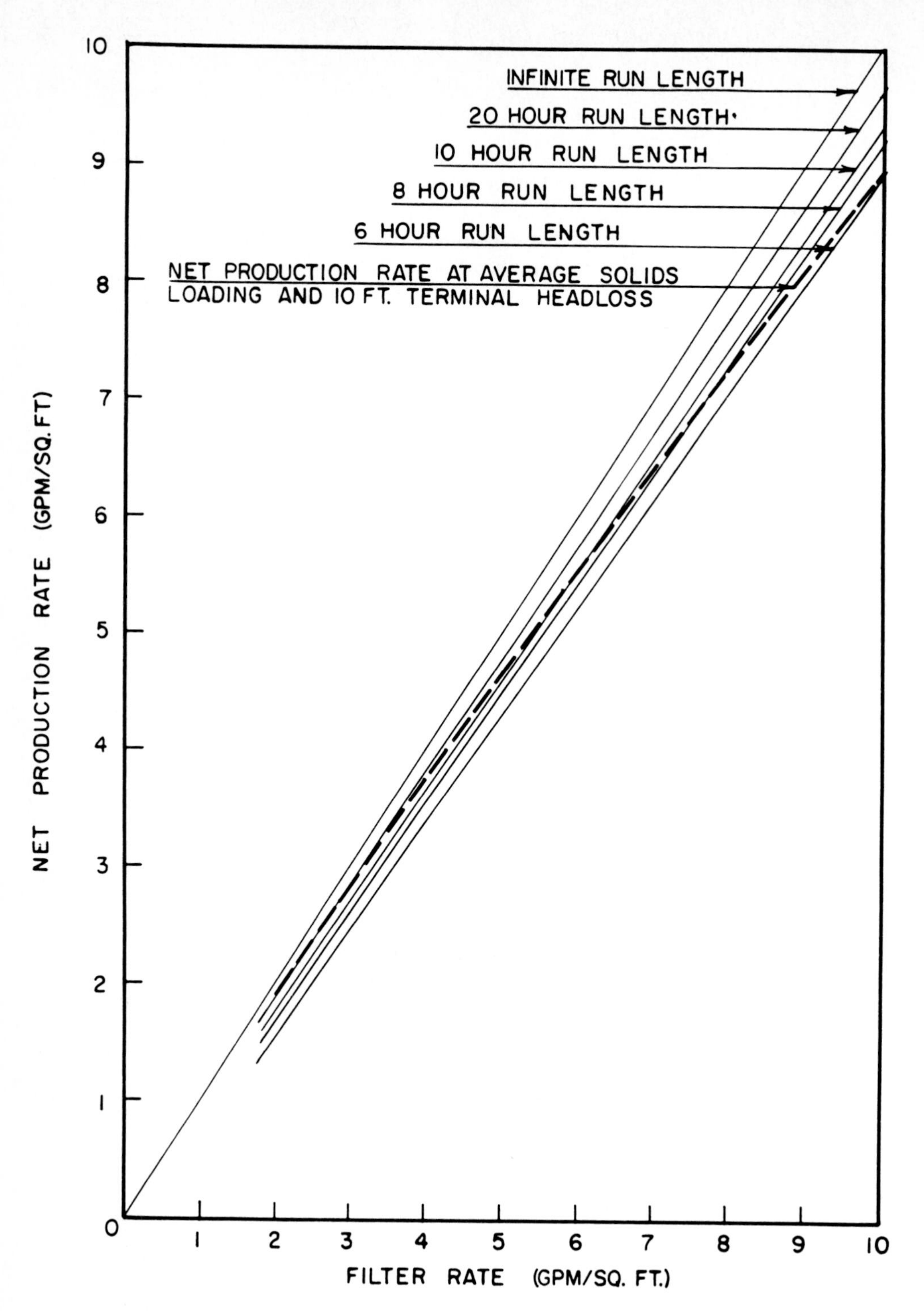

Figure 8.13 Illustrates net production rate vs filter rate for different run lengths.

4. For this filter rate and the assumed run length determine the net production rate from Figure 8.12.
5. Determine filter sizing based on this net production rate and the maximum design flow for a duration equal to the filter cycle time (run length plus downtime for cleaning).
6. Estimate capital costs for filters based on previous sizing and the design terminal head loss.
7. Determine average net production by dividing average flow by filter area.
8. Construct a plot of net production versus filter rate based on run lengths to reach the trial value of terminal head loss at various filter rates with average solids concentrations. (See Figure 8.13).
9. From the plot in step 8, determine filter rate and run length to provide average net production.
10. Calculate operating costs based on the assumed terminal head loss and the run length for average flow and solids loading.
11. Convert operating costs to present worth and add to capital cost to determine total present worth.
12. Repeat preceding analysis assuming different values for terminal head loss and filter runs. The objective is to find assumptions which minimize present worth, within technological constraints.

It is suggested that a conservative initial value of 8 ft be assumed for terminal head loss with a run length of at least 8 hours at maximum solids concentrations. Subsequent trials would explore use of higher head-loss values to permit either longer runs or higher filter rates, whichever appears more advantageous. Judgment must be applied to minimize amount of calculation required.

Where it is impossible to test proposed filter media on the actual influent, guidance may be obtained from results with the same media treating similar influents. In the absence of specifically applicable test results, filter rates, and head-loss allowances should be very conservatively selected, based on ample estimates of influent solids concentrations. To assure adequate capacity it is suggested that, as a minimum, sufficient filter area be provided to handle the 24-hour design flow at 4 gpm/sq ft or the 4-hour maximum design flow at 6 gpm/sq ft, whichever is more stringent. For predominantly chemical floc, the surface media should be no finer than 1 mm and allowance should be made for a terminal head loss of 10 feet. For filtration of biological solids in secondary effluent, the following procedures are suggested in selecting terminal head loss and final filter sizing:

1. For the minimum filter area as determined earlier, estimate head-loss

buildup based on expected solids removals and the following values of specific capture:

Minimum media size at influent surface (mm)	Specific capture (lb of solids removed/sq ft/ft if head-loss increase)
1.8	0.07
1.3	0.035

2. Avoid use of any finer surface media. Surface media coarser than 1.8 mm may permit higher specific captures, but problems of adequate cleaning must be considered.
3. For the minimum filter area, calculate the required head for 24-hour run length at average solids loading and for 8-hour run length at maximum (8-hours) solids loadings.
4. Provide for terminal head loss on the more critical basis determined earlier, or use more than the minimum filter size and recalculate solids loadings and head-loss requirements.

Designs based on the preceding criteria should be as flexible as possible to permit use of higher rates or lower heads if operating experience shows this is possible. Flexibility to increase rates is most valuable where capacity is to be increased in future stages. Flexibility in pumping and control systems will permit head to be reduced to what proves necessary in actual operation.

Filtration Media

Media commonly used in water and wastewater filtration include silica sand (sp gr 2.65), anthracite coal (sp gr 1.4 to 1.6) and in special multimedia designs garnet (sp gr 4.2) or ilmenite (sp gr 4.5).

As they occur in nature these materials are not of uniform size but instead typically have a grain-size, distribution such as that shown in Figure 8.14. Natural grain-size distributions frequently are close to geometrically normal, i.e., plot as a straight line on log probability paper. As shown in Figure 8.14, grain-size distributions are often characterized by two points, the 10 percent and 60 percent size (d_{10} and d_{60}). These are sizes such that the weight of all smaller particles constitutes respectively 10 percent or 60 percent of the whole. Media is frequently specified in terms of effective size (d_{10}) and the uniformity coefficient (d_{60}/d_{10}).

It is possible to change the characteristics of a given media material by removing certain size fractions. Coarser fractions may be seived out while finer fractions may be removed by *scalping* (removing surface layers) after hydraulically grading material during upflow washing.

The most important modification for most media is to remove any very fine particles—say those less than 80 percent of the effective size. Such fine

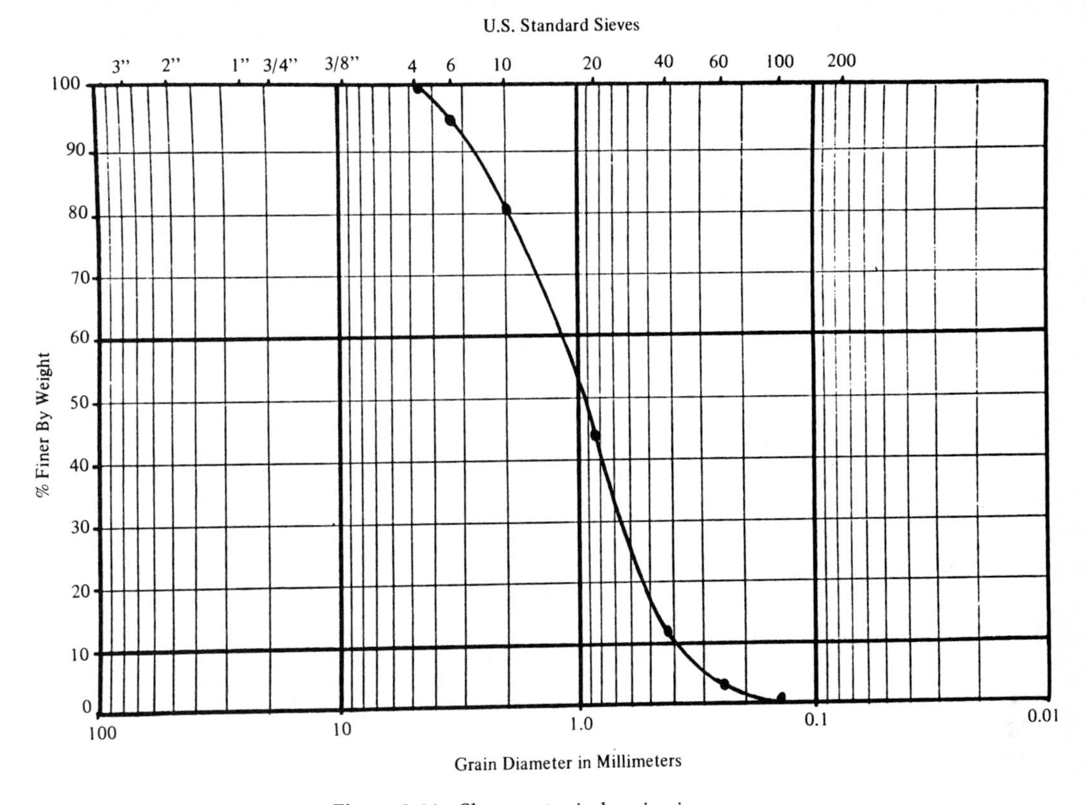

Figure 8.14 Shows a typical grain size curve.

material never constitutes more than a small fraction of the media volume but, if not removed, may cause head losses far greater than would be expected for the given effective size.

Dual and Multi Media

Upflow washing stratifies a bed in accordance with the settling velocities of the media particles as determined by their size, shape, and specific gravity. In a dual or tri-media bed, although each media component is still graded fine to coarse in a downward direction, lighter coarser media can be maintained above finer denser media. This makes it possible to approximate a coarse-to-fine gradation in down-flow filtration units. Another advantage of dual or multimedia over a single medium is that mud balls formed in the filter remain above the coal-sand interface where they are subject to auxiliary scrubbing action.

The maximum settling velocities of media particles also determine the minimum wash rate required for adequate fluidization of the bed during backwash. Hence, for a given media size at the top of the bed, lower wash rates can be used if each media component is more uniform and the top portion of the filtration is of anthracite rather than a heavier material. Dual or multimedia should be sized so the coarsest (d_{90}) sizes of each component have about the same minimum fluidization velocity.

The hydraulic behavior and filtration performance of any given media are more properly related to pore size than to grain size. For single-media component, pore size is directly proportional to grain size, and the porosity (percent of volume represented by pores) is a constant depending only on media shape. Coal which tends to be angular has a porosity of almost 0.5, whereas sand porosity is closer to 0.4. In water treatment applications, coal media, because of its greater porosity, has been found to give poorer removals but lower pressure losses than sand of the same grain size.

The pore size in multicomponent filter media depends on the degree of intermixing of the components. With no mixing, pore-size distribution simply follows that of the components. With intermixing, however, the finer layers of the denser material below are dispersed into the voids of the coarser layers of lighter material above. No precise methods have as yet been demonstrated for calculating actual pore size, or even the degree of intermixing, from the characteristics of media components. Where such information is of interest, it may be obtained from test columns or from experience with specific combinations of components in other installations. Limiting size ratios have been proposed to control intermixing and to avoid the extreme where lower-density coarse media is overtopped by very fine high-density media.

In tri-media designs the overall depths and the minimum depths of anthracite and of the combined finer media are in the same range as in dual-media designs. The single-media configurations employ depths of 60 in. or more. In downflow filtration this great depth is intended to improve efficiency, while in upflow units it has an additional purpose of adding weight to restrain the bed from uplift due to differential pressures during operation. Where uplift exceeds the submerged weight of the media, it will either fluidize the bed or lift it in a "prison" effect (small-diameter filters).

Some of the theoretical advantage of upflow coarse-to-fine filtration is lost because minimum grain sizes must be coarse enough to avoid excessive uplift.

Additional resistance to uplift is provided in many upflow designs by placing a restraining grid on top of the media. The spacing between bars of the grid must be large enough to prevent upward bed movement during filtration. Although these two requirements appear contradictory, arching of the grains takes place between the bars, allowing a reasonably large spacing, in the range of 100 to 150 times the diameter of the smallest grain size in the beds.

Pilot testing is indispensable to provide the information necessary for meaningful comparison of different media designs or to assure the effluent quality performance of any media design selected. Without pilot testing, the designer should select a media which, on the basis of experience with similar influents, may be expected to provide good solids removal with low head-loss buildup.

Pilot testing to guide media selection should define head-loss development versus time for each media design, under all test conditions. If one media design clearly gives lower head buildup at all times and under all test conditions, it may be selected directly provided its backwash requirements are not extraordinary. If different media provide essentially the same head-loss development over the range of test conditions, selection may be based on other factors. Where different media appear significantly better under different conditions, selection should be based on cost comparison of the alternative designs each at its most favorable rate, terminal head loss, and run length determined. Significant differences in backwash flows should be taken into account.

Control Systems

Major filter functions requiring monitoring and/or control are:

1. Head loss
2. Effluent quality
3. Initiation of backwash where automatic
4. Flow rate through the filters
5. Backwash sequence, rate, and duration

An important tool in performance control is the automatic turbidimeter which can continuously monitor the filter feed and product. This allows the operator to anticipate difficulties from changes in feed quality, and rapidly remedy process failures. In addition, these devices allow the operator to rapidly evaluate the effects of changes in process variables and provide a continuous record of plant performance. All turbidimeters operate on the principle of measurement of scattered or transmitted light. A variety of commercial instruments are available.

Filter installations should be equipped with appropriate loss of head and flow indicators. Individual filters should have multiple taps for pressure readings if full-scale experimental testing is desired.

Provisions for automatic or remote initiation of backwash by timers or based on head loss or turbidity monitoring may be justified to reduce the need for operating attention.

Three types of flow control systems are used for filters:

1. Effluent rate control
2. Influent flow splitting
3. Variable declining rate control

Features of these systems and of automatic backwash systems follow.

An effluent rate control system is common in traditional water treatment plant designs; maintaining a set flow for each filter by throttling the effluent. The throttle valve may be controlled directly by mechanical linkage to a venturi controller or indirectly by a set point controller linked to a pneumatic or hydraulic valve operator. The direct acting system is unsuitable if flows to individual filters must vary over the day. The indirect system is complex and both may be troublesome in maintenance. The system is also wasteful of head since available head not needed in a clean filter is lost in the controller. In addition, control valves may produce high-frequency surges in the filter bed with accompanying loss of efficiency.

Influent Flow Splitting

In this system flow is evenly divided among filters in a splitter box located at or above the level of the top of the filter boxes (see Figure 8.15A). The boxes themselves are made deep so that the water level in them can build up to provide the maximum operating head needed when the filter bed is dirty. A weir on the filter outlet maintains a constant back pressure or minimum water level to prevent accidentally dewatering of the bed.

Advantages of influent flow splitting include:

1. Rate controllers with attendant maintenance and surging problems are eliminated.
2. Flow variations are distributed to filters automatically.
3. Head loss may be read directly from water levels in filter boxes.
4. Only a single master flow meter is needed.
5. Changes in filter rate are gradual because of time required for head to build up in filter boxes.

Disadvantages are:

1. The head not needed for filter operation is lost in the drop between the splitter and the filters.
2. Capital cost of filter box construction is increased by the greater depth.

Declining Rate Filtration

This system requires multiple filters. All operate under the same head but at different flow rates depending on the degree of clogging. Under constant head the output from a single filter declines as the run progresses. The filter selected to be backwashed is always the one which has been on line the longest and is most clogged. Total output from all filters is controlled by varying the head applied. Figure 8.15B shows a variable declining rate filter.

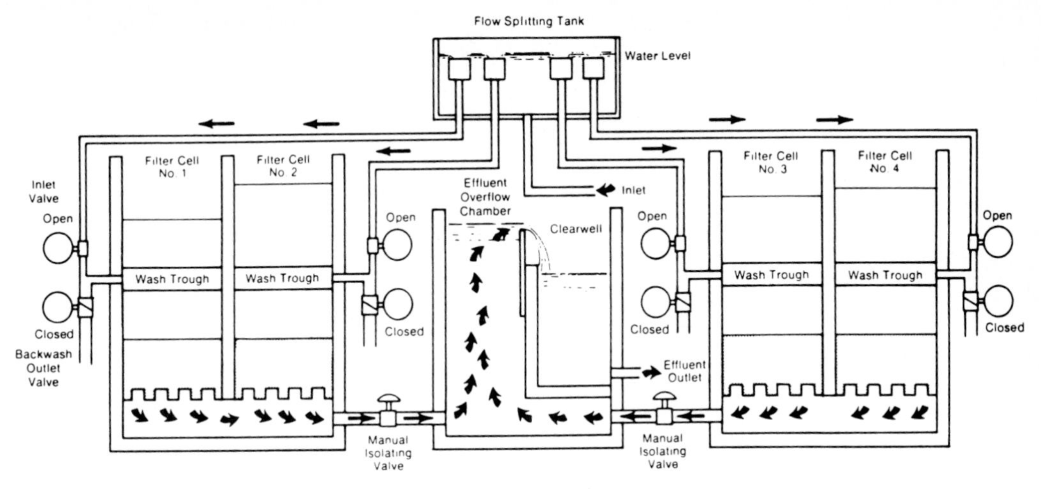

A. INFLUENT FLOW SPLITTING

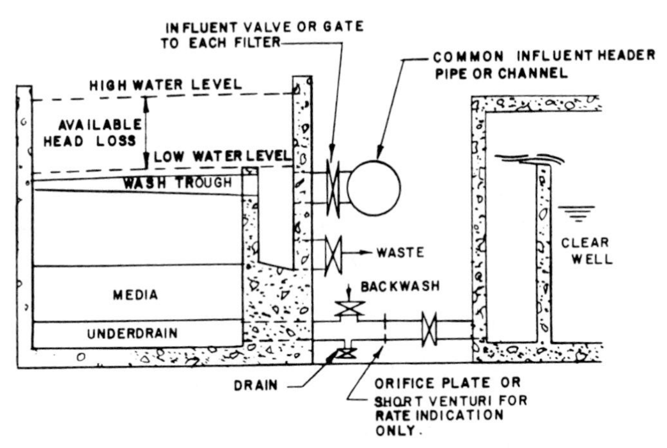

B. VARIABLE DECLINING RATE FILTRATION (1)

Figure 8.15 Illustrates common flow control systems.

The head on the filters may be controlled by varying either the upstream or downstream water level. With downstream water level control, an equalizing chamber must be provided to limit the rate of change of head and hence of flow, when filters are taken off line or restored to service. It is common to apply maximum design loadings to the filters as a group and to limit maximum rates on individual clean filters to from 20 percent to 40 percent above these design loadings.

Advantages cited for declining rate filtration include better effluent quality, absence of surges, and significantly lower total head requirements. Less head is needed because:

1. There is no loss due to throttling or due to free fall after flow splitting.
2. As rate declines turbulent head losses (underdrains, valves, and so on) reduce rapidly (in proportion to second power of flow), making head available to overcome resistance of clogged filter (proportional to flow).

For proper operating control, flow rates should be measured individually for each declining rate filter. Only single indicators are needed for inlet and outlet levels on head loss, since these are the same for all filters.

The chief disadvantage of this method of flow control is the need for a large volume of water storage upstream of the filter.

Backwash Control

Programmed backwash systems are widely used. Such systems consist of interlocked controllers and timers programmed to open and close valves, make or break siphons, start and stop pumps and blowers, and limit backwash flows to control the rate, duration, and sequence of activities during backwash. Even where backwash is manually initiated, the rest of the control system may be entirely automatic.

FILTER CLEANING SYSTEMS

Accumulated solids are removed from filters by a rapid upflow of washwater. The waste flow is then recycled to some prior treatment unit, usually primary settling. Washwater sources may include filter influent, filter effluent, or effluent from subsequent treatment units. Storage of washwater supply may be needed if rates required exceed the flow available. Recycled spent washwater flows should be equalized by storage so they do not disrupt prior treatment processes. Backwash rates for most effective cleaning vary with media size and density.

Figures 8.16 and 8.17 may be used to determine minimum upflow velocities or wash rates to fluidize coal, sand, and garnet media of various sizes. Rates should be variable to compensate for changes in temperature, viscosity, and hence bed expansion. Maximum hydrodynamic shear and most efficient cleaning have been shown to occur when the porosity of the expanded media is about 0.7. To reach this porosity in the surface layers of a nonuniform sand (effective size = 0.4, uniformity coefficient = 1.47) requires almost 50 percent expansion. Because of its higher unexpanded porosity coal requires only 20 percent to 25 percent expansion to reach a porosity of 0.7 in the surface layers. In practice the backwash rates generally used in filter designs range up to 20 gpm/sq ft, and do not provide more than 15 percent to 30 percent expansion except for very fine media. This means that higher backwash durations (5 to 10 or at the extreme 15 minutes) and somewhat higher washwater con-

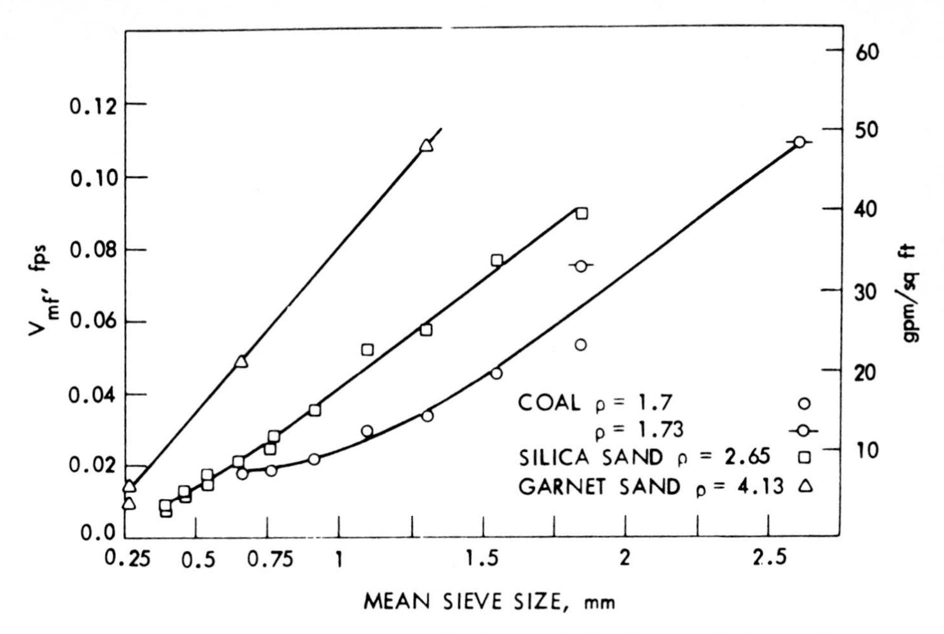

Figure 8.16 Shows minimum fluidization velocity for 10 percent bed expansion.

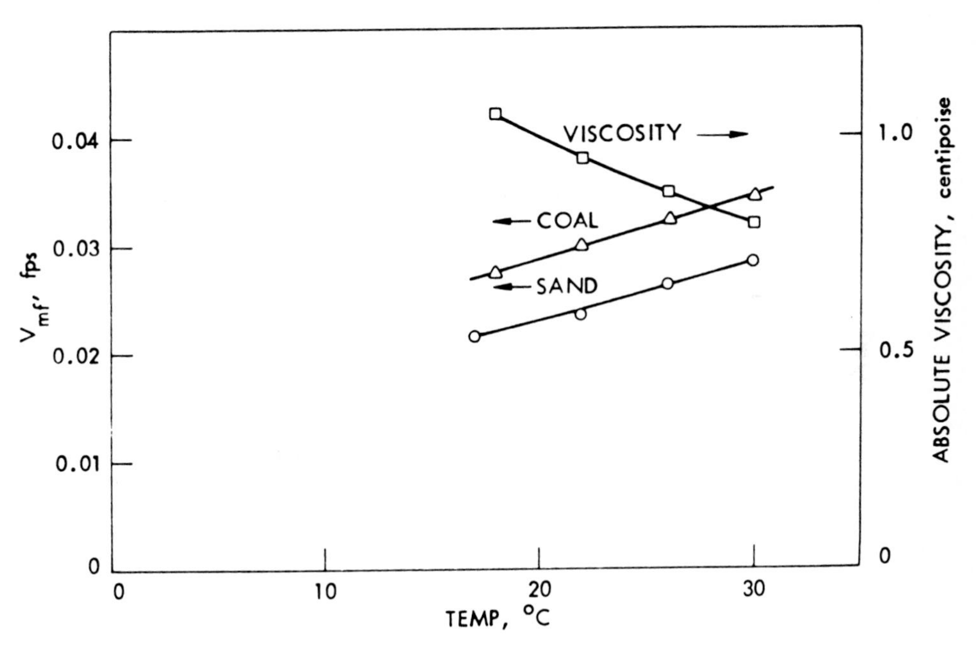

Figure 8.17 Shows effect of temperature on minimum fluidization velocity for sand and coal, and absolute viscosity of water.

sumption are required than at rates which would provide the most efficient washing.

Whatever wash rate and duration are expected, design of piping, valves, pumps, and storage tanks should provide extra capacity of at least 25 percent.

Methods are available for predicting expansion of sand beds accurately, but have not yet been satisfactorily extended to other media components. Expansion of multicomponent media can be rapidly obtained, however, from backwash tests in pilot columns.

Underdrains should distribute washwater as uniformly as possible over the area of the filter. Excessive variation in washwater rate results in uneven and ineffective cleaning. Moreover, the accompanying excessive jet action can lead to lateral displacement of gravel and clogging of the underdrains with filter media.

In general, underdrain systems developed for water filtration may also be used in wastewater applications. One of the first systems employed in later filtration consists of several layers of graded gravel surrounding manifold piping positioned on the filter floor. Orifices in the manifold piping provide preliminary distribution of the washwater as shown in Figure 8.18A. The final distribution is accomplished as the water moves upward through the gravel.

Air scour systems have been increasingly used in an attempt to reduce washwater requirements and to effect cleaning of the deeper portions of the bed. Some concern has been expressed concerning loss of finer lighter media particles when air washing is used. Where this is a problem, air scour should be applied separately from the backwash, with liquid in the filter box drawn down below the washwater overflow level so that no overflow occurs during air wash. Allowance must be made for 6 to 9 in. of water level rise due to air lift. Although some of the lighter media may remain on the surface of the water and subsequently be lost, the rate of such loss should generally be negligible. To prevent air scour from disrupting gravel placement, air is usually injected through a grid above the underdrain gravel. It may go directly into the underdrains where no gravel is used.

The cleaning cycle time (total downtime during one cleaning operation) includes time for valve openings and closings, time for drainage of inflow from the filter, and time for the actual upflow washing and auxiliary cleaning. Unless influent in the filter box is to be wasted, drainage time should be calculated at normal filter rates. Valve openings and closings to start and stop backwash and air scour should be gradual to keep from upsetting the media gradation and structure.

A typical sequence for backwash with auxiliary surface wash is:

1. Shut influent and permit water level to drain down to top of the washwater troughs or other washwater control weir.
2. Apply surface wash for 1 to 3 minutes.

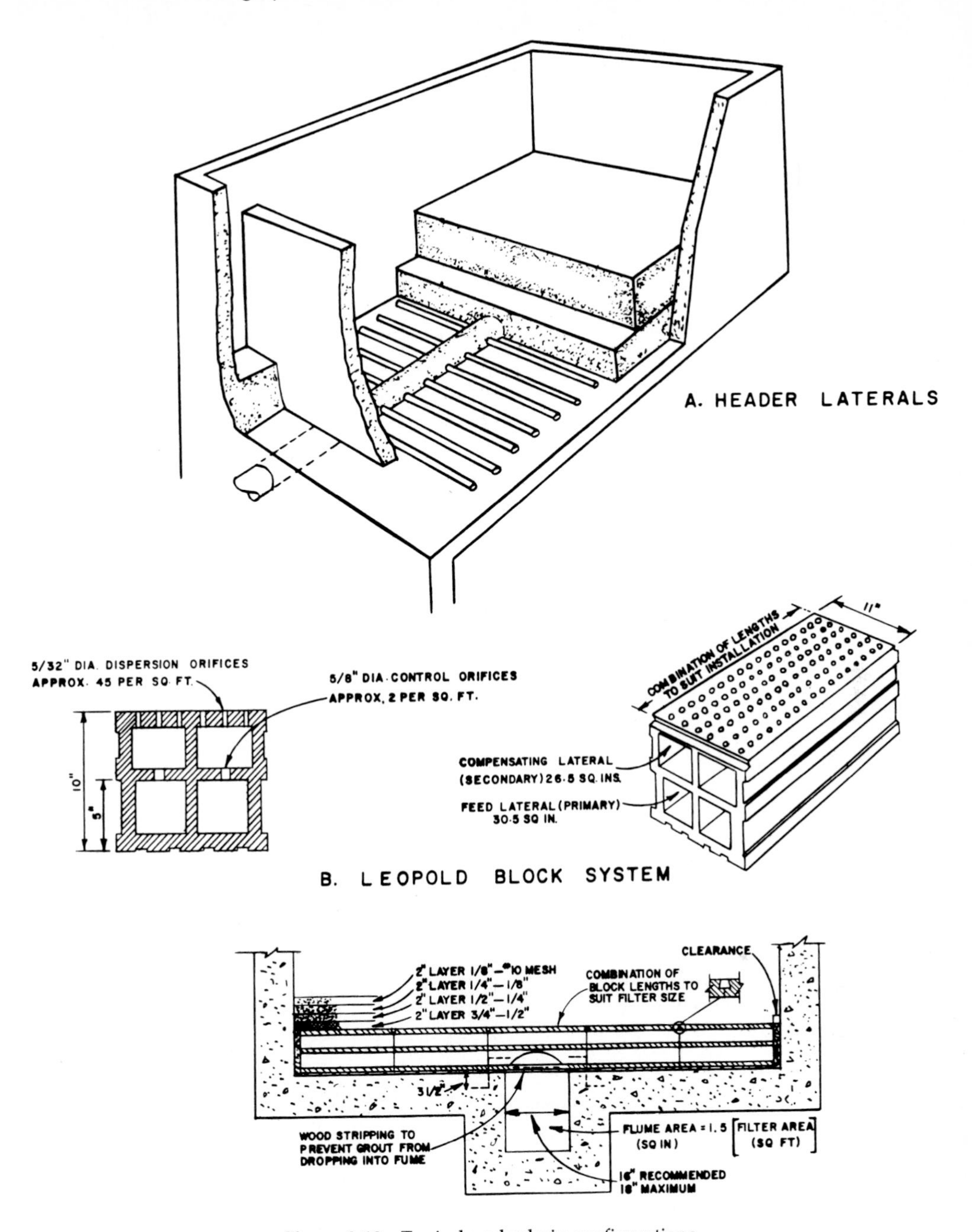

Figure 8.18 Typical underdrain configurations.

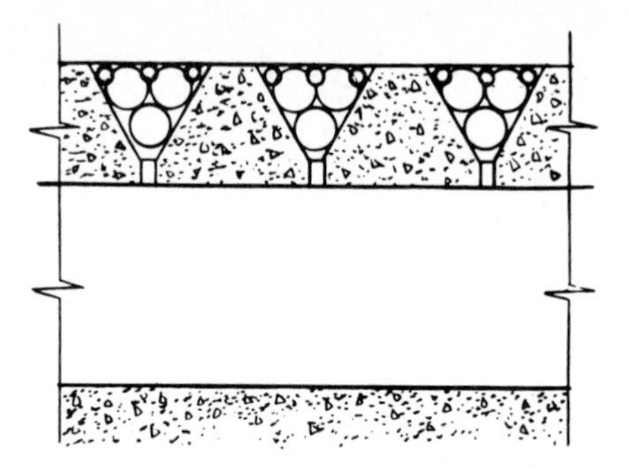

C. WHEELER FILTER BOTTOM

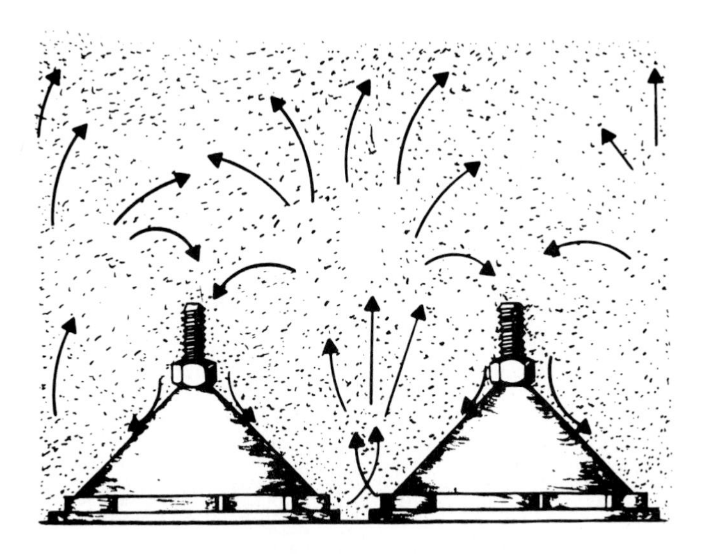

D. SUBFILL–LESS STRAINERS

Figure 8.18 *(Continued)*

3. Apply upflow wash and surface wash together 5 to 10 minutes as needed to flush out solids.
4. Shut off auxiliary wash and apply backwash alone for 1 to 2 minutes at rate needed to classify the bed.
5. Return bed to service.

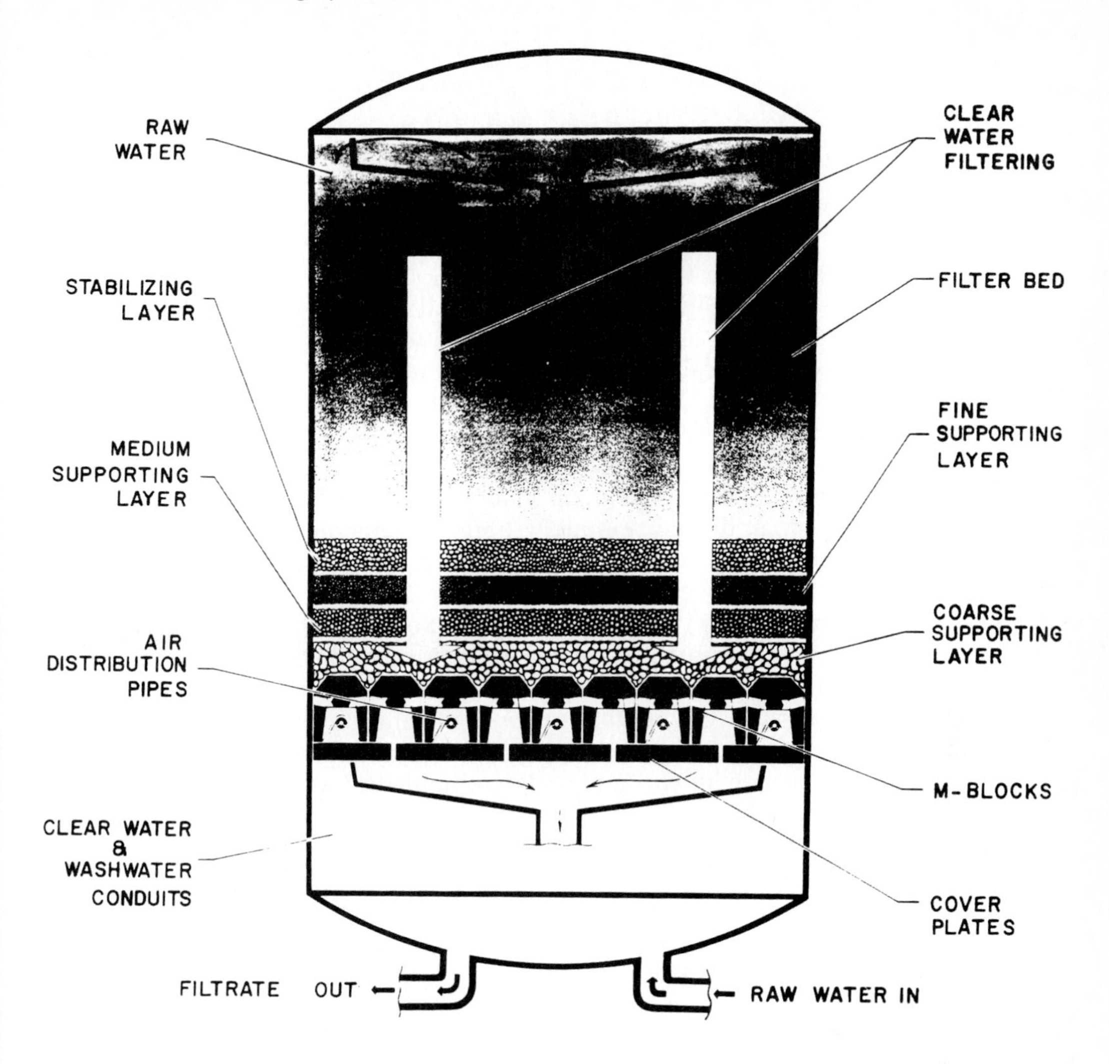

Figure 8.18 *(Continued)*

A typical sequence for cleaning using upflow wash and air scour is:

1. Stop influent and lower the water level to a few inches above bed.
2. Apply air alone at 2 to 5 cfm sq ft for 3 to 10 minutes.
3. Apply water backwash at 2 to 5 gpm/sq ft with air on until water is within one foot of washwater trough.
4. Shut air off.
5. Continue water backwash at normal rate for usual period of time.

6. Apply backwash for 1 to 2 minutes at a rate required to insure hydraulic classification of the filter media.
7. Return bed to service.

STRUCTURES AND GENERAL CONFIGURATION

A typical wastewater filter consists of a tank or filter box containing an underdrain system, media, and sufficient overall depth to contain media during backwash. In gravity units, the overall depth must also provide for operating submergence and freeboard. Influent, effluent, washwater, and waste connections are provided. In addition, all wastewater filters should have provisions for auxiliary cleaning.

Underdrains are designed to properly distribute the washwater during cleaning. During normal operation, underdrains collect filter effluent (downflow operation) or distribute influent (upflow operation). Washwater troughs and filter inlets (downflow) or effluent launders (upflow) are located in the submerged zone above the media.

For gravity filters of concrete construction, filter boxes are usually arranged in rows along one or two sides of a common pipe gallery, narrow side toward the gallery. This maximizes common wall construction and minimizes piping runs. Gravity filters may be of concrete or steel shell construction. Concrete units are generally square or rectangular and steel units circular. Sizes of gravity concrete units are limited to about 1,000 sq ft; steel units are generally smaller.

For filters using influent flow splitting, multiple filter boxes have been constructed as compartments in a single round or square tank (concrete or steel) with common influent and waste piping located above the center of the tank and common washwater and effluent piping around the outside base.

Steel shell package pressure filters are cylindrical units with either horizontal or vertical axes. To minimize piping runs, horizontal units are usually placed in rows with common piping along the ends. Vertical units are arranged in either rows or clusters. Horizontal pressure units are less restricted in size than the vertical pressure units and hence are normally used for plant capacities above 1 to 1.5 mgd. Where pressure units are used, it is essential that manholes be provided for interior access both above the bed and below the underdrains. Pressure filters should also be provided with a means for hydraulic removal of all the filter media, and with sight glasses for observation of the bed.

In municipal water filtration plant designs, it is common to totally enclose the pipe gallery and to locate controls in an enclosed superstructure above the gallery and overlooking the filters. In northern climates the filters themselves are usually included under the superstructure.

Wastewater effluent temperatures are generally somewhat higher than the local natural waters, so in a given locality there is less justification for housing wastewater filters than water filters. Piping and valves need to be protected in climates where freezing occurs either by housing or by insulation and heating. Controls need housing or weather-protected enclosures in any climate. Local controls for each filter should be placed in a location from which the backwashing filter can be observed.

9

Handling of Chemicals

This chapter surveys the chemicals most commonly used for suspended solids removal, with respect to their properties, availability, storage, transport, reactions, and feeding. The chemicals described include those used in coagulation and sedimentation-type applications.

ALUMINUM COMPOUNDS

The principal aluminum compounds that are commercially available and suitable for suspended solids removal are dry and liquid alum. Sodium aluminate has been used in activated sludge plants for phosphorus removal, but its applicability for suspended solids removal is limited.

Dry Alum

The commercial dry alum most often used in wastewater treatment is known as *filter alum,* and has the approximate chemical formula $Al_2(SO_4)_3 \cdot 14H_2O$ and a molecular weight of about 600. Alum is white to cream in color and a 1 percent solution has a pH of about 3.5. The commercially available grades of alum and their corresponding bulk densities and angles of repose are:

Grade	Angle of repose	Bulk density (lb./cubic feet)
Lump	—	62 to 68
Ground	43	60 to 71
Rice	38	57 to 71
Powdered	65	38 to 45

Each of these grades has a minimum aluminum content of 27 percent, expressed as Al_2O_3, and maximum Fe_2O_3 and soluble contents of 0.75 percent and 0.5 percent, respectively. Viscosity and solution crystallation temperatures are included in a later section on liquid alum.

Since dry alum is only partially hydrated, it is slightly hygroscopic. However, it is relatively stable when stored under the extremes of temperature and humidity encountered.

The solubility of commercial dry alum at various temperatures is as follows:

Temperature (°F)	Solubility (lb/gal)
32	6.03
50	6.56
68	7.28
86	8.45
104	10.16

Dry alum is not corrosive unless it absorbs moisture from the air, such as during prolonged exposure to humid atmospheres. Therefore, precautions should be taken to ensure that the storage space is free of moisture.

Alum is shipped in 100 lb bags, drums, or in bulk (minimum of 40,000 lb) by truck or rail. Bag shipments may be ordered on wood pallets if desired.

Ground and rice alum are the grades most commonly used by utilities because of their superior flow characteristics. These grades have less tendency to lump or arch in storage and therefore provide more consistent feeding qualities. Hopper agitation is seldom required with these grades, and in fact may be detrimental to feeding because of the possibility of packing the bin.

Alum dust is present in the ground grade and will cause minor irritation of the eyes and nose on breathing. A respirator may be worn for protection against alum dust. Gloves may be work to protect the hands. Because of minor irritation in handling and the possibility of alum dust causing rusting of adjacent machinery, dust removal equipment is desirable. Alum dust should be thoroughly flushed from the eyes immediately and washed from the skin with water.

A typical storage and feeding system for dry alum is shown in Figure 9.1. Bulk alum can be stored in mild steel or concrete bins with dust collector vents located in, above, or adjacent to the equipment room. Recommended

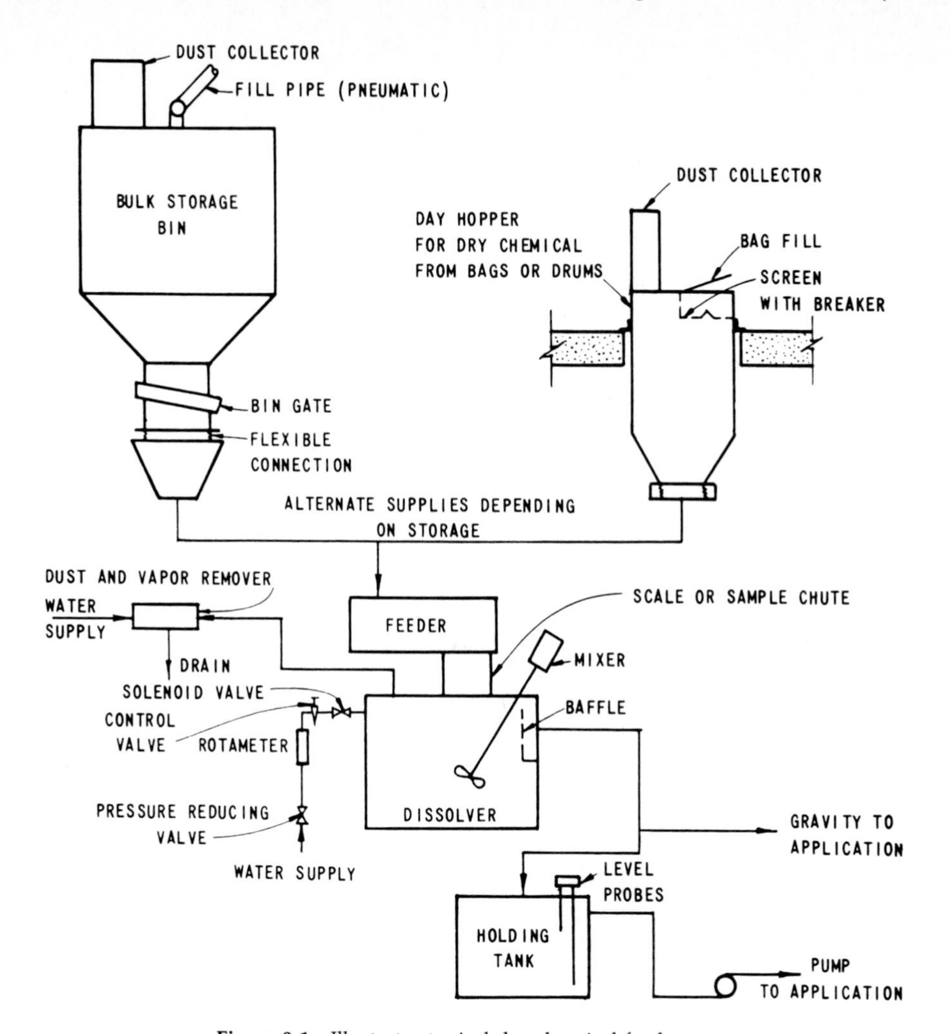

Figure 9.1 Illustrates typical dry chemical feed system.

storage capacity is about 30 days. Dry alum in bulk can be transferred with screw conveyors, pneumatic conveyors, or bucket elevators made of mild steel. Pneumatic conveyor elbows should have a reinforced backing as the alum can contain abrasive impurities.

Bags and drums of alum should be stored in a dry location to avoid caking. Bag or drum-loaded hoppers should have a nominal storage capacity for eight hours at the nominal maximum feed rate so that personnel are not required to charge the hopper more than once per shift. Converging hopper sections should have a minimum slope of 60 degrees to prevent arching.

Bulk storage hoppers should terminate at a bin gate so that the feeding equipment may be isolated for servicing. The bin gate should be followed by a flexible connection, and a transition hopper chute or hopper which acts as a conditioning chamber over the feeder.

The feed system includes all of the components required for the proper preparation of the chemical solution. Capacities and assemblies should be selected to fulfill individual system requirements. Three basic types of chemical feed equipment are used: volumetric, belt gravimetric, and loss-in-weight gravimetric. Volumetric feeders are usually used where initial low cost and usually lower capacities are the basis of selection. Volumetric feeder mechanisms are usually exposed to the corrosive dissolving chamber vapors which can cause corrosion of discharge areas. Manufacturers usually control this problem by use of an electric heater to keep the feeder housing dry or by using plastic components in the exposed areas.

Volumetric dry feeders presently in general use are of the screw type. Two designs of screw-feed mechanisms are available. Both allow even withdrawal across the bottom of the feeder hopper to prevent hopper dead zones. One screw design is the variable-pitch type with the pitch expanding unevenly to the discharge point. The second screw design is the constant-pitch type expanding evenly to the discharge point. This type of screw design is the constant-pitch-reciprocating type. This type has each half of the screw turned in opposite directions so that the turning and reciprocating motion alternately fills one half of the screw while the other half of the screw is discharging. The variable-pitch screw has one point of discharge, while the constant-pitch-reciprocating screw has two points of discharge, one at each end of the screw. The accuracy of volumetric feeders is influenced by the character of the material being fed and ranges between ±1 percent for free-flowing materials and ±7 percent for cohesive materials. This accuracy is volumetric and should not be related to accuracy by weight (gravimetric).

Where the greatest accuracy and the most economical use of chemicals is desired, the loss-in-weight-type feeder should be selected. This feeder is limited to the low and intermediate feed rates up to a maximum rate of approximately 4,000 lb/hr. The loss-in-weight-type feeder consists of a material hopper and feeding mechanism mounted on enclosed scales. The feed-rate controller retracts the scale poise weight to deliver the dry chemical at the desired rate. The feeding mechanism must feed at this rate to maintain the balance of the scale. Any unbalance of the scale beam causes a corrective change in the output of the feeding mechanism. Continuous comparison of actual hopper weight with set hopper weight prevents cumulative errors. Accuracy of the loss-in-weight feeder is ±1 percent by weight of the set rate.

Belt-type gravimetric feeders span the capacity ranges of volumetric and loss-in-weight feeders and can usually be sized for all applications encountered in wastewater treatment applications. Initial expense is greater than for the volumetric feeder and slightly less than for the loss-in-weight feeder. Belt-type gravimetric feeders consist of a basic belt feeder incorporating a weighing and control system. Feed rates can be varied by changing either the weight per foot of belt, or the belt speed, or both. Controllers in general use are mechanical, pneumatic, electric, and mechanical-vibrating. Accuracy specified for belt-type gravimetric feeders should be within ± 1 percent of set rate. Materials of construction of feed equipment normally include mild steel hoppers, stainless steel mechanism components, and rubber-surfaced feed belts.

Because alum solution is corrosive, dissolving or solution chambers should be constructed of type 316 stainless steel, fiberglass reinforced plastic (FRP), or plastics. Dissolvers should be sized for preparation of the desired solution strength. The solution strength usually recommended is 0.5 lb of alum to 1 gal. of water, or a 6 percent solution. The dissolving chamber is designed for a minimum detention time of 5 minutes at the maximum feed rate. Because excessive dilution may be detrimental to coagulation, eductors, or float valves that would ordinarily be used ahead of centrifugal pumps are not recommended. Dissolvers should be equipped with water meters and mechanical mixers so that the water-to-alum ratio may be properly established and controlled.

FRP, plastics (polyvinyl chloride, polyethylene, polypropylene, and other similar materials), and rubber are general use and are recommended for alum solutions. Care must be taken to provide adequate support for these piping systems, with close attention given to spans between supports so that objectionable deflection will not be experienced. The alum solution should be injected into a zone of rapid mixing or turbulent flow.

Solution flow by gravity to the point of discharge is desirable. When gravity flow is not possible, transfer components should be selected that require little or no dilution. When metering pumps or proportioning weir tanks are used, return of excess flow to a holding tank should be considered. Metering pumps are discussed further in the section on liquid alum.

Valves used in solution lines should be plastic, type 316 stainless steel, or rubber-lined iron or steel.

Standard instrument control and pacing signals are generally acceptable for common feeder system operation. Volumetric and gravimetric feeders are usually adaptable to operation from any standard instrument signals.

When solution must be pumped, consideration should be given to use of holding tanks between the dry feed system and feed pumps, and the solution water supply should be controlled to prevent excessive dilution. The dry feeders may be started and stopped by tank level probes. Variable-control metering pumps can then transfer the alum stock solution to the point of application without further dilution.

Means should be provided for calibration of the chemical feeders. Volumetric feeders may be mounted on platform scales. Belt feeders should include a sample chute and box to catch samples for checking actual delivery with set delivery.

Gravimetric feeders are usually furnished with totalizers only. Remote instrumentation is frequently used with gravimetric equipment, but seldom used with volumetric equipment.

Liquid Alum

Liquid alum is shipped in rubber-lined or stainless steel insulated tank cars or trucks. Alum shipped during the winter is heated prior to shipment so that crystallization will not occur during transit. Liquid alum is shipped at a solution strength of about 8.3 percent as Al_2O_3 or about 49 percent as $Al_2(SO_4)_3 \cdot 14H_2O$. The latter solution weighs about 11 lb/gal at 60°F and contains about 5.4 lb dry alum (17 percent Al_2O_3) per gal of liquid. This solution will begin to crystallize at 30°F and freezes at about 18°F.

Crystallization temperatures of various solution strengths are shown in Figure 9.2. The viscosity of various alum solutions is given in Figure 9.3.

Bulk unloading facilities usually must be provided at the treatment plant. Rail cars are constructed for top unloading and therefore require an air supply system and flexible connectors to pneumatically displace the alum from the car. U.S. Department of Transportation regulations concerning chemical tank car unloading should be observed. Tank truck unloading is usually accomplished by gravity or by a truck mounted pump.

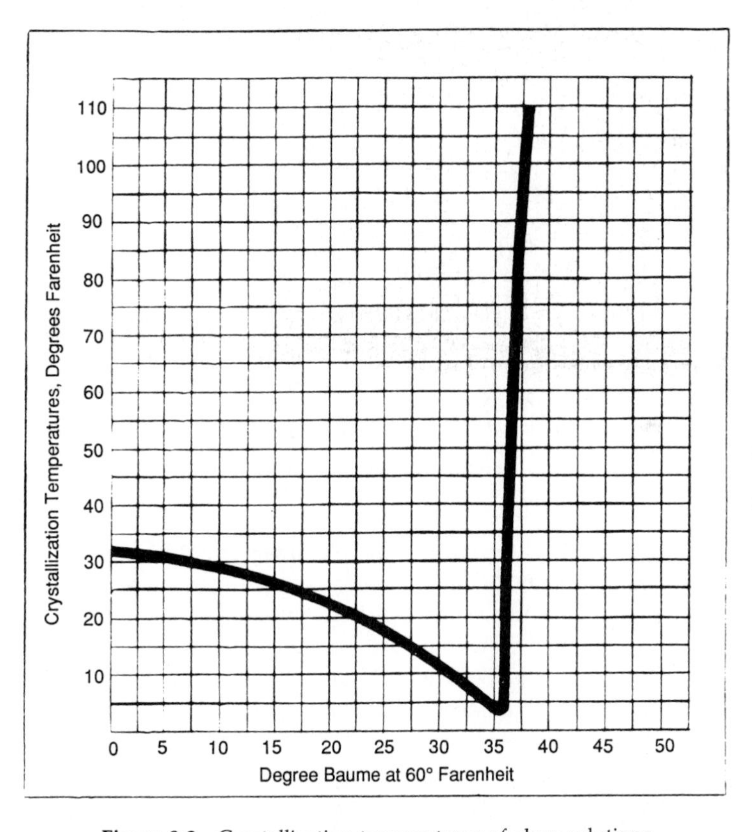

Figure 9.2 Crystallization temperatures of alum solutions

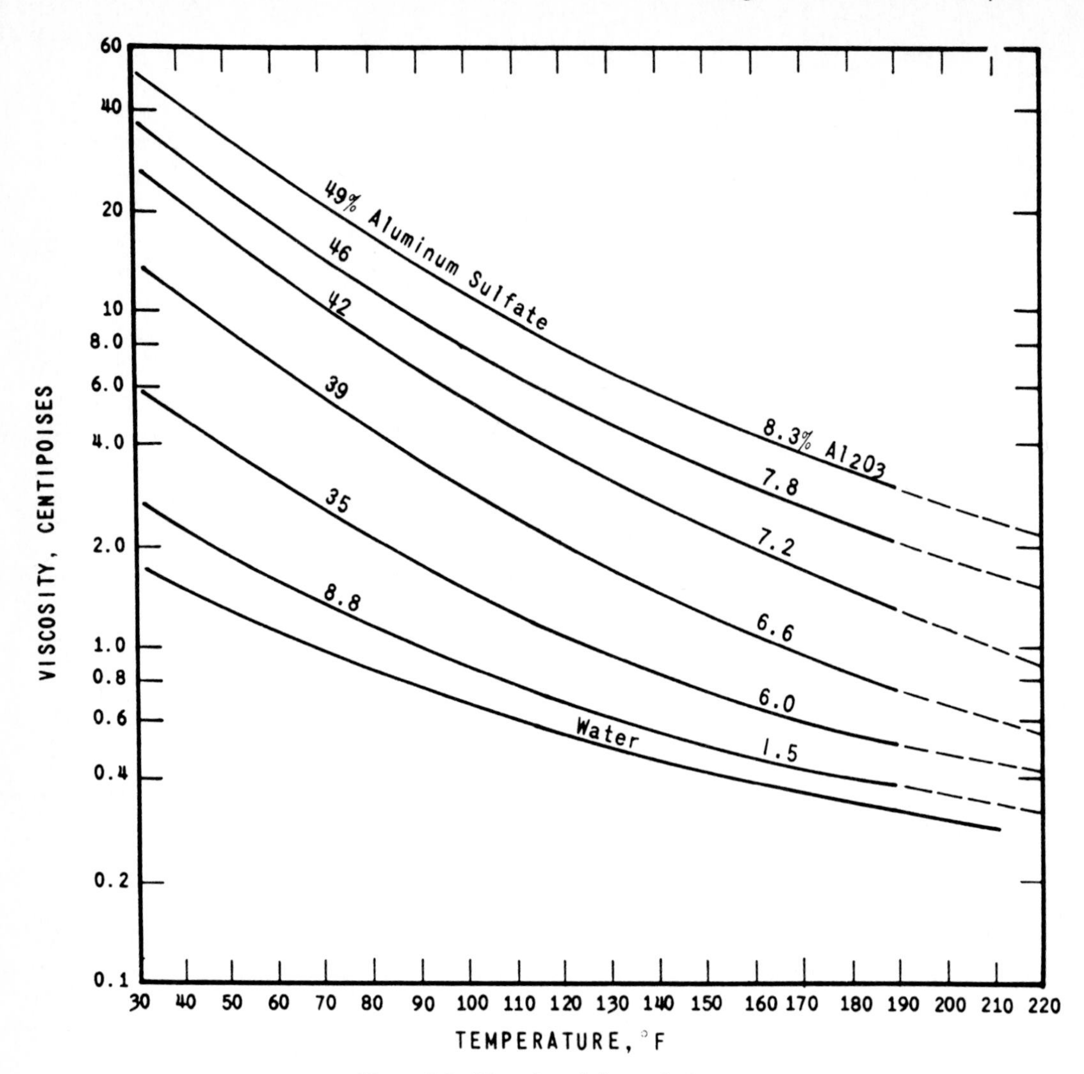

Figure 9.3 Viscosity of alum solutions.

Established practice in the treatment field has been to dilute liquid alum prior to application. However, recent studies have shown that feeding undiluted liquid alum results in better coagulation and settling. This is reportedly due to prevention of hydrolysis of the alum.

No particular industrial hazards are encountered in handling liquid alum. However, a face shield and gloves should be worn around leaking equipment. The eyes or skin should be flushed and washed upon contact with liquid alum. Liquid alum becomes very sick upon evaporation and therefore spillage should be avoided.

Liquid alum is stored without dilution at the shipping concentration.

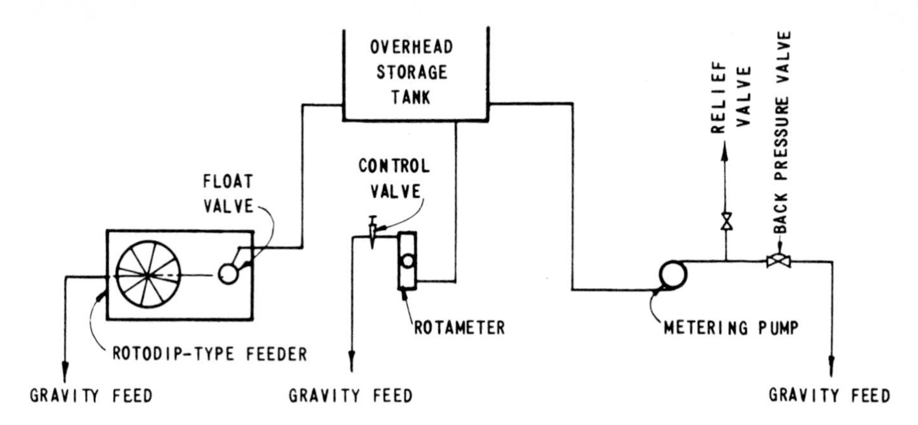

Figure 9.4 Illustrates alternative liquid feed systems for overhead storage.

Storage tanks may be open if indoors but must be closed and vented if outdoors. Outdoor tanks should also be heated, if necessary, to keep the temperature above 45°F to prevent crystallization. Storage tanks should be constructed of type 316 stainless steel, FRP, steel lined with rubber, polyvinyl chloride, or lead. Liquid alum can be stored indefinitely without deterioration.

Storage tanks should be sized according to maximum feed rate, shipping time required, and quantity of shipment. Tanks should generally be sized for 1.5 times the quantity of shipments. A ten-day to two-week supply should be provided to allow for unforeseen shipping delays.

Various types of gravity or pressure feeding and metering units are available. Figures 9.4 and 9.5 illustrate commonly used feed systems. The rotodip-type feeder or rotameter is often used for gravity feed and the metering pump for pressure feed systems.

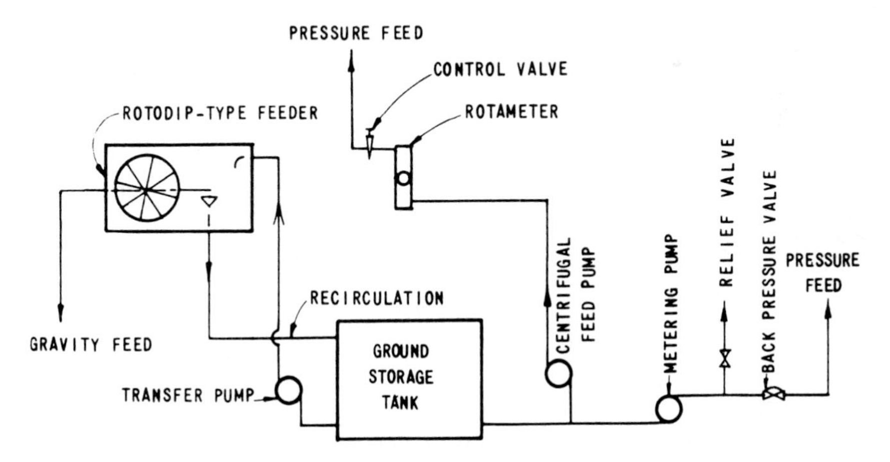

Figure 9.5 Shows alternative liquid feed systems for ground storage.

The pressure or head available at the point of application frequently determines the feeding system to be used. The rotodip feeder can be supplied from overhead storage by gravity with the use of an internal level control value, as shown by Figure 9.4. It may also be supplied by a centrifugal pump. The latter arrangement requires an excess flow return line to the storage tank, as shown by Figure 9.5. Centrifugal pumps should be direct connected but not close coupled because of possible leakage into the motor, and should be constructed of type 316 stainless steel, FRP, and plastics.

Metering pumps currently available allow a wide range of capacity compared with the rotodip and rotameter systems. Hydraulic diaphragm-type pumps are preferable to other type pumps and should be protected with an internal or external relief valve. A back pressure valve is usually required in the pump discharge to provide efficient check valve action. Materials of construction for feeding equipment should be as recommended by the manufacturer for the service, but depending on the type of system, will generally include type 316 stainless steel, FRP, plastics and rubber.

Piping systems for alum should be FRP, plastics (subject to temperature limits), type 316 stainless steel, or lead. Piping and valves used for alum solutions are also discussed in the preceding section on dry alum.

The feeding systems described previously are volumetric, and the feeders generally available can be adapted to receive standard instrument pacing signals. The signals can be used to vary motor speed, variable-speed transmission setting, stroke speed, and stroke length where applicable. A totalizer is usually furnished with a rotodip-type feeder, and remote instruments are available. Instrumentation is rarely used with rotameters and metering pumps.

Reactions between alum and the normal constituents of wastewaters are influenced by many factors; hence, it is impossible to predict accurately the amount of alum that will react with a given amount of alkalinity, lime, or soda ash which may have been added to the wastewater. Theoretical reactions can be written which will serve as a general guide, but in general the optimum dosage in each case must be determined by laboratory jar tests.

The simplest case is the reaction of Al^{3+} with OH^- ions made available by the ionization of water or by the alkalinity of the water.

Solution of alum in water produces:

$$Al_2(SO_4)_3 \rightleftarrows 2Al^{3+} + 3(SO_4)^{2-}$$

Hydroxyl ions become available from ionization of water:

$$H_2O \rightleftarrows H^+ + OH^-$$

The aluminum ions (Al^{3+}) then react:

$$2\ Al^{3+} + 6\ OH^- \rightleftarrows 2\ Al(OH)_3$$

Consumption of hydroxyl ions will result in a decrease in the alkalinity. Where the alkalinity of the wastewater is inadequate for the alum dosage, the pH

TABLE 9.1 REACTIONS OF ALUMINUM SULFATE

$Al_2(SO_4)_3 + 3\ Ca(HCO_3)_2 \rightarrow 2\ Al(OH)_3 \downarrow + 3\ CaSO_4 + 6\ CO_2 \uparrow$
$Al_2(SO_4)_3 + 3\ Na_2CO_3 + 3\ H_2O \rightarrow 2\ AL(OH)_3 \downarrow + 3\ CO_2 \uparrow$
$Al_2(SO_4)_3 + 3\ Ca(OH)_2 \rightarrow 2\ Al(OH)_3 \downarrow + 3\ CaSO_4$

must be increased by the addition of hydrated lime, soda ash, or caustic soda. The reactions of alum with the common alkaline reagents are shown in Table 9.1. While these reactions are an oversimplification of what actually takes place, they do serve to indicate orders of magnitude and some by-products of alum treatment.

These approximate amounts of alkali when added to wastewater will maintain the alkalinity of the water unchanged when 1 mg/1 of alum is added. For example, if no alkalinity is added, 1 mg/1 of alum will reduce the alkalinity of 0.50 mg/1 as $CaCO_3$ but alkalinity can be maintained unchanged if 0.39 mg/1 of hydrated lime is added. This lowering of natural alkalinity is desirable in many cases to attain the pH range for optimum coagulation.

For each mg/1 of alum dosage, the sulfate (SO_4) content of the water will be increased approximately 0.49 mg/1 and the CO_2 content of the water will be increased approximately 0.44 mg/1.

Iron Compounds

Iron compounds have pH coagulation ranges and floc characteristics similar to aluminum sulfate. The cost of iron compounds may often be less than the cost of alum. However, the iron compounds are generally corrosive and often present difficulties in dissolving, and their use may result in high soluble iron concentrations in process effluents.

Liquid Ferric Chloride

Liquid ferric chloride is a corrosive, dark brown oily-appearing solution having a weight as shipped and stored of 11.2 to 12.4 lb/gal (35 percent to 45 percent $FeCl_3$). The ferric chloride content of these solutions, as $FeCl_3$, is 3.95 to 5.58 lb/gal. Shipping concentrations vary from summer to winter due to the relatively high crystallization temperature of the more concentrated solutions as shown by Figure 9.6. The pH of a 1 percent solution is 2.0.

The molecular weight of ferric chloride is 162.22. Viscosities of ferric chloride solutions at various temperatures are presented in Figure 9.7.

Ferric chloride solutions are corrosive to many common materials and cause stains which are difficult to remove. Areas which are subject to staining should be protected with resistant paint or rubber mats.

Normal precautions should be employed when cleaning ferric chloride handling equipment. Workers should wear rubber gloves, rubber apron, and goggles or a face shield. If ferric chloride comes in contact with the eyes or

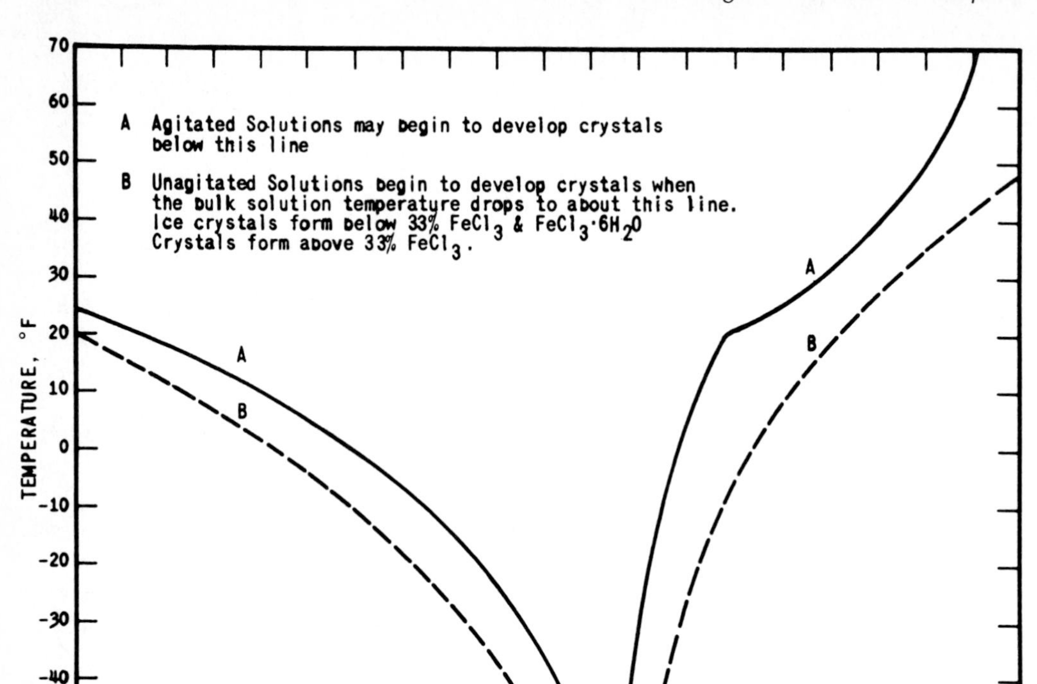

Figure 9.6 Freezing point curves for ferric chloride solutions.

skin, flush with copious quantities of running water and call a physician. If ferric chloride is ingested, induce vomiting and call a physician.

Ferric chloride solution can be stored as shipped. Storage tanks should have a free vent or vacuum relief valve. Tanks may be constructed of FRP, rubber-lined steel, or plastic-lined steel. Resin-impregnated carbon or graphite are also suitable materials for storage containers.

It may be necessary in most instances to house liquid ferric chloride tanks in heated areas or provide tank heaters or insulation to prevent crystallization. Ferric chloride can be stored for long periods of time without deterioration. The total storage capacity should be 1.5 times the largest anticipated shipment, and should provide at least a ten-day to two-week supply of the chemical at the design average dosage.

Feeding equipment and systems describe for liquid alum generally apply to ferric chloride except for materials of construction and the use of glass tube rotameters.

It may not be desirable to dilute the ferric chloride solution from its shipping concentration to a weaker feed solution because of possible hydrolysis. Ferric chloride solutions may be transferred from underground storage to day tanks with impervious graphite or rubber-lined self-priming centrifugal pumps having Teflon rotary and stationary seals. Because of the tendency for liquid ferric chloride to stain or deposit, glass-tube rotameters should not be used for metering this solution. Rotodip feeders and diaphragm metering pumps are often used for ferric chloride, and should be constructed of materials such as rubber-lined steel and plastics.

Materials for piping and transporting ferric chloride should be rubber or Saran-lined steel, hard rubber, FRP, or plastics. Valving should consist of

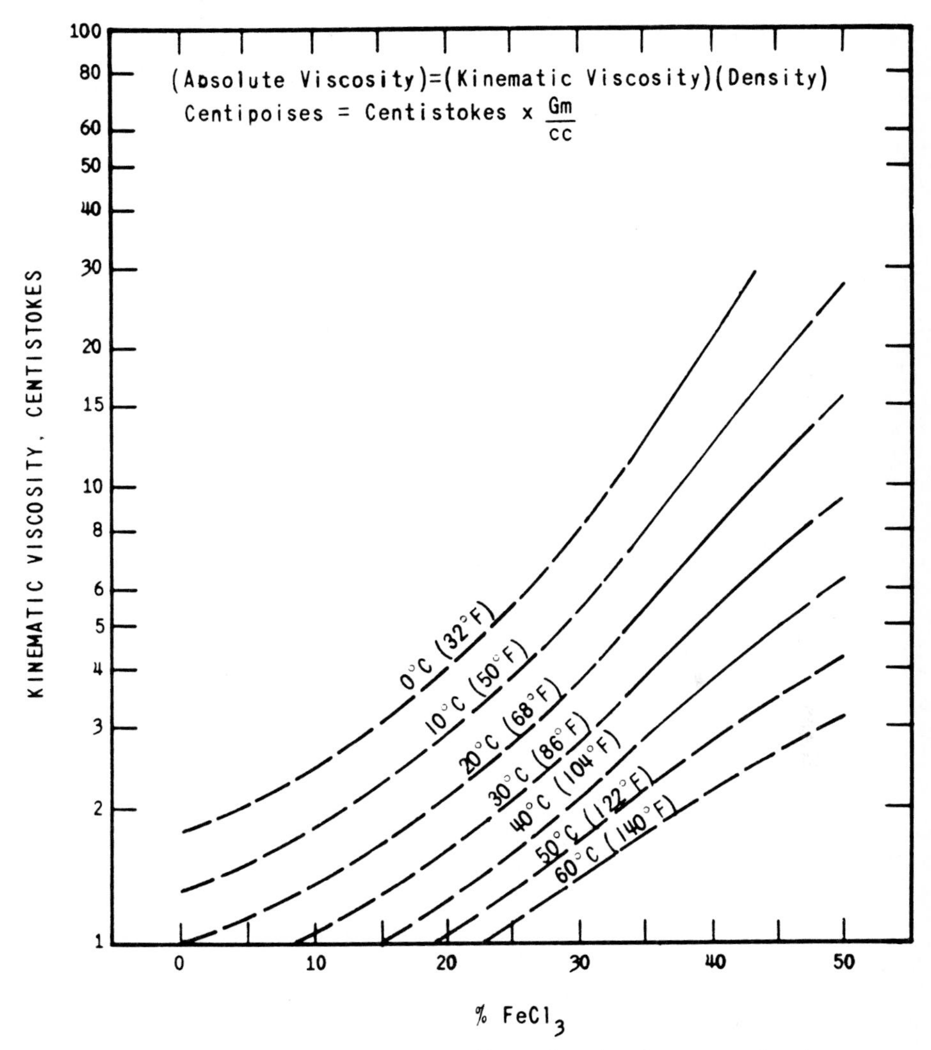

Figure 9.7 Viscosity vs. composition of ferric chloride solutions as a function of temperature.

rubber or resin-lined diaphragm valves. Saran-lined valves with Teflon diaphragms, rubber-sleeved pinch-type valves, or plastic ball valves. Gasket material for large openings such as manholes in storage tanks should be soft rubber; all other gaskets should be graphite-impregnated blue asbestos, Teflon, or vinyl. System pacing and control requirements are similar to those discussed previously for liquid alum.

Ferrous Chloride (Waste Pickle Liquor)

Ferrous chloride, $FeCl_2$, as a liquid is available in the form of waste pickle liquor from steel processing. The liquor weighs between 9.9 and 10.4 lb/gal and contains 20 percent to 25 percent $FeCl_2$ or about 10 percent available Fe^{2+}. A 22 percent solution of $FeCl_2$ will crystallize at a temperature of $-4°F$. The molecular weight of $FeCl_2$ is 126.76. Free acid in waste pickle liquor can vary from 1 percent to 10 percent and usually averages about 1.5 percent to 2.0 percent. Ferrous chloride is slightly less corrosive than ferric chloride.

Waste pickle liquor is available in 4,000 gal truckload lots and a variety of carload lots. In most instances the availability of waste pickle liquor will depend on the proximity to steel processing plants.

Since ferrous chloride or waste pickle liquor may not be available on a continuous basis, storage and feeding equipment should be suitable for handling ferric chloride. Therefore, the ferric chloride section should be referred to for storage and handling details.

Ferric Sulfate

Ferric sulfate is marketed as dry, partially-hydrated granules with the formula $Fe_2(SO_4)_3 \cdot X\ H_2O$, where X is approximately 7. Typical properties of a commercial product are presented as follows:

Molecular Weight	526
Bulk Density	56–60 lb/cu ft
Water Soluble Iron Expressed as Fe	21.5 percent
Water Soluble $Fe+^3$	19.5 percent
Water Soluble $Fe+^2$	2.0 percent
Insolubles Total	4.0 percent
Free Acid	2.5 percent
Moisture @ 105°C	2.0 percent

Ferric sulfate is shipped in car and truck load lots of 50 lb and 100 lb moisture-proof paper bags and 200 lb and 400 lb fiber drums.

General precautions should be observed when handling ferric sulfate, such as wearing goggles and dust masks, and areas of the body that come in contact with the dust or vapor should be washed promptly.

Aeration of ferric sulfate should be held to a minimum because of the hygroscopic nature of the material, particularly in damp atmospheres. Mixing of ferric sulfate and quicklime in conveying and dust vent systems should be avoided as caking and excessive heating can result. The presence of ferric sulfate and lime in combination has been known to destroy cloth bags in pneumatic unloading devices. Because ferric sulfate in the presence of moisture will stain, precautions similar to those discussed for ferric chloride should be observed.

Ferric sulfate is usually stored in the dry state either in the shipping bags or in bulk in concrete or steel bins. Bulk storage bins should be as tight as possible to avoid moisture absorption, but dust collector vents are permissible and desirable. Hoppers on bulk storage bins should have a minimum slope of 36°; however, a greater angle is preferred.

Bins may be located inside or outside and the material transferred by bucket elevator, screw, or air conveyors. Ferric sulfate stored in bins usually absorbs some moisture and forms a thin protective crust which retards further absorption until the crust is broken.

Feed solutions are usually made up at a water to chemical ratio of 2:1 to 8:1 (on a weight basis) with the usual ratio being 4:1 with a 20-minute detention time. Care must be taken not to dilute ferric sulfate solutions to less than 1 percent to prevent hydrolysis and deposition of ferric hydroxide. Ferric sulfate is actively corrosive in solution, and dissolving and transporting equipment should be fabricated of type 316 stainless steel, rubber, plastics, ceramics, or lead.

Dry feeding requirements are similar to those for dry alum except that belt-type feeders are rarely used because of their open type of construction. Closed construction, as found in the volumetric and loss-in-weight-type feeders, generally exposes a minimum of operating components to the vapor, and thereby minimizes maintenance. A water jet vapor remover should be provided at the dissolver to protect both the machinery and operator.

Piping systems for ferric sulfate should be FRP, plastics, type 316 stainless steel, rubber, or glass. System pacing and control are the same as discussed for dry alum.

Ferrous Sulfate

Ferrous sulfate or copperas is a by-product of pickling steel and is produced as granules, crystals, powder, and lumps. The most common commercial form of ferrous sulfate is $FeSO_4 \cdot 7H_2O$, with a molecular weight of 278, and containing 55 percent to 58 percent $FeSO_4$ and 20 percent to 21 percent Fe. The product has a bulk density of 62 to 66 lb/cu ft. When dissolved, ferrous sulfate is acidic. The composition of ferrous sulfate may be quite variable and should be established by consulting the nearest manufacturers.

Ferrous sulfate is also available in a wet state in bulk form from some plants. This form is likely to be difficult to handle and the manufacturer should be consulted for specific information and instructions.

TABLE 9.2A REACTIONS OF FERRIC SULFATE

$Fe_2(SO_4)_3 + 3\ Ca(HCO_3)_2 \rightarrow 2\ Fe(OH)_3 \downarrow + 3\ CaSO_4 + 6\ CO_2 \uparrow$
$Fe_2(SO_4)_3 + 3\ Na_2CO_3 + 3\ H_2O \rightarrow 2\ Fe(OH)_3 \downarrow + 3\ Na_2SO_4 + 3\ CO_2 \uparrow$
$Fe_2(SO_4)_3 + 3\ Ca(OH)_2 \rightarrow 2\ Fe(OH)_3 \downarrow + 3\ CaSO_4$

Dry ferrous sulfate cakes at storage temperatures above 68°F, is efflorescent in dry air, and oxidizes and hydrates further in moist air.

General precautions similar to those for ferric sulfate, with respect to dust and handling acidic solutions, should be observed when working with ferrous sulfate. Mixing quicklime and ferrous sulfate produces high temperatures and the possibility of fire.

The granular form of ferrous sulfate has the best feeding characteristics and gravimetric or volumetric feeding equipment may be used.

The optimum chemical-to-water ratio for continuous dissolving is 0.5 lb/gal. of 6 percent with a detention time of 5 minutes in the dissolver. Mechanical agitation should be provided in the dissolver to assure complete solution. Lead, rubber, iron, plastics, and type 304 stainless steel can be used as construction materials for handling solutions of ferrous sulfate.

Storage, feeding, and transporting systems probably should be suitable for handling ferric sulfate as an alternative to ferrous sulfate.

Ferric sulfate and ferric chloride react with the alkalinity of wastewater or with the added alkaline materials such as lime or soda ash. The reactions may be written to show precipitation of ferric hydroxide, although in practice, as with alum, the reactions are more complicated than this. The reactions are shown in Table 9.2A using ferric sulfate.

Storage, feeding, and transporting systems probably should be suitable for handling ferric sulfate as an alternative to ferrous sulfate.

Ferrous sulfate and ferrous chloride react with the alkalinity of wastewater or with the added alkaline materials such as lime to precipitate ferrous hydroxide. The ferrous hydroxide is oxidized to ferric hydroxide by dissolved oxygen in wastewater. Typical reactions using ferrous sulfate are shown in the following table (Table 9.2B).

TABLE 9.2B REACTIONS OF FERROUS SULFATE

$FeSO_4 + Ca(HCO_3)_2 \rightarrow Fe(OH)_2 \downarrow + Ca\ SO_4 + 2CO_2 \uparrow$
$FeSO_4 + Ca(OH)_2 \rightarrow Fe(OH)_2 \downarrow + Ca\ SO_4$
$4\ Fe(OH)_2 + O_2 + 2H_2 \rightarrow 4\ Fe(OH)_3 \downarrow$

Ferrous hydroxide is rather soluble and oxidation to the more insoluble ferric hydroxide is necessary if high iron residuals in effluents are to be avoided. Flocculation with ferrous iron is improved by addition of lime or caustic soda at a rate of 1 to 2 mg/mg Fe to serve as a floc-conditioning agent. Polymers are also generally required to produce a clear effluent.

LIME

The term *lime* applies to a variety of chemicals which are alkaline in nature and contain principally calcium, oxygen and, in some cases, magnesium. In this grouping are included quicklime, dolomitic lime, hydrated lime, dolomitic hydrated lime, limestone, and dolomite. This section is restricted to discussion of quicklime and hydrated lime, but the dolomitic counterparts of these chemicals (i.e., the high-magnesium forms) are quite applicable for wastewater treatment and are generally similar in physical requirements.

Quicklime

Quicklime, CaO, has a density range of approximately 55 to 75 lb/cu ft, and a molecular weight of 56.08. A slurry for feeding, called milk of lime, can be prepared with up to 45 percent solids. Lime is only slightly soluble, and both lime dust and slurries are caustic in nature. A saturated solution of lime has a pH of about 12.4.

Lime can be purchased in bulk in both carload and truckload lots. It is also shipped in 80 lb and 100 lb multiwall "moisture-proof" paper bags.

The CaO content of commercially available quicklime can vary quite widely over an approximate range of 70 percent to 96 percent. Content below 88 percent is generally considered below standard in the municipal use field. Purchase contracts are often based on 90 percent CaO content with provisions for payment of a bonus for each 1 percent over and a penalty for each 1 percent under the standard. A CaO content less than 75 percent probably should be rejected because of excessive grit and difficulties in slaking.

Workers should wear protective clothing and goggles to protect the skin and eyes, as lime dust and hot slurry can cause severe burns. Areas contacted by lime should be washed immediately. Lime should not be mixed with chemicals which have water of hydration. The lime will be slaked by the water of hydration causing excessive temperature rise and possibly explosive conditions. Conveyors and bins used for more than one chemical should be thoroughly cleaned before switching chemicals.

Pebble quicklime, all passing a 34-in. screen and not more than 5 percent passing a No. 100 screen, is normally specified because of easier handling and less dust. Hopper agitation is generally not required with the pebble form. Published slaker capacity ratings require "soft or normally burned" limes which provide fast slaking and temperature rise, but poorer grades of limes may also be satisfactorily slaked by selection of the appropriate slaker retention time and capacity.

Storage of bagged lime should be in a dry place, and preferably elevated on pallets to avoid absorption of moisture. System capacities often make the use of bagged quicklime impractical. Maximum storage period is about 60 days.

Bulk lime is stored in airtight concrete or steel bins having a 55-degree to 60-degree slope on the bin outlet. Bulk lime can be conveyed by conventional bucket elevators and screw, belt, apron, drag-chain, and bulk conveyors of mild steel construction. Pneumatic conveyors subject the lime to air slaking and particle sizes may be reduced by attrition. Dust collectors should be provided on manually and pneumatically-filled bins.

A typical lime storage and feed system is illustrated in Figure 9.8. Quicklime feeders are usually limited to the belt or loss-in-weight gravimetric types because of the wide variation of the bulk density. Feed equipment should have an adjustable feed range of at least 20:1 to match the operating range

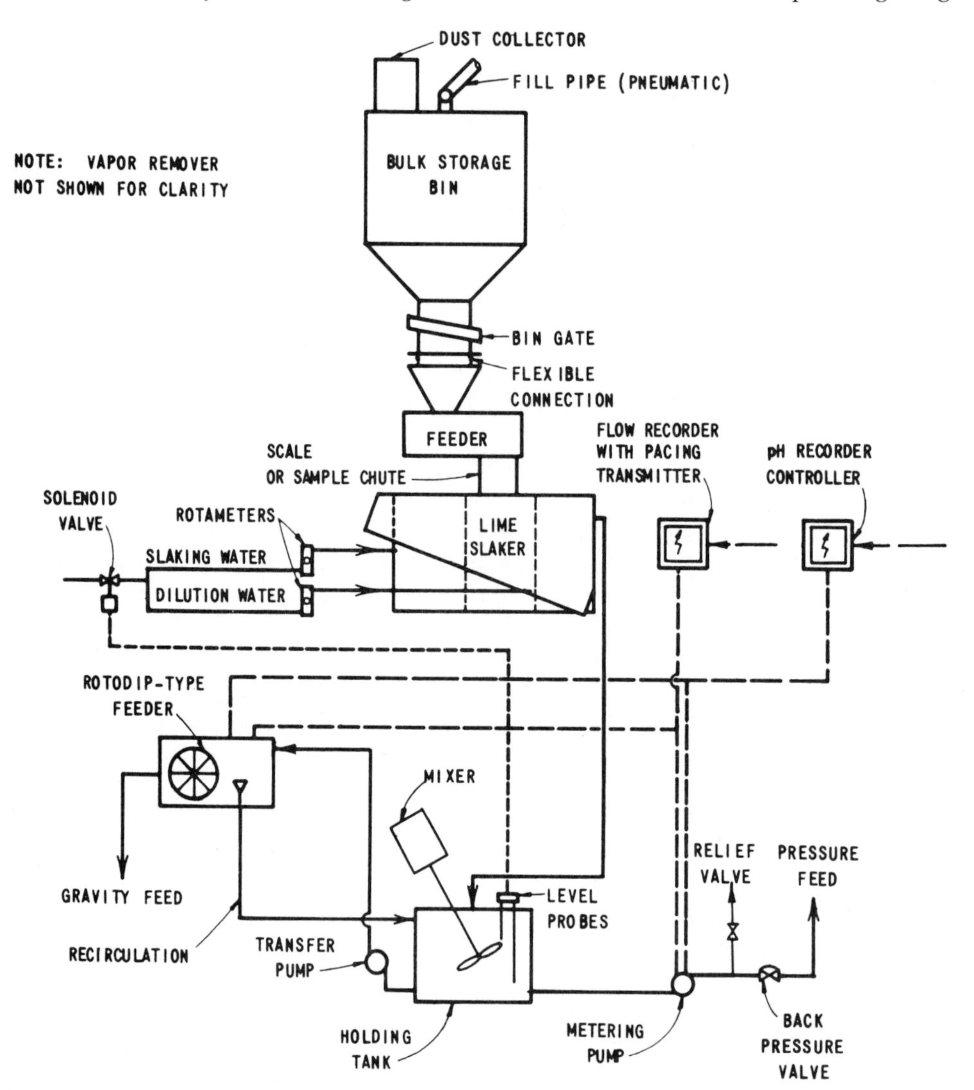

Figure 9.8 Illustrates a typical line feed system.

of the associated slaker. The feeders should have an over-under feed rate alarm to immediately warn of operation beyond set limits of control. The feeder drive should be instrumented to be interrupted in the event of excessive temperature in the slaker compartment.

Lime slakers for wastewater treatment should be of the continuous type, and the major components should include one or more slaking compartments, a dilution compartment, a grit separation compartment, and a continuous grit remover. Commercial designs vary in regard to the combination of water to lime, slaking temperature, and slaking time in obtaining the "milk of lime" suspensions.

The *paste-type* slaker admits water as required to maintain a desired mixing viscosity. This viscosity therefore sets the operating retention time of the slaker. The paste slaker usually operates with a low water-to-lime ratio (approximately 2:1 by weight), elevated temperature, and 5-minute slaking time at maximum capacity.

The *detention-type* slaker admits water to maintain a desired ratio with the lime, and therefore the lime feed rate sets the retention time of the slaker. The detention slaker operates with a wide range of water-to-lime ratios (2.5:1 and 6:1), moderate temperature, and a 10-minute slaking time at maximum capacity. A water-to-lime ratio of from 3.5:1 to 4:1 is most often used. The operating temperature in lime slakers is a function of the water-to-lime ratio, lime quality, heat transfer, and water temperature. Lime slaking evolves heat in hydrating the CaO to $Ca(OH)_2$ and therefore vapor removers are required for feeder protection.

Lime slurry should be transported by gravity in open channels wherever possible. Piping channels and accessories may be rubber, iron, steel, concrete, and plastics. Glass tubing, such as that in rotameters, will cloud rapidly and therefore should not be used. Any abrupt directional changes in piping should include plugged tees or crosses to allow rodding out of deposits. Long sweep elbows should be provided to allow the piping to be cleaned by the use of a cleaning "pig." Daily cleaning is desirable.

Milk-of-lime transfer pumps should be of the open impeller centrifugal type. Pumps having an iron body and impeller with bronze trim are suitable for this purpose. Rubber-lined pumps with rubber-covered impellers are also frequently used. Makeup tanks are usually provided ahead of centrifugal pumps to ensure a flooded suction at all times. *Plating out* of lime is minimized by the use of soft water in the makeup tank and slurry recirculation. Turbine pumps and eductors should be avoided in transferring milk of lime because of scaling problems.

Lime slaker water proportioning is integrally controlled or paced from the feeder. Therefore, the feeder-slaker system will follow pacing controls applied to the feeder only. As discussed previously, gravimetric feeders are adaptable to receive most standard instrumentation pacing signals. Systems can be instrumented to allow remote pacing with telemetering of temperature and feed rate to a central panel for control purposes.

The lime feeding system may be controlled by an instrumentation system integrating both plant flow and pH of the wastewater after lime addition. However, it should be recognized that pH probes require daily maintenance in this application to monitor the pH accurately. Deposits tend to build up on the probe and necessitate frequent maintenance. The low pH lime treatment systems (pH 9.5 to 10.0) can be more readily adapted to this method of control than high-lime treatment systems (pH 11.0 or greater) because less maintenance of the pH equipment is required. In a close-loop pH-flow control system, milk of lime is prepared on a batch basis and transferred to a holding tank with variable output feeders set by the flow and pH meters to proportion the feed rate. Figure 9.8 illustrates such a control system.

Hydrated Lime

Hydrated lime, $Ca(OH)_2$, is usually a white powder (200 to 400 mesh); has a bulk density of 20 to 50 lb/cu ft, contains 82 percent to 98 percent $Ca(OH)_2$, is slightly hydroscopic, tends to flood the feeder, and will arch in storage bins if packed. The modular weight is 74.08. The dust and slurry of hydrated lime are caustic in nature. The pH of hydrated lime solution is the same as that given for quicklime.

Hydrated lime is slaked lime and needs only enough water added to form milk of lime. Wetting or dissolving chambers are usually designed to provide 5-minutes detention with a ratio of 0.5 lb/gal of water or 6 percent slurry at the maximum feed rate. Hydrated lime is usually used where maximum feed rates do not exceed 250 lb/hr., i.e., in smaller plants. Hydrated lime and milk of lime will irritate the eyes, nose, and respiratory system and will dry the skin. Affected areas should be washed with water.

Information given for quicklime also applies to hydrated lime except that bin agitation must be provided. Bulk bin outlets should be provided with nonflooding rotary feeders. Hopper slopes vary from 60 degrees to 66 degrees.

Volumetric or gravimetric feeders may be used, but volumetric feeders are usually selected only for installations where comparatively low feed rates are required. Dilution does not appear to be important, therefore, control of the amount of water used in the feeding operation is not considered necessary. Inexpensive hydraulic jet agitation may be furnished in the wetting chamber of the feeder as an alternative to mechanical agitation. The jets should be sized for the available water supply pressure to obtain proper mixing.

Piping and accessories as described for quicklime are also appropriate for hydrated lime.

Controls as listed for dry alum apply to hydrated lime. Hydraulic jets should operate continuously and only shut off when the feeder is taken out of service. Control of the feed rate with pH as well as pacing with the plant flow may be used with hydrated lime as well as quicklime.

Lime is somewhat different from the hydrolyzing coagulants. When added to wastewater it increases pH and reacts with the carbonate alkalinity to precipitate calcium carbonate. If sufficient lime is added to reach a high pH, approximately 10.5, magnesium hydroxide is also precipitated. This latter precipitation enhances clarification due to the flocculant nature of the $Mg(OH)_2$. Excess calcium ions at high pH levels may be precipitated by the

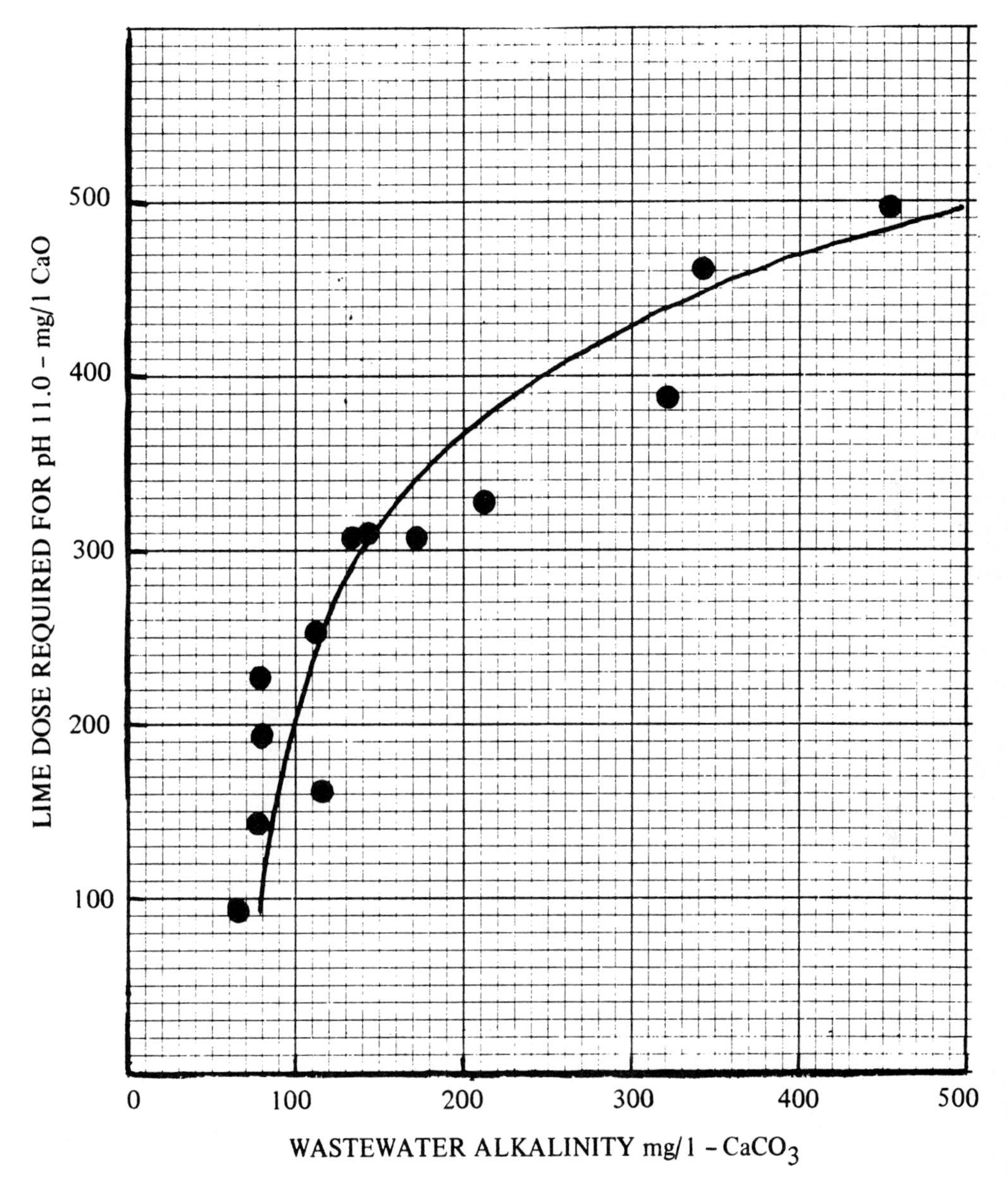

Figure 9.9 Lime requirement for pH 11.0 as a function of wastewater alkalinity.

addition of soda ash. The preceding reactions are shown as follows:

$$Ca(OH)_2 + Ca(HCO_3)_2 \rightarrow 2\ CaCO_3\downarrow + 2H_2O$$

$$2\ Ca(OH)_2 + Mg(HCO_3)_2 \rightarrow 2\ CaCO_3\downarrow + Mg(OH)_2\downarrow + 2H_2O$$

$$Ca(OH)_2 + Na_2CO_3 \rightarrow CaCO_3\downarrow + 2\ NaOH$$

Reduction of the resulting high pH levels may be accomplished in one or two stages. The first stage of the two-stage method results in the precipitation of calcium carbonate through the addition of carbon dioxide according to the following reaction:

$$Ca(OH)_2 + CO_2 \rightarrow CaCO_3\downarrow + H_2O$$

Single-stage pH reduction is generally accomplished by the addition of carbon dioxide, although acids have been employed. This reaction, which also represents the second stage of the two-stage method, is as follows:

$$Ca(OH)_2 + 2\ CO_2 \rightarrow Ca(HCO_3)_2$$

As noted for the other chemicals, the preceding reactions are merely approximations to the more complex interactions which actually occur in wastewaters.

The lime demand of a given wastewater is a function of the buffer capacity or alkalinity of the wastewater. Figure 9.9 shows this relationship for a number of different wastewaters.

OTHER INORGANIC CHEMICALS

In addition to aluminum and iron salts and lime, a number of other inorganic chemicals have been used in wastewater treatment. The only three discussed in this section are, soda ash, caustic soda, and carbon dioxide, but others have been and will be employed. Mineral and other acids are prime examples. For information on any of these chemicals, the local supplier or manufacturer should be contacted.

Soda Ash

Soda ash, Na_2CO_3, is available in two forms. Light soda ash has a bulk density range of 35 to 50 lb/cu ft and a working density of 41 lb/cu ft. Dense soda ash has a density range of 60 to 76 lb/cu ft and a working density of 63 lb/cu ft. The pH of a 1 percent solution of soda ash as 11.2. It is used for pH control and in lime treatment.

The molecular weight of soda ash is 106. Commercial purity ranges from 98 percent to greater than 99 percent Na_2CO_3. The viscosities of sodium carbonate solutions are given in Figure 9.10. Soda ash by itself is not particularly

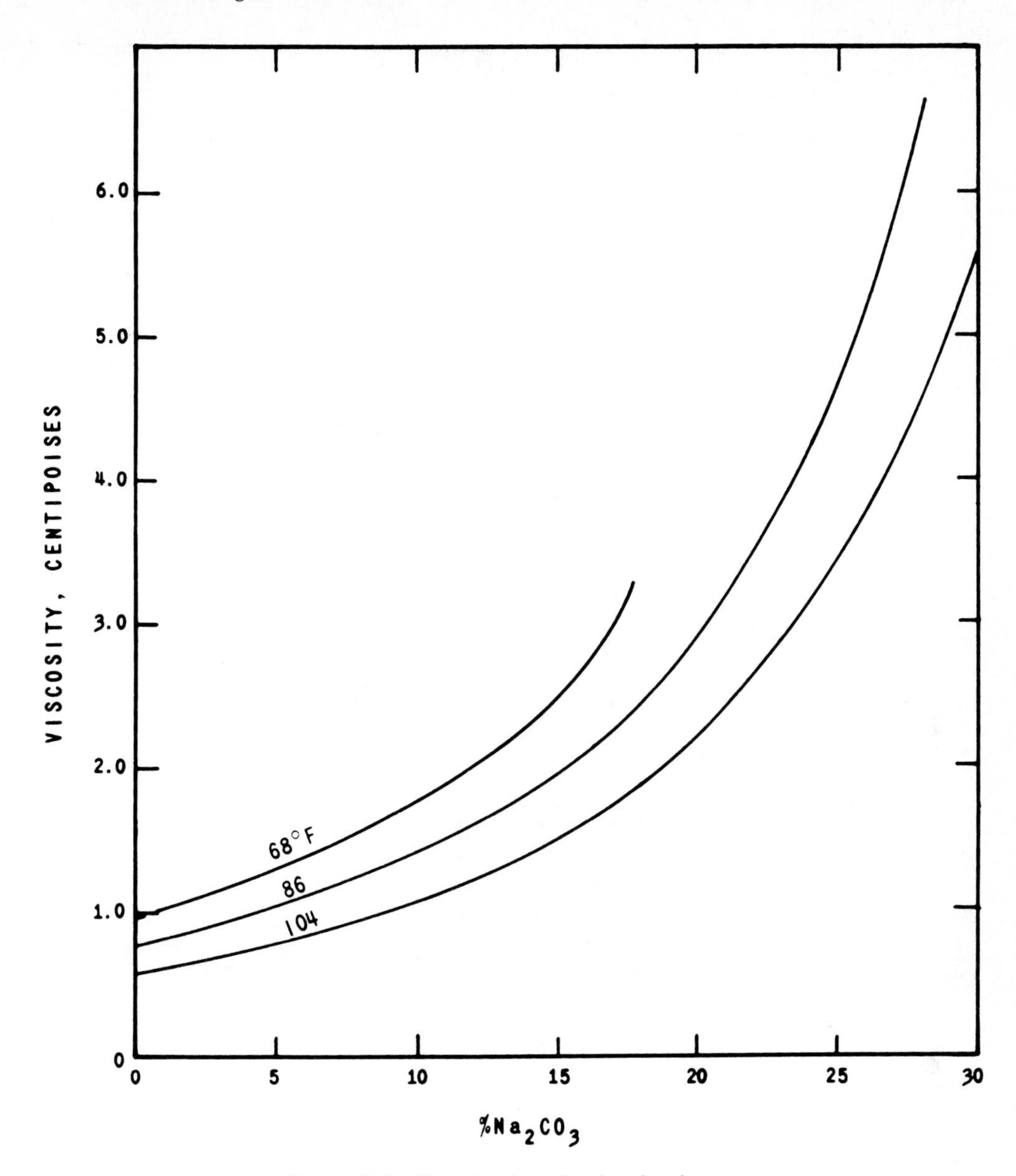

Figure 9.10 Viscosity data of soda ash solutions.

corrosive, but in the presence of lime and water, caustic soda is formed which is quite corrosive.

Dense soda ash is generally used in municipal applications because of superior handling characteristics. It has little dust, good flow characteristics, and will not arch in the bin or flood and feeder. It is relatively hard to dissolve and ample dissolver capacity must be provided. Normal practice calls for 0.5

lb of dense soda ash per gallon of water or a 6 percent solution retained for 20 minutes in the dissolver.

The dust and solution are irritating to the eyes, nose, lungs, and skin and therefore general precautions should be observed and the affected areas should be washed promptly with water.

Soda ash is usually stored in steel bins and where pneumatic-filling equipment is used, bins should be provided with dust collectors. Bulk and bagged soda ash tend to absorb atmospheric CO_2 and water, forming the less active sodium bicarbonate ($NaHCO_3$). Material recommended for unloading facilities is steel.

Feed equipment as described for dry alum is suitable for soda ash. Dissolving of soda ash may be hastened by the use of warm dissolving water. Mechanical or hydraulic jet mixing should be provided in the dissolver.

Materials of construction for piping and accessories should be iron, steel, rubber, and plastics.

Liquid Caustic Soda

Anhydrous caustic soda (NaOH) is available but its use is generally not considered practical in water and wastewater treatment applications. Consequently, only liquid caustic soda is discussed here.

Liquid caustic soda is generally shipped at two concentrations, 50 percent and 73 percent NaOH. The densities of the solutions as shipped are 12.76 lb/gal for the 50 percent solution and 14.18 lb/gal for the 73 percent solution. These solutions contain 6.38 lb/gal NaOH and 10.34 lb/gal NaOH, respectively. The crystallization temperature is 53°F for the 50 percent solution and 165°F for the 73 percent solution. The molecular weight of NaOH is 40. Viscosities of various caustic soda solutions are presented in Figure 9.11. The pH of a 1 percent solution of caustic soda is 12.9.

Truckload lots of 1,000 to 4,000 gallons are available in the 50 percent concentration only. Both shipping concentrations can be obtained in 8,000, 10,000 and 16,000 gal carload lots. Tank cars can be unloaded through the dome eduction pipe using air pressure or through the bottom valve by gravity or by using air pressure or a pump. Trucks are usually unloaded by gravity or with air pressure or a truck-mounted pump.

Liquid caustic soda is received in bulk shipments, transferred to storage, and diluted as necessary for feeding to the points of application. Caustic soda is poisonous and is dangerous to handle. U.S. Department of Transportation Regulations for "White Label" materials must be observed. However, if handled properly caustic soda poses no particular industrial hazard. To avoid accidental spills, all pumps, valves, and lines should be checked regularly for leaks. Workers should be thoroughly instructed in the precautions related to the handling of caustic soda. The eyes should be protected by goggles at all times when exposure to mist or splashing is possible. Other parts of the body should be protected as necessary to prevent alkali burns. Areas exposed to

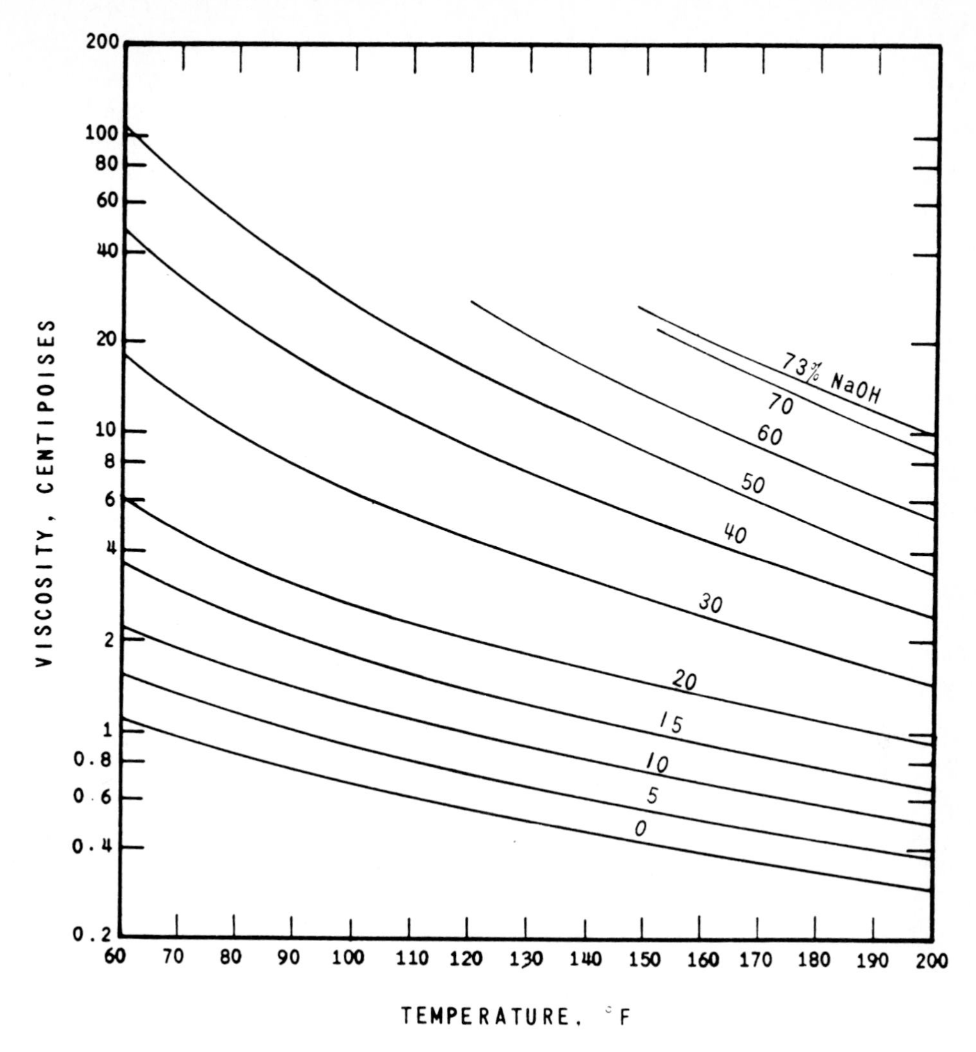

Figure 9.11 Viscosity data of caustic soda solutions.

caustic soda should be washed with copious amounts of water for 15 minutes to 2 hours. A physician should be called when exposure is severe. Caustic soda taken internally should be diluted with water or milk and then neutralized with dilute vinegar or fruit juice. Vomiting may occur spontaneously but should not be induced except on the advice of a physician.

Liquid caustic soda may be stored at the 50 percent concentration. However, at this solution strength, it crystallizes at 53°F. Therefore, storage tanks must be located indoors or provided with heating and suitable insulation if outdoors. Because of its relatively high crystallization temperature, liquid caustic soda is often diluted to a concentration of about 20 percent NaOH for storage. A 20 percent solution of NaOH has a crystallization temperature of about −20°F. Recommendations for dilution of both 73 percent and 50 percent solutions should be obtained from the manufacturer because special considerations are necessary.

Storage tanks for liquid caustic soda should be provided with an air vent for gravity flow. The storage capacity should be equal to 1.5 times the largest expected delivery, with an allowance for dilution water, if used, or two-weeks supply at the anticipated feed rate, whichever is greater. Tanks for storing 50 percent solution at a temperature between 75°F and 140°F may be constructed of mild steel. Storage temperatures above 140°F require more elab-

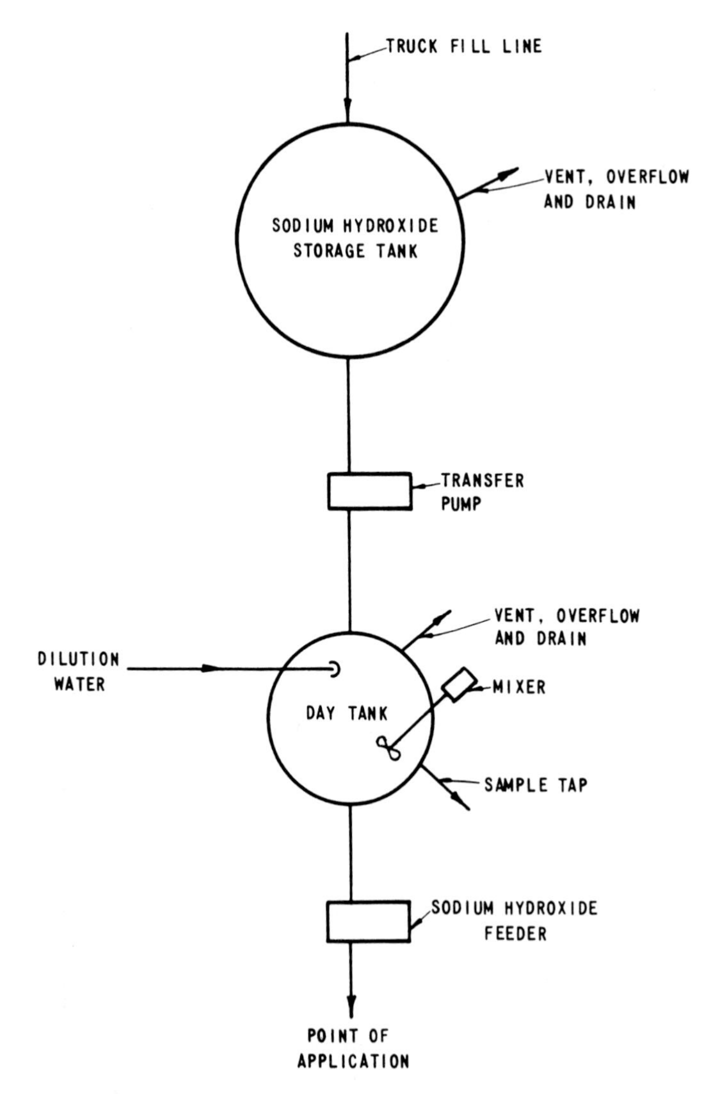

Figure 9.12 Illustrates a caustic soda feed system.

orate materials selection and are not recommended. Caustic soda will tend to pick up iron when stored in steel vessels for extended periods. Subject to temperature and solution-strength limitations, rubber, 316 stainless steel, nickel, nickel alloys, or plastics may be used when iron contamination must be avoided.

Further dilution of liquid caustic soda below the storage strength may be desirable for feeding by volumetric feeders. Feeding systems as described for liquid alum generally apply to caustic soda with appropriate selection of materials of construction. A typical system schematic is shown in Figure 9.12. Feeders will usually include materials such as ductile iron, stainless steels, rubber, and plastics.

Transfer lines from the shipping unit to the storage tank should be spiral-wire-bound neoprene or rubber hose, solid steel pipe with swivel joints, or steel hose. Because caustic soda attacks glass, use of glass materials should be avoided. Other miscellaneous materials for use with liquid caustic soda feeding and handling equipment follow.

Components	Recommended materials for use with 50% NaOH up to 140°F
Rigid Pipe	Standard Weight Black Iron
Flexible Connections	Rigid Pipe with Ells or Swing Joints Stainless Steel or Rubber Hose
Diluting Tees	Type 304 Stainless Steel
Fittings	Steel
Permanent Joints	Welded or Screwed Fittings
Unions	Screwed Steel
Valves—Nonleaking (Plug)	
Body	Steel
Plug	Type 304 Stainless Steel
Pumps (Centrifugal)	
Body	Steel
Impeller	Ni-Resist
Packing	Blue Asbestos
Storage Tanks	Steel

Carbon Dioxide

Carbon dioxide, CO_2, is available for use in wastewater treatment plants in gas and liquid form. The molecular weight of CO_2 is 44. Dry CO_2 is not chemically active at normal temperatures and is a nontoxic safe chemical; however, the gas displaces oxygen and adequate ventilation of closed areas should be provided. Solutions of CO_2 in water are very reactive chemically and form carbonic acid. Saturated solutions of CO_2 have a pH of 4.0 at 68°F.

The gas form may be produced at the treatment plant site by scrubbing and compressing the combustion product of lime recalcining furnaces, sludge furnaces, or generators used principally for the production of CO_2 gas only.

These generators are usually fired with combustible gases, fuel oil, or coke and have CO_2 yields as shown in Table 9.3.

The gas forms, as generated at the plant site, usually have a CO_2 content of between 6 percent and 18 percent depending on the source and efficiency of the producing system.

The liquid form is available from commercial suppliers in 20 to 50 lb cylinders, 10 to 20 ton trucks, and 30 to 50 ton rail cars. The commercial liquid form has a minimum CO_2 content of 99.5 percent.

Recovery of CO_2 from recalcining furnaces or incinerators is the least expensive source, but maintenance of stack gas systems is likely to be extensive because of the corrosive nature of the wet gas and the presence of particulate matter. Scrubber systems are required to clean the stack gas and specially designed gas compressors are necessary to provide the process injection pressure.

Pressure generators and submerged burners require less maintenance because the system pressure is established by compressors or blowers handling dry air or gas. On-site generating units have a limited range of CO_2 production as compared with the liquid storage and feed system, and therefore may require multiple units.

The liquid CO_2 storage and feed system generally includes a temperature-pressure controlled, bulk storage tank, an evaporation unit, and a gas feeder to meter the gas. Solution feeders, similar in construction to chlorinators, may also be used to feed CO_2.

This section applies only to use of commercial liquid CO_2. Liquid system capacities encountered in wastewater treatment usually require on-site bulk storage units. Standard prepackaged units are available, ranging in size from $\frac{3}{8}$ to 50 tons capacity, and are furnished with temperature-pressure controls to maintain approximately 300 psi at 0°F conditions. The typical package unit contains refrigeration, vaporization, safety, and control equipment. The units are well insulated and protected for outdoor location. The gas from the evaporation unit usually passes through two stages of pressure reduction before entering the gas feeder to prevent the formation of dry ice.

Feeding systems for the stack gas source of CO_2 consist of simple valving

TABLE 9.3 CO_2 YIELDS OF COMMON FUELS

Fuel	Quantity	CO_2 yield (lb)
Natural Gas	1,000 cu ft	115
Coke	1 lb	3
Kerosene	1 gal	20
Fuel Oil (No. 2)	1 gal	23
Propane	1,000 cu ft	141
Butane	1,000 cu ft	142

arrangements, for admitting varying quantities of makeup air to the suction side of the constant volume compressors, or for venting excess gas on the compressor discharge.

Pressure generators and submerged burners are regulated by valving arrangements on the fuel and air supply. Generation of CO_2 by combustion is usually difficult to control, requires frequent operator attention, and demands considerable maintenance over the life of the equipment, when compared with liquid CO_2 systems.

Commercial liquid carbon dioxide is becoming more generally used because of its high purity, the simplicity and range of feeding equipment, ease of control, and smaller, less expensive piping systems. After vaporization, the CO_2 with suitable metering and pressure reduction may be fed directly to the point of application as a gas. However, vacuum-operated, solution-type gas feeders are often preferred. Such feeders generally include safety devices and operating controls in a compact panel housing, with materials of construction suitable for CO_2 service. Absorption of CO_2 in the injector water supply approaches 100 percent when a ratio of 1.0 lb of gas to 60 gal of water is maintained.

Mild steel piping and accessories are suitable for use with cool, dry, carbon dioxide. Hot, moist gases, however, require the use of type 316 stainless steel or plastic materials. Plastics or FRP pipe are generally used for solution piping and diffusers. Diffusers should be submerged at least 8 ft, and preferably deeper, to assure complete absorption of the gas.

Standard instrument signals and control components can be used to pace or control carbon dioxide feed systems.

Using stack gas as the source of CO_2, the feed rate can be controlled by proper selection and operation of compressors, by manual control of vent or bleed valves, or by automatic control of such valves by a pH meter-controller system.

In commercial CO_2 feed systems, solution feeders may function as controllers and can be paced by instrument signals from pH monitors and plant flow meters.

In feeding commercial CO_2 directly to the point of application as a gas, a differential pressure transmitter and a control valve may function as the primary elements of a control system. Standard instrument signals may be used to pace or control the rate of CO_2 feed.

CO_2 generators are difficult to pace or control other than by manual or automatic operation of vent or bleed valves that waste a portion of the produced gas according to the plant requirements.

Polymers

Polymeric flocculants are high molecular weight organic chains with ionic or other functional groups incorporated at intervals along the chains. Because these compounds have characteristics of both polymers and electrolytes, they are frequently called *polyelectrolytes*. They may be of natural or synthetic origin.

All synthetic polyelectrolytes can be classified on the basis of the type of charge on the polymer chain. Thus, polymers possessing negative charges are called *anionic* while those carrying positive charges are *cationic*. Certain compounds carry no electrical charge and are called *nonionic* polyelectrolytes.

Because of the great variety of monomers available as starting material and the additional variety that can be obtained by varying the molecular weight, charge density, and ionizable groups, it is not surprising that a great assortment of polyelectrolytes are available to the wastewater plant operator.

Extensive use of any specific polymer as a flocculant is of necessity determined by the size, density, and ionic charge of the colloids to be coagulated. As other factors need to be considered (such as the coagulants used, pH of the system, techniques and equipment for dissolution of the polyelectrolyte, and so on), it is mandatory that extensive jar testing be performed to determine the specific polymer that will perform its function most efficiently. These results should be verified by plant-scale testing.

Dry Polymers

Types of polymers vary widely in characteristics. Manufacturers should be consulted for properties, availability, and cost of the polymer being considered.

Dry polymer and water must be blended and mixed to obtain a recommended solution for efficient action. Solution concentrations vary from fractions of a percent up. Preparation of the stock solution involves wetting of the dry material and usually an aging period prior to application. Solutions can be very viscous, and close attention should be paid to piping size and length and pump selections. Metered solution is usually diluted just prior to injection to the process to obtain better dispersion at the point of application.

Two types of systems are frequently combined to feed polymers. The solution preparation system includes a manual or automatic blending system with the polymer dispensed by hand or by a dry feeder to a wetting jet and then to a mixing-aging tank at a controlled ratio. The aged polymer is transported to a holding tank where metering pumps or rotodip feeders dispense the polymer to the process. A schematic of such a system is shown by Figure 9.13. It is generally advisable to keep the holding or storage time of polymer solutions to a minimum, one to three days or less, to prevent deterioration of the product.

Selection must be made after determination of the polymer; however, type 316 stainless steel or plastics are generally used.

The solution preparation system may be an automatic batching system, as shown by the schematic in Figure 9.14, that fills the holding tank with aged polymer as required by level probes. Such as system is usually provided

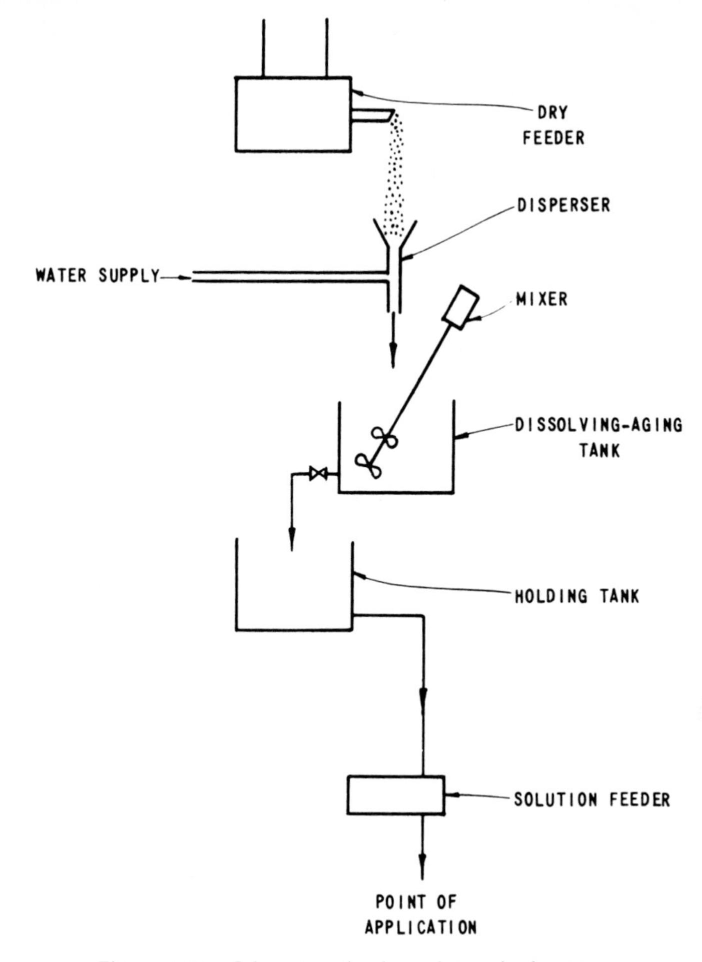

Figure 9.13. Schematic of a dry polymer feed system.

only at large plants. Prepackaged solution preparation units are available but have a limited capacity.

Liquid Polymers

As with dry polymers, there is a wide variety of products, and manufacturers should be consulted for specific information.

Liquid systems differ from the dry systems only in the equipment used to blend the polymer with water to prepare the solution. Liquid solution preparation is usually a hand-batching operation with manual filling of a mixing-aging tank with water and polymer.

Liquid polymers need no aging and simple dilution is the only require-

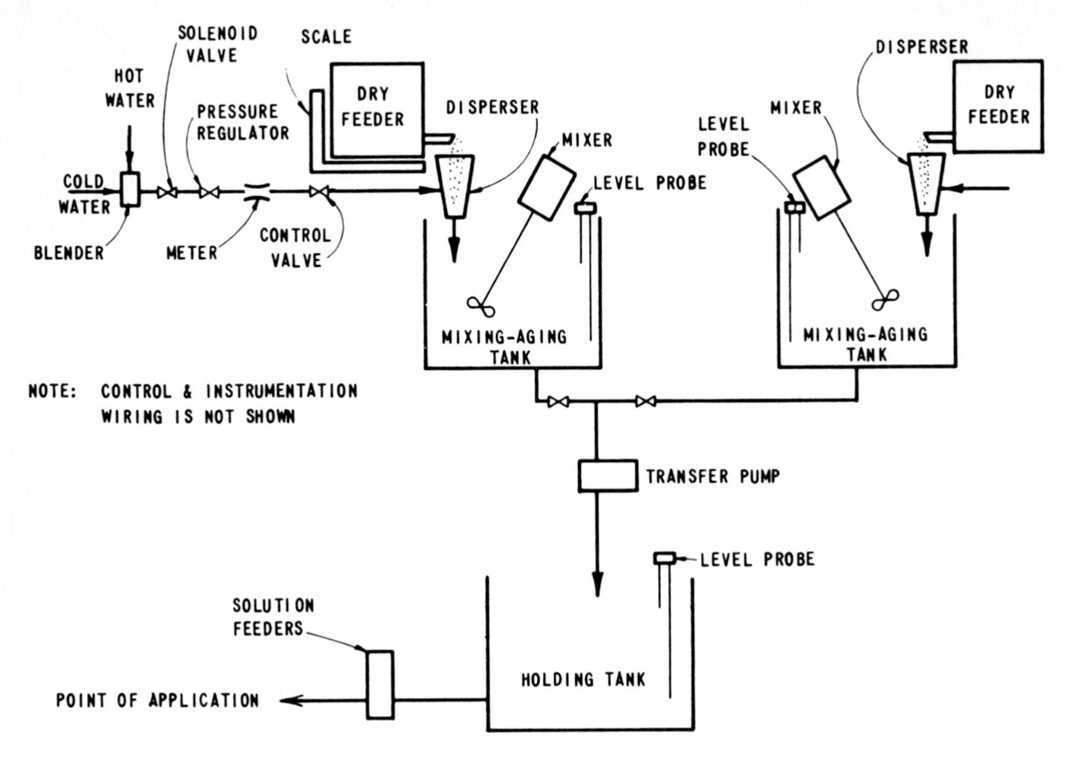

Figure 9.14. Shows an automatic dry polymer feed system.

ment for feeding. The dosage of liquid polymers may be accurately controlled by metering pumps or rotodip feeders.

The balance of the process is generally the same as described for dry polymers.

CHEMICAL FEEDERS

Chemical feed systems must be flexibly designed to provide for a high degree of reliability in light of the many contingencies which may affect their operation. Thorough waste characterization in terms of flow extremes and chemical requirements should precede the design of the chemical feed system. The design of the chemical feed system must take into account the form of each chemical desired for feeding, the particular physical and chemical characteristics of the chemical, maximum waste flows, and the reliability of the feeding devices.

In suspended and colloidal solids removal from wastewaters, the chemicals employed are generally in liquid or solid form. Those in solid form are generally converted to solution or slurry form prior to introduction to the wastewater stream; however, some chemicals are fed in a dry form. In any case, some type of solids feeder is usually required. This type of feeder has numerous different forms due to wide ranges in chemical characteristics, feed rates, and degree of accuracy required. Liquid feeding is somewhat more restrictive, depending mainly on liquid volume and viscosity.

The capacity of a chemical feed system is an important consideration in both storage and feeding. Storage capacity design must take into account the advantage of quantity purchase versus the disadvantage of construction cost and chemical deterioration with time. Potential delivery delays and chemical use rates are necessary factors in the total picture. Storage tanks or bins for solid chemicals must be designed with proper consideration of the angle of repose of the chemical and its necessary environmental requirements, such as temperature and humidity. Size and slope of feeding lines are important along with their materials of construction with respect to the corrosiveness of the chemicals.

Chemical feeders must accommodate the minimum and maximum feeding rates required. Manually-controlled feeders have a common range of 20:1, but this range can be increased to about 100:1 with dual-control systems. Chemical feeder control can be manual, automatically proportioned to flow, dependent on some form of process feedback, or a combination of any two of these. More sophisticated control systems are feasible if proper sensors are available. If manual-control systems are specified with the possibility of future automation, the feeders selected should be amenable to this conversion with a minimum of expense. An example would be a feeder with an external motor which could easily be replaced with a variable-speed motor or drive when automation is installed.

Standby or backup units should be included for each type of feeder used. Reliability calculations will be necessary in larger plants with a greater multiplicity of these units. Points of chemical addition and piping to them should be capable of handling all possible changes in dosing patterns in order to have proper flexibility of operation. Designed flexibility in hoppers, tanks, chemical feeders, and solution lines is the key to maximum benefits at least cost.

Liquid feeders are generally in the form of metering pumps or orifices. Usually these metering pumps are of the positive-displacement variety, plunger, or diaphragm type. The choice of liquid feeder is highly dependent on the viscosity, corrosivity, solubility, suction and discharge heads, and internal pressure-relief requirements. Some examples are shown in Figure 9.15. In some cases control valves and rotameters may be all that is required. In other cases, such as lime slurry feeding, centrifugal pumps with open impellers are used with appropriate controls. More complete descriptions of liquid feeder requirements can be found in the literature and elsewhere.

Solids characteristics vary to a great degree and the choice of feeder must be considered carefully, particularly in the smaller-sized facility where a single feeder may be used for more than one chemical. Generally, provisions should be made to keep all chemicals cool and dry. Dryness is very important,

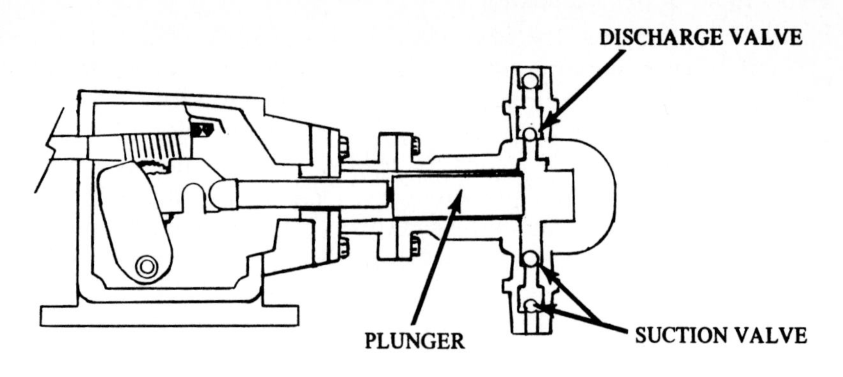

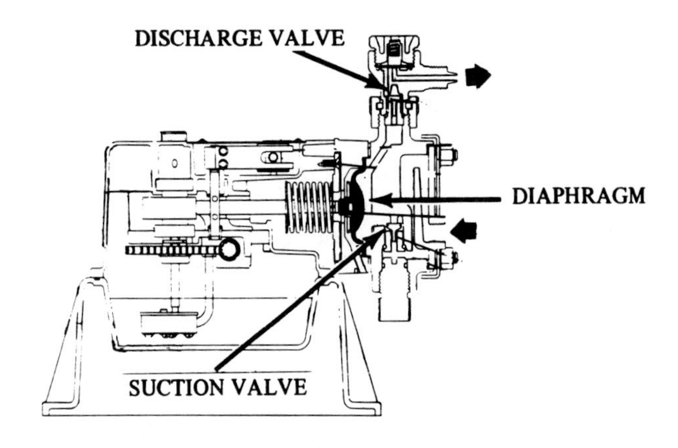

Figure 9.15. Shows positive displacement pumps.

as hygroscopic (water-absorbing) chemicals may become lumpy, viscous, or even rock hard; other chemicals with less affinity for water may become sticky from moisture on the particulate surfaces, causing increased arching in hoppers. In either case, moisture will affect the density of the chemical and may result in underfeed. Dust-removal equipment should be used at shoveling locations, bucket elevators, hoppers, and feeders for neatness, corrosion prevention, and safety reasons. Collected chemical dust may often be used.

The simplest method for feeding solid chemicals is by hand. Chemicals may be preweighed or simply shoveled or poured by the bagful into a dissolving tank. This method is of economic necessity limited to very small operations, or to chemicals used in very weak solutions.

Because of the many factors, such as moisture content, different grades, and compressibility, which can affect chemical density (weight-to-volume ratio), volumetric feeding of solids is normally restricted to smaller plants, specific types of chemicals which are reliably constant in composition and low rates of feed. Within these restrictions several volumetric types are available. Accuracy of feed is usually limited to ±2 percent by weight but may be as high as ±15 percent.

One type of volumetric dry feeder uses a continuous belt of specific width moving from under the hopper to the dissolving tank. A mechanical gate mechanism regulates the depth of material on the belt, and the rate of feed is governed by the speed of the belt and or the height of the gate opening. The hopper normally is equipped with a vibratory mechanism to reduce arching. This type of feeder is not suited for easily fluidized materials.

Another type employs a screw or helix from the bottom of the hopper through a tube opening slightly larger than the diameter of the screw or helix. Rate of feed is governed by the speed of screw or helix rotation. Some screw-type designs are self-cleaning, while others are subject to clogging. Figure 9.16 shows a typical screw feeder.

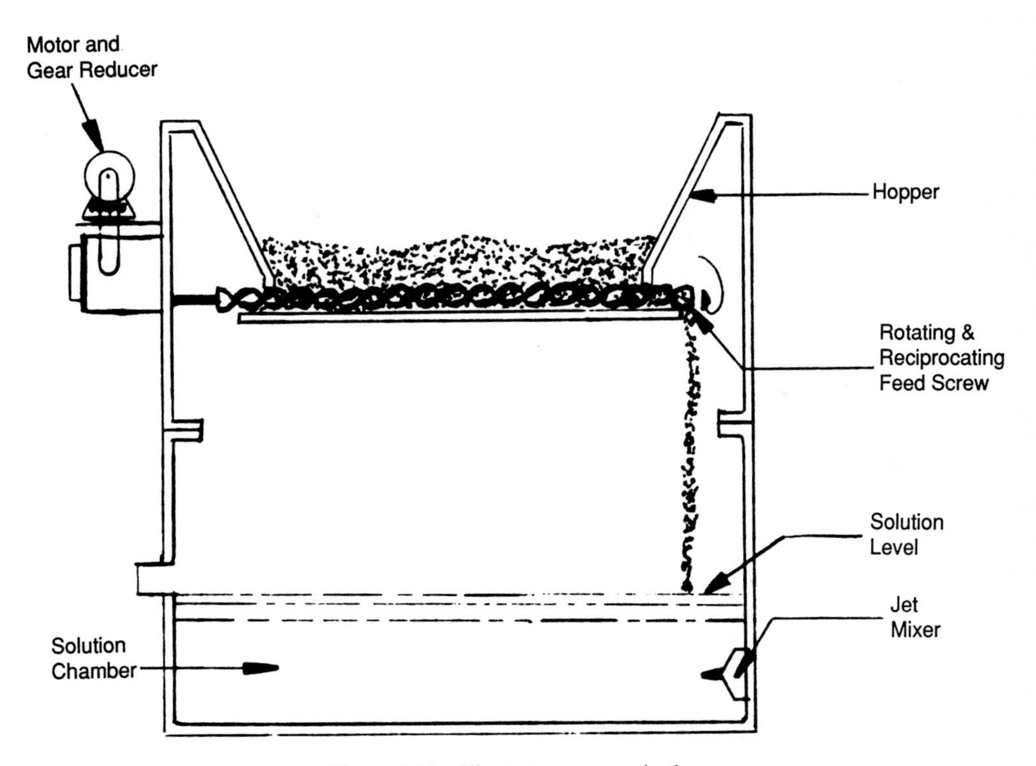

Figure 9.16. Illustrates a screw feeder.

Most remaining types of volumetric feeders generally fall into the positive-displacement category. All designs of this type incorporate some form of moving cavity of a specific or variable size. In operation, the chemical falls by gravity into the cavity and is more or less fully enclosed and separated from the hopper's feed. The size of the cavity and the rate at which the cavity moves and is discharged govern the amount of material fed. The positive control of the chemical may place a low limit on rates of feed. One unique design is the progressive cavity-metering pump, a nonreciprocating type. Positive-displacement feeders often utilize air injection to improve the flow of the material. Some examples of positive-displacement units are illustrated in Figure 9.17.

The basic drawback of volumetric feeder design (i.e., its inability to compensate for changes in the density of materials) is overcome by modifying the volumetric design to include a gravimetric or loss-in-weight controller. This modification allows for weighing of the material as it is fed. The beam-balance type measures the actual mass of material. This is considerably more accurate, particularly over a long period of time, than the less common spring-loaded gravimetric designs. Gravimetric feeders are used where feed accuracy of about 1 percent is required for economy, as in large-scale operations and for materials which are used in small, precise quantities. It should be noted, however, that even gravimetric feeders cannot compensate for weight added to the chemical by excess moisture. Many volumetric feeders may be con-

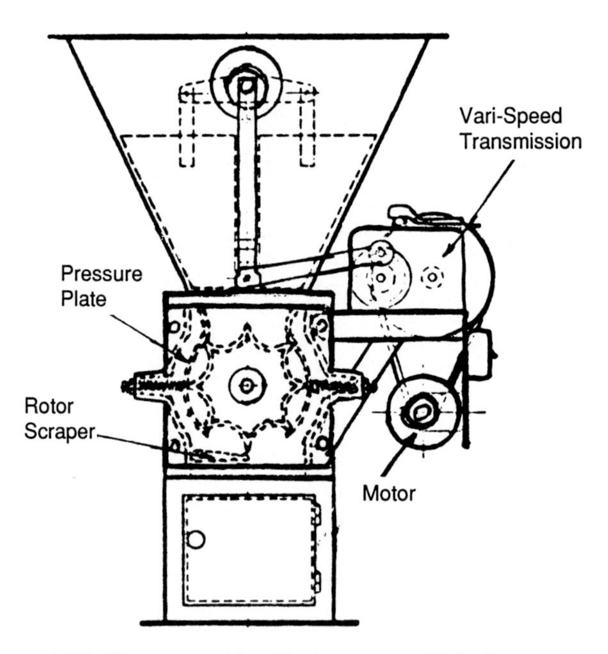

Figure 9.17 Shows a positive displacement solid feeder-rotary type.

verted to loss-in-weight function by placing the entire feeder on a platform scale which is tared to neutralize the weight of the feeder.

Good housekeeping and need for accurate feed rates dictate that the gravimetric feeder be shut down and thoroughly cleaned on a regular basis. Although many of these feeders have automatic or semiautomatic devices which compensate to some degree for accumulated solids on the weighing mechanism, accuracy is affected, particularly on humid days when hygroscopic materials are fed. In some cases, built-up chemicals can actually jam the equipment.

No discussion of feeders is complete without at least passing reference to dissolvers, as any metered material must be mixed with water to provide a chemical solution of desired strength. Most feeders, regardless of type, discharge their material to a small dissolving tank which is equipped with a nozzle system and/or mechanical agitator depending on the solubility of the chemical being fed. Solid materials, such as polyelectrolytes, may be carefully spread into a vortex spray or washdown jet of water immediately before entering the dissolver. It is essential that the surface of each particle become thoroughly wetted before entering the feed tank to ensure accurate dispersal and to avoid clumping, settling, or floating.

A dissolver for a dry chemical feeder is unlike a chemical feeding mechanism, which by simple adjustment and change of speed can vary its output tenfold. The dissolver must be designed for the job to be done. A dissolver suitable for a rate of 10 lb/hr may not be suitable for dissolving at a rate of 100 lb/hr. As a general rule, dissolvers may be oversized, but dissolvers for commercial ferric sulfate or lime slakers do not perform well if greatly oversized.

It is essential that specifications for dry chemical feeders include specifications on dissolver capacity. A number of factors need to be considered in designing dissolvers of proper capacity. These include detention times and water requirements, as well as other factors specific to individual chemicals.

The capacity of a dissolver is based on detention time, which is directly related to the wettability or rate of solution of the chemical. Therefore, the dissolver must be large enough to provide the necessary detention for both the chemical and the water at the maximum rate of feed. At lower rates of feed, the strength of solution or suspension leaving the dissolver will be less, but the detention time will be approximately the same unless the water supply to the dissolver is reduced. When the water supply to any dissolver is controlled for the purpose of forming a constant-strength solution, mixing within the dissolver must be accomplished by mechanical means, because sufficient power will not be available from the mixing jets at low rates of flow. Hot-water dissolvers are also available in order to minimize the required tankage.

The foregoing descriptions give some indication of the wide variety of materials which may be handled. Because of this variety, a modern facility may contain any number of a variety of feeders with combined or multiple materials capability. Ancillary equipment to the feeder also varies according

TABLE 9.4 TYPES OF CHEMICAL FEEDERS

Type of feeder	Use	Limitations		
		General	Capacity (cu ft/hr)	Range
Dry feeder:				
Volumetric:				
Oscillating plate	Any material, granules or powder.		0.01 to 35	40 to 1
Oscillating throat (universal)	Any material, any particle size.		0.02 to 100	40 to 1
Rotating disk	Most materials including NaF, granules or powder	Use disk unloader for arching.	0.01 to 1.0	20 to 1
Rotating cylinder (star)	Any material, granules or powder.		8 to 2,000 or 7.2 to 300	10 to 1 or 100 to 1
Screw	Dry, free-flowing material, powder or granular.		0.05 to 18	20 to 1
Ribbon	Dry, free-flowing material, powder, granular, or lumps.		0.002 to 0.16	10 to 1
Belt	Dry, free-flowing material up to 1½-inch size, power or granular.		0.1 to 3,000	10 to 1 or 100 to 1
Gravimetric				
Continuous-belt and scale	Dry, free-flowing, granular material, or floodable material.	Use hopper agitator to maintain constant density.	0.02 to 2	100 to 1
Loss in weight	Most materials, powder, granular or lumps.		0.02 to 80	100 to 1
Solution feeder:				
Nonpositive displacement:				
Decanter (lowering pipe)	Most solutions or light slurries		0.01 to 10	100 to 1
Orifice	Most solutions	No slurries	0.16 to 5	10 to 1
Rotameter (calibrated valve)	Clear solutions	No slurries	0.005 to 0.16 or 0.01 to 20	10 to 1
Loss in weight (tank with control valve)	Most solutions	No slurries	0.002 to 0.20	30 to 1
Positive displacement:				
Rotating dipper	Most solutions or slurries		0.1 to 30	100 to 1
Proportioning pump:				
Diaphragm	Most solutions. Special unit for 5% slurries.[a]		0.004 to 0.15	100 to 1
Piston	Most solutions, light slurries		0.01 to 170	20 to 1

TABLE 9.4 *(Continued)*

Type of feeder	Use	Limitations		
		General	Capacity (cu ft/hr)	Range
Gas feeders:				
Solution feed	Chlorine		8,000 lb/day max	20 to 1
	Ammonia		2,000 lb/day max	20 to 1
	Sulfur dioxide		7,600 lb/day max	20 to 1
	Carbon dioxide		6,000 lb/day max	20 to 1
Direct feed	Chlorine		300 lb/day max	10 to 1
	Ammonia		120 lb/day max	7 to 1
	Carbon dioxide		10,000 lb/day max	20 to 1

[a] Use special heads and valves for slurries.

to the material to be handled. Liquid feeders encompass a limited number of design principles which account for density and viscosity ranges. Solids feeders, relatively speaking, vary considerably due to the wide range of physical and chemical characteristics, feed rates, and the degree of precision and repeatability required.

Table 9.4 describes several types of chemical feeders commonly used in wastewater treatment.

Index

C

D

diatomite, 3, 179, 186

E

F

G

H

T

U